Tectonic Modeling: A Volume in Honor of Hans Ramberg

Edited by

Hemin A. Koyi
Hans Ramberg Tectonic Laboratory
Department of Earth Sciences
Uppsala University
Villavägen 16
SE-752 36 Uppsala
SWEDEN

and

Neil S. Mancktelow
Geologisches Institut
ETH-Zentrum
CH-8092 Zurich
SWITZERLAND

Geological Society of America
3300 Penrose Place
P.O. Box 9140
Boulder, Colorado 80301-9140
2001

Copyright © 2001, The Geological Society of America, Inc. (GSA). All rights reserved. GSA grants permission to individual scientists to make unlimited photocopies of one or more items from this volume for noncommercial purposes advancing science or education, including classroom use. Permission is granted to individuals to make photocopies of any item in this volume for other noncommercial, nonprofit purposes provided that the appropriate fee ($0.25 per page) is paid directly to the Copyright Clearance Center, 222 Rosewood Drive, Danvers, MA 01923, USA, phone (978) 750-8400, http://www.copyright.com (include title and ISBN when paying). Written permission is required from the Copyright Clearance Center for all other forms of capture or reproduction of any item in the volume including, but not limited to, all types of electronic or digital scanning or other digital or manual transformation of articles or any portion thereof, such as abstracts, into computer-readable and/or transmittable form for personal or corporate use, either noncommercial or commercial, for-profit or otherwise. Send permission requests to Copyright Clearance Center.

Copyright is not claimed on any material prepared wholly by government employees within the scope of their employment.

Published by The Geological Society of America, Inc.
3300 Penrose Place, P.O. Box 9140, Boulder, Colorado 80301
www.geosociety.org

Printed in U.S.A.

GSA Books Science Editor Abhijit Basu
GSA Books Editor Rebecca Herr
Cover design by Margo Good

Library of Congress Cataloging-in-Publication Data

Tectonic modeling : a volume in honor of Hans Ramberg / edited by Hemin A. Koyi and
Neil S. Mancktelow.
    p. cm. -- (Memoir ; 193)
  Includes bibliographical references and index.
  ISBN 0-8137-1193-2
    1.Geology, Structural--Mathematical models. I. Ramberg, Hans, 1917- II. Koyi,
Hemin A., 1960- III. Mancktelow, Neil S. IV. Memoir (Geological Society of America) ;
  193.

QE601 .T385 2000
551.8--dc21

                                                00-058686

Cover: Image assemblage of some of the geologic structures and processes described in this volume. Illustrations courtesy of Neil Mancktelow, Guido Schreurs, Hans Peter Steyrer, who provided originals of their model photographs. Prepared by H.A. Koyi.

Contents

Preface v

Introduction: Hans Ramberg vii
    Christopher J. Talbot

1. Types of transpressional and transtensional deformation 1
    Subir Kumar Ghosh

2. Experimental modeling of strike-slip faults and the self-similar behavior 21
    Martin P.J. Schöpfer and Hans Peter Steyrer

3. Finite-element approach to deformation with diffusive mass transfer 29
    Brian Bayly and D.H. Minkel

4. Modeling of anisotropic grain growth in minerals 39
    Paul D. Bons, Mark W. Jessell, Lynn Evans, Terrence Barr, and Kurt Stüwe

5. Modifications of early lineations during later folding in simple shear 51
    Sudipta Sengupta and Hemin A. Koyi

*6. Single-layer folds developed from initial random perturbations: The effects
of probability distribution, fractal dimension, phase, and amplitude* 69
    Neil S. Mancktelow

7. Experimental study of single layer folding in nonlinear materials 89
    Tatiana Tentler

*8. Sheath fold development in bulk simple shear: Analogue modeling
of natural examples from the Southern Iberian Variscan Fold belt* 101
    Filipe M. Rosas, Fernando O. Marques, Sara Coelho, and Paulo Fonseca

9. Modeling the role of erosion in diapir development in contractional settings 111
    Maura Sans and Hemin A. Koyi

*10. Diapirism in convergent settings triggered by hinterland pinch-out
of viscous decollement: A hypothesis from modeling* 123
    Elisabetta Costa and Bruno Vendeville

*11. Salt tectonics and sedimentation along Atlantic margins: Insights from seismic
interpretation and physical models* 131
    Webster Ueipass Mohriak and Peter Szatmari

12. *Compressional structures in a multilayered mechanical stratigraphy: Insights from sandbox modeling with three-dimensional variations in basal geometry and friction* 153
    Claudio Turrini, Antonio Ravaglia, and Cesare Perotti

13. *Four-dimensional analysis of analog models: Experiments on transfer zones in fold and thrust belts* 179
    Guido Schreurs, Reto Hänni, and Peter Vock

14. *Effective indenters and the development of double-vergent orogens: Insights from analogue sand models* 191
    Katarina S. Persson

15. *Horses and duplexes in extensional regimes: A scale-modeling contribution* 207
    Roy H. Gabrielsen and Jill A. Clausen

16. *Crustal rheology and its effect on rift basin styles* 221
    Anthony P. Gartrell

17. *New apparatus for controlled general flow modeling of analog materials* 235
    Sandra Piazolo, Saskia M. ten Grotenhuis, and Cees W. Passchier

18. *New apparatus for thermomechanical analogue modeling* 245
    Elmar M. Wosnitza, Djordje Grujic, Robert Hoffman, and Jan H. Behrmann

19. *Modeling of temperature-dependent strength in orogenic wedges: First results from a new thermomechanical apparatus* 253
    Federico Rossetti, Claudio Faccenna, Giorgio Rannelli, Renato Funiciello, and Fabrizio Storti

20. *Flow and fracturing of clay: Analogue experiments in bulk pure shear* 261
    Fernando O. Marques

Index 271

Preface

Preparing a volume in honor of any well-known scientist is not an easy task. In our case, it was even more difficult than usual because Hans Ramberg was an international scientist with a very broad field of interest.

The idea of this special volume developed during the preparation of the Hans Ramberg Symposium on tectonic modeling, which was organized as part of European Union of Geosciences biennial meeting (EUG-10) in Strasbourg in 1998. We thank the attendees for contributing so much to the success of the symposium. The contributors we approached during this symposium strongly supported the idea of a volume dedicated to Hans Ramberg, who made such a major contribution to developing and promoting modeling of tectonic structures as a way to better understanding the geologic record.

Earth sciences is by nature an international discipline that is not (or should not be) restricted by national boundaries. Hans Ramberg lived up fully to this criterion. He was an international scientist who had worked in many parts of the world. After receiving his first degrees in Scandinavia, he moved to the United States and worked as a professor at the University of Chicago. He spent couple of years in Brazil before returning to Sweden and building his well-known tectonic lab at Uppsala. The tectonics laboratory, named after him, has over the years hosted many scientists from all over the world, and some of these alumni are represented in this volume. Hans Ramberg strongly promoted such cooperation at the international level. He was a member of various geological societies; he was made an honorary member of the Geological Society of America and was twice recognized with important awards (Arthur L. Day Medal [1976] and Career Contribution Award [1998]). It is therefore particularly fitting that this society offered to publish a commemorative volume.

Many people helped us prepare the volume. We thank the authors for their contributions and for helping us to stick to the deadlines: we had to be generally quite strict in keeping to our timetable, both through consideration for the other authors and for the Geological Society of America. Each manuscript was reviewed by two or three experts who were not contributors to the volume. This was a Geological Society of America policy that we followed strictly. Special thanks are due to the following reviewers, some of whom reviewed more than one manuscript more than once. Their invaluable help was essential for preparing a volume worthy of honoring Hans Ramberg.

Jesus Aller	Tim Dooley	Robert Krantz
Ian Alsop	Bruce Douglas	Nina Kukowski
Wayne Bailey	Gregg Erickson	Richard Lisle
Kevin Burke	Magnus Friberg	Jacques Malavieille
Martin Burkhard	Marc-André Gutscher	Tom Mauduit
Jordi Carreras	Chris Harrison	Ken McClay
Alexandre Chemenda	Mike Hudec	Win Means
Mike Coward	Peter Hudleston	Alan Geoffery Milnes
Alexander Cruden	Benoit Ildefonse	Chris Morley
Kenneth Cruikshank	William Jamison	Alan Morris
Ian Davison	Dazi Jiang	Gene Mulugeta
Philippe Davy	Richard Jones	Francis Odonne

Youndo Park	Deborah Spratt	Frederick Vollmer
Adrian Pfiffner	Simon Stewart	Thomas Walter
Josep Poblet	Christopher J. Talbot	Ruud Weijermars
Bertram Schott	Mark Taylor	
Lilian Skjernaa	Antonio Teixell	

We also thank Abhijit Basu, the Books Science Editor at the Geological Society of America, for his advice and guidance during preparation of the volume. Last but not least, we thank the Geological Society of America for offering to publish this volume.

We have tried to include contributions covering three fields (analytical, numerical, and physical modeling) in which Hans Ramberg worked during the last 30 years of his career. This modeling approach reflects Hans Ramberg's interest in understanding the principles and processes that govern the development of deformation structures in natural field examples. We hope that this selection gives some idea of the range of modeling techniques currently available and the results that can be obtained, and is a fitting tribute to the pioneering work of Hans Ramberg.

Hemin A. Koyi and Neil Mancktelow

Introduction: Hans Ramberg and this volume

Christopher J. Talbot

Hans Ramberg Tectonic Laboratory, Department of Earth Sciences, Uppsala University, SE-752 36 Uppsala, Sweden

Hans Ramberg died of cancer in Uppsala on 7th June 1998, at the age of 81, soon after the pleasure of adding the Career Contribution Award of the Structural Geology and Tectonics Division of the Geological Society of America (GSA) to his collection of major awards. These included the Arthur L. Day Medal from the GSA (1976), the Celsius Medal from the Royal Society of Sciences at Uppsala (1969), the Wollaston Medal from the Geological Society of London (1972), the Grand Prize from the Royal Academy for Natural Sciences in Sweden (1973), and the Arthur Holmes Medal from the European Union of Geoscientists (1983). Such awards recognized the achievements of an extraordinary career, which can be divided into three phases.

Hans Ramberg was born in Trondheim on the west coast of Norway (15 March 1917), and earned a strong grounding in chemistry during his first degree at Oslo University (1943). The first phase of his career involved meticulous field observations of metamorphic rocks and the structures in them, first in the Norwegian Caledonides for his Doctor of Philosophy degree (1946), and for the next five summers in west Greenland. There he distinguished the Nagsuggtoqidian orogen using the deformation of Paleoproterozoic dikes that were undeformed in the adjoining Archean craton. His observations in the Norwegian Caledonides and Greenland led to a deep personal drive to understand the fundamental thermodynamics behind mineral assemblages with long histories of deep burial. This phase was documented in 30 papers analyzing the fundamental thermodynamics behind mineral assemblages and his first highly influential book, *The origin of metamorphic and metasomatic rocks* (University of Chicago Press, 1952).

Hans moved from professor of geology at the University of Chicago (1948–1961) to research associate with the Carnegie Geophysical Laboratory in Washington (1952–1955), where he began the second phase by turning to mechanics. He used engineering theory to attribute natural and experimental boudinage to extension along thin sheets induced by compression across them. His picture of tensile stresses concentrating midway along successive generations of boudins that repeatedly halve in length was developed in a single paper (1955) that has been improved on remarkably little since.

Hans then used a combination of engineering and fluid dynamic theory to explain the ratios of wavelength to thickness of ptygmatic folds in terms of buckling of thin viscous sheets. This led to a remarkable suite of about 13 papers (1959–1964), one of which appeared in the new journal *Tectonophysics*, which he coedited for more than two decades. These works differentiated passive from active folds and accounted for folds in multilayers that could be harmonic and disharmonic and major or minor on scales at which the influence of gravity is significant or not. After a session as visiting professor at Universidade do Brasil, Ouro Preto (1960–1961), Hans returned to Scandinavia as professor of mineralogy and petrology at Uppsala University (1961–1982) while maintaining his links with the United States. On arrival at Uppsala his papers on theoretical petrogenesis dwindled to a trickle while he began establishing the tectonic laboratory now named after him. Having equipped himself, he then began testing his ideas, first in pure and simple shear boxes and then, after they had been installed, two centrifuges.

Until then, experimental rock mechanics was only about squeezing and pulling real rocks in real time. Hans Ramberg's new approach was rigorously based on the principles of geometric, kinematic, and dynamic scaling. By increasing the body force exerted in nature by gravity, complicated 20×20 cm models carefully constructed with relatively weak ductile materials at 1 g are then deformed at as much as 2000 g in a centrifuge for about 10 min to simulate natural crustal structures that took tens of millions of years to develop.

Within a decade of start-up, the centrifuges had become a cornucopia. Models produced by Hans's assistants and students simulated crustal isostasy, rift valleys, opening oceans, and the growth of continents (1964), as well as individual structures such as glaciers (1964), plutons (1970), and salt structures (1970). Such models simulated the structural patterns then being mapped in sedimentary basins and orogens of all ages in every continent, from Archean granite-greenstone terrains to nappe piles in the Alps. Profuse illustrations of these beautiful models in his second book *Gravity, deformation and the Earth's crust* (Academic Press, London 1967; second edition 1981) helped generations of geoscientists to reach new levels of understanding of the dynamics behind the phenomena they saw in the field or literature. Hans's introduction of scaled analogue experiments had a profound influence on structural geologists and tectonicians at a time when the concept of orogeny was

Talbot, C.J., 2001, Introduction: Hans Ramberg and this volume, *in* Koyi, H.A., and Mancktelow, N.S., eds., Tectonic Modeling: A Volume in Honor of Hans Ramberg: Boulder, Colorado, Geological Society of America Memoir 193, p. vii–ix.

floundering in countless categories of geosynclines. However, the almost simultaneous advent of plate tectonics changed the focus of continental geologists toward the lateral forces that open and close oceans due to gravity on a larger scale, and intracrustal gravity overturns tended to be relegated to the old fixist view. The sinking of ocean floor from between colliding continents results in enormous lateral displacements, which leads to the lateral escape and/or vertical extrusion and gravity spreading or sliding of ocean fills in proportions controlled by their boundaries between continent- and ocean-capped lithosphere. Hans's analogue experimental approach, controversial when it was introduced in a laboratory informally known as "the baking shop," can now be considered as one of mainstreams of the geosciences.

While analogue modeling laboratories were proliferating throughout industry and academia (many of the latter under the direction of former students or visitors to Uppsala), Hans was becoming addicted to computers. Thereafter, he left analogue modeling to others while he explored the potential of numerical modeling in the third phase of his career. His first paper in which numerical models joined theory and analogue experiments (with Harold Berner and Ové Stephnasson in 1972) included the application of finite elements to the widening of the Atlantic. Hans went on to develop analytical theories for particle paths, displacement, and progressive strains (1975), and opened up another field, the spectrum of pure through subsimple and simple shear to oscillating supershears. On retirement, Hans was appointed emeritus professor (1982) and his position was replaced by two chairs.

As an undergraduate in 1942, Hans married one of his first school friends, Marie Louise (Lillemor), and for the next 30 years they maintained open house at their summer cottage in Westranden in Norway to a stream of friends prepared to help Hans fish from his boat. In 1947, Hans acknowledged his former mentor at Oslo, Tom Barth, for introducing him to the American concept of open and informal communications between teachers and students (which were often still lamentably formal in Europe at the time). Social evenings with Hans and Lillemor invariably involved discussions on the implications of major advances in current science, from the DNA molecule to black holes and quasars. Long after his official retirement, when his early students and visitors were becoming professors throughout the world, Hans would appear in the laboratory, peer over shoulders, and ask a few pertinent questions that would often change the course of the experiments. Hans celebrated his 70th birthday by completing his unique world view by adding the thermodynamics of rock deformation (1989) to his early studies on the thermodynamics of rock formation. His last papers advocated the application of stream functions and work efficiency (1986) and developed numerical models to analyze the gravity spreading and sliding of nappes (1991).

Including the two books, Hans Ramberg's publications number no more than 90, but their extraordinary influence (and continued citation) emphasises their penetrating quality. In essence, Hans Ramberg almost single-handedly dragged first metamorphic petrology, and then structural geology, from descriptive exercises to theoretical and experimental sciences.

THIS VOLUME

Hans did not live to know that this volume was in preparation to honor his memory. However, he would have enjoyed the demonstration that so many of the subjects he first raised have developed so far, and would have appreciated the advances in equipment and materials. This collection of 19 papers by authors from 12 countries celebrates the health and vigor of subjects to which Hans Ramberg devoted much of his career.

He would have particularly appreciated contributions by former friends and colleagues: Bayly and Minkel, with their FEM treatment of mass diffusion in terms of energy dissipation; Ghosh's subdivision of transpressive and transtensional deformations into 18 types (some with unexpected deformation paths), and the analysis of how early lineations are modified during later folding with different amounts of simple shear by Sengupta and Koyi. Hans would have beamed had he read Mancktelow's FEM study demonstrating that it is the amplitude, rather than the shape or location, of the initial irregularities that has most influence of the amplification of fold trains along single layers.

Hans would doubtless have been excited by the new hardware for analogue modeling announced here. There are two new hot machines; one in Rome (Rosetti et al.), the other in Frieburg (Wosnitza et al.). The latter is sufficiently large (35 L) to dynamically scale the effect of gravity during lateral shortening and extension of a model lithospheric in which viscosities can vary over 11 orders of magnitude and in which the isotherms are mapped by an infrared camera. He would have admired the time-lapse spiral X-ray computed tomographic images of fold-and-thrust belts (Schreurs et al.), and would have enjoyed designing experiments for the new general flow machine at Mainz (Piazolo et al.), with its six computer-driven flexible step motors. Rosas et al. use an elegant new simple shear box to simulate sheath folds from noncylindrical perturbations in the foliations around stiff inclusions, and Gabrielsen and Clausen explore in great detail the development of arrays of extensional faults using spectacular models in wet plaster (the scaling of which would Hans would have read very carefully). Hans would also have appreciated the proliferation of new model materials, e.g., strain-rate softening and hardening materials in folded single layers (Tentler), and a back-to-Riedel use of strain hardening clay, but here in pure shear (Marques). He would have followed with great interest every lesson learned from such crystalline rock analogues as octachloropropane, particularly when, matched with numerical modeling, experiments exploring the role of surface energy anisotropy during polycrystalline grain growth raise a theoretical problem (Bons et al.).

Several authors use sand boxes to explore aspects of thin-skinned compression. Turrini et al. report the influence of various mechanical stratigraphies on patterns of thrusts and folds and Persson discovers a second category of effective indenter while simulating orogenic wedges and illustrates seismic reflection profiles of natural examples both with and without such an effective indenter. Costa and Vendeville explore diapirism in laterally shortened cover sequences and compare their results with structures in the eastern Mediterranean, while Sans and Koyi elucidate how erosion can trigger the growth of diapirs in regions subjected to thin-skinned buckling, as in Spain. Mohriak and Szatmari use results of analogue models and seismic data to give a short review about salt tectonics and sedimentation. Schöpfer and Steyrer report intriguing details about the compaction of sand packs and the relation of grain size to self similarity in their modeling of strike-slip faults. Gartrell also used sand-box models, but for systematic studies of how crustal rheology, slab movement, and synrift sedimentation affect the styles of rift valleys during lateral extension of the crust.

Although this volume is about modeling, it is important to remember that a famous saying of Hans's was *"without fieldwork there is no geology."*

BIBLIOGRAPHY: Some of Hans Ramberg's Landmark Publications

Ramberg, H., 1952, The origin of metamorphic and metasomatic rocks: Chicago, University of Chicago Press, 317 p.

Ramberg, H., 1955, Natural and experimental boudinage and pinch-and-swell structures: Journal of Geology, v. 63, p. 512–526.

Ramberg, H., 1959, Evolution of ptygmatic folding: Nørsk Geologisk Tidskriffter, v. 39, p. 99–152.

Ramberg, H., 1960, Relationship between length of arc and thickness of ptygmatically folded veins: American Journal of Science, v. 258, p. 36–46.

Ramberg, H., 1961a, Relationship between concentric longitudinal strain and concentric shearing strain during folding of homogeneous sheets of rocks: American Journal of Science, v. 259, p. 382–390.

Ramberg, H., 1961b, Contact strain and folding instability of a multilayered body under compression: Geologische Rundschau, v. 51, p. 405–493.

Ramberg, H., 1963a, Fluid dynamics of viscous buckling applicable to folding of layered rocks: Bulletin of the American Association of Petroleum Geologists, v. 47, p. 484–505.

Ramberg, H., 1963b, Evolution of drag folds: Geological Magazine, v. 100, p. 97–106.

Ramberg, H., 1963c, Strain distribution and geometry of folds: Bulletin of the Geological Institutions, University of Uppsala, v. 47, p. 484–505.

Ramberg, H., 1964a, Notes on model studies of folding of moraines in piedmont glaciers: Journal of Glaciology, v. 5, p. 209–218.

Ramberg, H., 1964b, A model for the evolution of continents, oceans and orogens: Tectonophysics, v. 1, p. 207–341.

Ramberg, H., 1967 (second edition 1981), Gravity, deformation, and the Earth's crust: Academic Press, London.

Ramberg, H., 1970, Model studies in relation to intrusion of plutonic bodies, *in* Mechanism of igneous intrusions: Liverpool, Geological Journal, Special Issue No. 2, p. 261–286.

Ramberg, H., 1970, Experimental and theoretical study of salt-dome evolution: Third Symposium on Salt: Northern Ohio Geological Society, p. 261–270.

Berner, H., Stephansson, O., and Ramberg, H., 1972, Diapirism in theory and experiment: Tectonophysics, v. 15, p. 197–218.

Ramberg, H., 1975, Particle paths, displacement and progressive strain application to rocks: Tectonophysics, v. 28, p. 1–37.

Ramberg, H., 1986, Particle paths, displacement and progressive strain application to rocks: A correction: Tectonophysics, v. 121, p. 355.

Ramberg, H., 1989, Thermodynamics applied to deformation structures: Bulletin of the Geological Institutions, University of Uppsala, New Series 14, p. 1–12.

Ramberg, H., 1986, The stream function and Gauss' principle of least constraint: Two useful concepts for structural geology: Tectonophysics, v. 131, p. 205–246.

Ramberg, H., 1988, A new numerical simulation method applied to spreading nappes: Tectonophysics, v. 162, p. 173–192.

Ramberg, H., 1991, Numerical simulation of spreading nappes, sliding against basal friction: Tectonophysics, v. 188, p. 159–186.

Types of transpressional and transtensional deformation

Subir Kumar Ghosh*
Department of Geological Sciences, Jadavpur University, Calcutta 700 032, India

ABSTRACT

The transpressional (or transtensional) model of Sanderson and Marchini considered a horizontal displacement to be oblique to the walls of a vertical tabular shear zone that has no extrusion along the strike direction of the shear zone. In a more general model that includes lateral extrusion, the nature of deformation becomes much more complex. This chapter analyzes such a tabular shear zone in terms of simultaneous simple shearing and coaxial straining. In such a deformation, one of the principal axes of the strain ellipsoid is always vertical and parallel to the vorticity vector and the other two lie on the horizontal plane. I present a new approach for analyzing and classifying transpressional (and transtensional) tectonism. The classification enables us to compare the degree of similarity of the deformation types and to visualize their relation with the kinematic vorticity number. It has been shown that a transpression does not always produce a flattening. Similarly, a transtension does not always produce a constriction. The history of progressive deformation may be complex, involving switching of the direction of maximum stretching, development of a cleavage and its folding in the same continuous deformation, and folding of early boudin lines in later stages of deformation.

INTRODUCTION

Ramberg's (1975) paper on particle path and progressive deformation led to many studies (e.g., Means et al., 1980; Passchier, 1986) of deformation types more complex than pure shear and simple shear. Ramberg considered the case of plane strain by superposing pure shearing and simple shearing. As suggested by Means (1994), the ending with "ing" is used to specify that the deformation is instantaneous. Ramberg distinguished two types of plane strain for nonpulsating and pulsating strain ellipsoids. For the nonpulsating type, he considered the case in which the shortening direction of pure shearing is perpendicular to the direction of simple shearing. This type of deformation was described as general shear by Simpson and De Paor (1997).

The main concern of this chapter is the classification of deformations generally described as transpression and transtension (e.g., Harland, 1971; Sanderson and Marchini, 1984; Hudleston et al., 1988; Robin and Cruden, 1994; Fossen and Tikoff, 1993, 1998; Greene and Schweickert, 1995; Tikoff and Fossen, 1993; Tikoff and Teyssier, 1994; Tikoff and Greene, 1997; Jones et al., 1997; Tikoff and Peterson, 1998). Simultaneous simple shearing and coaxial straining may give rise to very complex types of deformation. Tikoff and Fossen (1993), for example, combined coaxial straining with three orthogonal simple shears. This chapter is concerned with a simpler type of deformation of a tabular shear zone with wall-parallel simple shearing. The strain rates of the component of coaxial deformation are parallel to the shear direction x (ε_x), parallel to the vorticity vector of simple shearing (ε_z) and perpendicular to the tabular zone (ε_y) (Fig. 1). The simple shear strain rate parallel to the shear zone is γ'. If ε_y is negative, causing a shortening across the tabular zone, the deformation is transpression (Fig. 1). If ε_y is

*E-mail: subir@jugeo.clib0.ernet.in

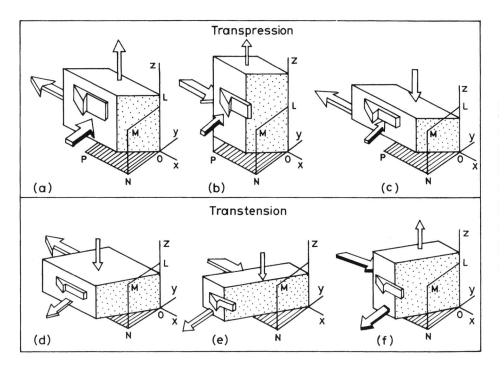

Figure 1. Transpression and transtension by simple shearing and coaxial straining, with reference to coordinate axes x, y, and z. Shear zone is parallel to xz plane or plane containing lines MN and PN. In transpression y axis (normal to shear zone) is direction of instantaneous contraction. In transtension it is direction of extension. For both transpression and transtension strain rate along vertical z axis or along horizontal x axis may be either positive (extension) or negative (contraction).

positive, causing extension across the zone, it is transtension (Fig. 1).

Transpression or transtension can be classified in different ways. Passchier (1991), for example, classified the types of instantaneous deformation in terms of two parameters W and A, with $W = \cos \alpha$ and $A = (\varepsilon_1 + \varepsilon_2)/(\varepsilon_1 - \varepsilon_2) = \cos \beta$, where ε_i is the principal strain rate, α is the angle between the two directions of zero rotation rate, and β is the angle between the two directions of zero strain rate of lines on the xy plane. Passchier recognized three basic types of instantaneous deformation depending on whether $(W^2 + A^2)$ is less than, equal to, or greater than 1. The classification proposed here is closer to that of Fossen and Tikoff (1998), who combined simple shearing and coaxial straining to represent transpressional and transtensional deformations. They recognized five reference types of constant volume transpression and another five types of transtension. The classification proposed here is based on two parameters, a and b, with $a = \varepsilon_x/\gamma'$ and $b = \varepsilon_y/\gamma'$. It is assumed that the volume remains constant, i.e., $a + b + c = 0$, where $c = \varepsilon_z/\gamma'$. Hence $c = -(a + b)$. In comparison with the classification of Fossen and Tikoff (1998), in which the ratios of strain rates are fixed for the five reference types, the classification proposed here has a broader basis. The objective of this study is to classify and analyze the characteristic features of different deformation types; the classification is such that, for any pair of values of the strain-rate ratios a and b, the type of deformation and its degree of similarity with other deformation types are immediately established. This method of geometrical representation of the deformation as a point on the *ab* plane also enables us to represent other parameters of deformation as functions of a and b on the *ab* plane, so that their relation with the type of deformation can be visualized.

This method of classification enables us to group together deformations that are closely similar and to separate them from other groups.

RATE OF DISPLACEMENT EQUATION AND PARTICLE PATHS

Transpression and transtension

The types of deformation considered here are produced by simultaneous superposition of simple shear and a coaxial deformation with constant strain rates: ε_x, ε_y, and ε_z are the strain rates along the x, y, and z coordinate axes and $\gamma' = \gamma'_{xy}$ is the simple shear strain rate with reference to the x and y axes. The deformation is considered to be transpression when the strain rate along the y axis is negative, i.e., the strain is contraction. The deformation is transtension if the strain rate along the y axis is positive, i.e., the strain is extension (Fig. 1). For both types of deformation the strain rates along x and the z axes can be positive, negative, or zero.

Rate of displacement equation

Let us start with the rate of displacement equations,

$$dx/dt = \varepsilon_x x + \gamma' y, \quad (1)$$

$$dy/dt = \varepsilon_y y, \quad (2)$$

$$dz/dt = \varepsilon_z z. \quad (3)$$

The ratios ε_x/γ' and ε_y/γ' are kept constant during progressive deformation. Let

$$a = \varepsilon_x/\gamma', \quad (4)$$

$$b = \varepsilon_y/\gamma', \quad (5)$$

$$c = \varepsilon_z/\gamma' = -(a + b), \quad (6)$$

because, for constant-volume deformation, $\varepsilon_x + \varepsilon_y + \varepsilon_z = 0$.

The coefficient matrix of the system of differential equations 1–3 is

$$A = \begin{pmatrix} \varepsilon_x & \gamma' & 0 \\ 0 & \varepsilon_y & 0 \\ 0 & 0 & \varepsilon_z \end{pmatrix} \quad (7)$$

The characteristic polynomial equation of the coefficient matrix can be expressed as

$$(\varepsilon_x - k)(\varepsilon_y - k)(\varepsilon_z - k) = 0. \quad (8)$$

The three roots or eigenvalues of this equation are $k_1 = \varepsilon_x$, $k_2 = \varepsilon_y$, $k_3 = \varepsilon_z$.

The solution of the system of differential equations 1–3 will depend on whether the three roots k_1, k_2, and k_3 are unequal, or among the three, k_1 and k_2 are equal.

Unequal roots of the characteristic equation

If the three roots or eigenvalues k_1, k_2, and k_3 are distinct, the corresponding eigenvectors are

$$\begin{pmatrix} 1 \\ 0 \\ 0 \end{pmatrix}, \begin{pmatrix} \gamma'/(\varepsilon_y - \varepsilon_x) \\ 1 \\ 0 \end{pmatrix}, \begin{pmatrix} 0 \\ 0 \\ 1 \end{pmatrix}. \quad (9)$$

Then,

$$\begin{pmatrix} x \\ y \\ z \end{pmatrix} = c_1 \exp(\varepsilon_x t) \begin{pmatrix} 1 \\ 0 \\ 0 \end{pmatrix} + c_2 \exp(\varepsilon_y t) \begin{pmatrix} \dfrac{\gamma'}{\varepsilon_y - \varepsilon_x} \\ 1 \\ 0 \end{pmatrix}$$

$$+ c_3 \exp(\varepsilon_z t) \begin{pmatrix} 0 \\ 0 \\ 1 \end{pmatrix}. \quad (10)$$

The constants c_1, c_2, and c_3 are determined from the initial values. At $t = 0$, $x = x_o$, $y = y_o$, and $z = z_o$, so that

$$c_1 + c_2 \gamma'/(\varepsilon_y - \varepsilon_x) = x_o, \quad (11)$$

$$c_2 = y_o, \quad (12)$$

$$c_3 = z_o. \quad (13)$$

Therefore,

$$x = \exp(\varepsilon_x t) x_o + \{[\exp(\varepsilon_y t) - \exp(\varepsilon_x t)]/[(\varepsilon_y - \varepsilon_x)/\gamma']\} y_o, \quad (14)$$

$$y = \exp(\varepsilon_y t) y_o, \quad (15)$$

and

$$z = \exp(\varepsilon_z t) z_o. \quad (16)$$

Because

$$a = \varepsilon_x/\gamma', b = \varepsilon_y/\gamma', c = \varepsilon_z/\gamma', \quad (17)$$

$$\varepsilon_x t = a\gamma' t = a\gamma, \quad (18)$$

$$\varepsilon_y t = b\gamma' t = b\gamma, \quad (19)$$

and

$$\varepsilon_z t = c\gamma' t = c\gamma. \quad (20)$$

In constant volume deformation,

$$\varepsilon_z = -(\varepsilon_x + \varepsilon_y),$$
$$c = -(a + b), \quad (21)$$

$$x = e^{a\gamma} x_o + \{(e^{b\gamma} - e^{a\gamma})/(b - a)\} y_o, \quad (22)$$

$$y = e^{b\gamma} y_o, \quad (23)$$

and

$$z = e^{-(a+b)\gamma} z_o, \quad (24)$$

where $e^{a\gamma} = \exp(a\gamma)$ and $e^{b\gamma} = \exp(b\gamma)$.

The path of a particle represented by the point (x, y, z) can be determined by progressively increasing γ in the equation.

Roots k_1 and k_2 are equal

The case when the roots are equal represents the situation when, for instantaneous deformation, simple shear is superposed on either a pure constriction, with $\varepsilon_x = \varepsilon_y < 0$, i.e., $a = b < 0$, or pure flattening, with $\varepsilon_x = \varepsilon_y > 0$, i.e., $a = b > 0$.

If the roots k_1 and k_2 of the characteristic polynomial equation are equal, i.e., $\varepsilon_x = \varepsilon_y$, the solution of the differential equation is of the form

$$\begin{pmatrix} x \\ y \\ z \end{pmatrix} = c_1 \exp(\varepsilon_x t) \begin{pmatrix} 1 \\ 0 \\ 0 \end{pmatrix} + c_2 \exp(\varepsilon_y t) \begin{pmatrix} \gamma' t \\ 1 \\ 0 \end{pmatrix} + c_3 \exp(\varepsilon_z t) \begin{pmatrix} 0 \\ 0 \\ 1 \end{pmatrix} \quad (25)$$

At $t = 0$, $x = x_o, y = y_o$, and $z = z_o$, so that $c_1 = x_o$, $c_2 = y_o$, and $c_3 = z_o$.

Therefore,

$$x = \exp(\varepsilon_x t)x_o + \exp(\varepsilon_y t)\gamma y_o, \quad (26)$$

$$y = \exp(\varepsilon_y t)y_o, \quad (27)$$

and

$$z = \exp(\varepsilon_z t)z_o. \quad (28)$$

For $\varepsilon_x = \varepsilon_y$, or $a = b$, and for constant volume deformation, $\varepsilon_x + \varepsilon_y + \varepsilon_z = 0$, we have $\varepsilon_z t = -(a+b)\gamma' t = -2b\gamma$, so that

$$x = \exp(b\gamma)x_o + \gamma \exp(b\gamma)y_o, \quad (29)$$

$$y = \exp(b\gamma)y_o, \quad (30)$$

and

$$z = \exp(-2b\gamma)z_o. \quad (31)$$

Equations 21–24 and 29–31 follow from the two cases of unequal and equal roots of the characteristic equation 8. The special case of combined simple shear and either pure constriction or pure flattening cannot be obtained by putting $a = b$ in equations 21–24.

For both cases of equal and unequal roots, the deformation is transpression when $b < 0$; it is transtension when $b > 0$ (Fig. 1).

Apophyses of the particle paths

Equations 21–24 and 29–31 show how the position of a particle represented by a point (x, y, z) changes in the course of progressive deformation. Consequently, these equations also show how the orientation of a material line changes. For given values of a and b, successive stages of progressive deformation are obtained by increasing the value of γ. The three-dimensional forms of the particle paths can be represented in stereographic projection (Fossen and Tikoff, 1998). The two-dimensional forms of the particle paths on the xy plane are also very informative.

The physical significance of the eigenvectors is clearly brought out by the particle paths. Among the three eigenvectors of equation 9, one is parallel to the x axis, another is on the xy plane and is oblique to the x and y axes, and the third is parallel to the z axis. The eigenvectors give the directions of the apophyses of particle paths. Thus, for example, it is evident from equation 9 that the oblique eigenvector makes an angle of θ_1 with the x axis, with $\tan \theta_1 = y/x = 1/[\gamma'/(\varepsilon_y - \varepsilon_x)] = 1/[1/(b-a)\} = b - a$.

An apophysis of the particle path on the xy plane coincides with the directions along which the particle paths are straight and pass through the origin, and hence $(dx/dt)/(dy/dt) = x/y$ (Ramberg, 1975). From equations 1–3:

$$x/y = (\varepsilon_x x + \gamma' y)/\varepsilon_y y, \quad (32)$$

or,

$$x/y = \cot \theta_1 = \gamma'/(\varepsilon_y - \varepsilon_x), \quad (33)$$

and

$$\tan \theta_1 = b - a. \quad (34)$$

This orientation of the oblique apophysis is the same as that of the oblique eigenvector lying on the xy plane.

STRAIN ELLIPSE AND PRINCIPAL AXES ON xy PLANE

Strain ellipse for $a = b$

After deriving the expressions for x_o and y_o from equations 29 and 30 and substituting these expressions in the equation of the unit circle $x_o^2 + y_o^2 = 1$, it is found that for $a = b$, the equation of the strain ellipse on the xy plane is:

$$Ax^2 + By^2 + 2hxy = 1, \quad (35)$$

where

$$A = e^{-2b\gamma}, B = (\gamma^2 + 1)e^{-2b\gamma} \text{ and } h = -\gamma e^{-2b\gamma}. \quad (36)$$

The principal axes, R_1 and R_2, of the strain ellipse are:

$$R_1 = (e^{b\gamma})/[(s^2 + 1)^{1/2} - s], \quad (37)$$

$$R_2 = (e^{b\gamma})/[(s^2 + 1)^{1/2} + s], \quad (38)$$

where $s = \gamma/2$.

Because R_1 and R_2 are also principal axes of the strain ellipsoid, and for constant volume deformation $R_1 R_2 R_3 = 1$,

$$R_3 = 1/(R_1 R_2) = e^{-2b\gamma}. \quad (39)$$

The representation of the expressions for R_1, R_2, and R_3 in this way is for easy comparison with the corresponding equations of simple shear, which are obtained by putting $b = 0$, and in which case $R_1 R_2 = 1$.

$$R_1 = X = (s^2 + 1)^{1/2} + s, \quad (40)$$

$$R_2 = Z = (s^2 + 1)^{1/2} - s, \quad (41)$$

$$R_3 = Y = 1 \quad (42)$$

(Jaeger, 1964, p. 33; Ghosh, 1993, p. 148). R_1 and R_2 lie on the xy coordinate plane, i.e., on the plane perpendicular to the vorticity vector along the z axis (Fig. 2). Whether any one of the principal axes coincides with the X or Y or Z axis of the strain ellipsoid depends on the values of a and b and on the stage of progressive deformation (i.e., on γ).

Strain ellipse for $a \neq b$

In a similar manner for $a \neq b$, the equation of the strain ellipse on the xy plane can be obtained from equations 21–24:

$$Ax^2 + By^2 + 2hxy = 1, \quad (43)$$

where

$$A = e^{-2a\gamma}, \quad (44)$$

$$B = [(e^{a\gamma} - e^{b\gamma})^2 + (b-a)^2 e^{2a\gamma}]/[(b-a)^2 e^{2\gamma(a+b)}], \quad (45)$$

$$h = (b-a)^{-1}[e^{-a\gamma}e^{-b\gamma} - e^{-2a\gamma}]. \quad (46)$$

With invariants

$$I_1 = A + B, \quad (47)$$

$$I_2 = AB - h^2, \quad (48)$$

and

$$s_i (i = 1, 2) = \tfrac{1}{2}[I_1 \pm (I_1^2 - 4I_2)^{1/2}], \quad (49)$$

we have

$$R_1 = 1/(s_2)^{1/2}, \quad (50)$$

$$R_2 = 1/(s_1)^{1/2}. \quad (51)$$

R_1 and R_2 are also the two principal axes of the strain ellipsoid, and because for constant-volume deformation $R_1 R_2 R_3 = 1$, the other principal axis R_3 parallel to the z axis is found from the relation

$$R_3 = 1/(R_1 R_2) = (s_1 s_2)^{1/2}. \quad (52)$$

STRAIN RATES IN DIFFERENT DIRECTIONS IN THE xy PLANE

Strain rates in different directions

A large amount of information regarding the nature of deformation can be obtained from the equation of strain rates (see Jaeger, 1964, p. 40, for the corresponding equation for infinitesimal strain) along lines on the xy plane:

$$\varepsilon = \varepsilon_x \cos^2\theta + \gamma' \sin\theta \cos\theta + \varepsilon_y \sin^2\theta \quad (53)$$

or, dividing by γ',

$$\varepsilon/\gamma' = a \cos^2\theta + \sin\theta \cos\theta + b \sin^2\theta, \quad (54)$$

where a and b are given by equations 4–6 and where θ is the angle between the x axis and a line lying on the xy plane. These simple equations can be the basis of separating the different types of transpression and transtension. In this chapter I do not consider the case for which the roots of the characteristic equation 8 are complex conjugates, a situation in which we may obtain the pulsating strain ellipsoids (Ramberg, 1975).

Directions in which the strain rate vanishes

To obtain the directions in which the strain rate is zero, let us put $\varepsilon = 0$ in equation 54:

$$a \cos^2\theta + \sin\theta \cos\theta + b \sin^2\theta = 0 \quad (55)$$

Dividing by $\cos^2\theta$, we find $b \tan^2\theta + \tan\theta + a = 0$. Hence

$$\tan\theta = (2b)^{-1}[-1 \pm (1 - 4ab)^{1/2}]. \quad (56)$$

Equation 56 gives two directions in which $\varepsilon/\gamma' = 0$.

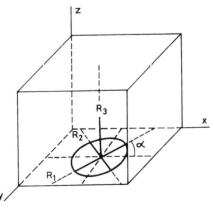

Figure 2. Orientation of principal axes of strain ellipsoid, R_1, R_2, and R_3, with respect to x, y, z coordinate axes. Strain ellipse on horizontal xy plane has principal axes R_1 and R_2, with R_1 at angle α with x axis.

Condition for which the strain rate does not vanish in any direction

Equations 53 and 56 imply that when $\varepsilon = 0$, the infinitesimal strain ellipse on the xy plane intersects the unit circle. Equation 56 then gives two directions along which the unit circle and the strain ellipse intersect. However, there is a situation in which the strain ellipse lies entirely within or entirely outside the unit circle. The strain rate can vanish only if $(1 - 4ab) > 0$ in equation 56. Thus, the condition

$$ab = 1/4 \tag{57}$$

separates fields for $\varepsilon/\gamma' = 0$ and $\varepsilon/\gamma' \neq 0$. For the present purpose, it is important to note that equation 57, in the form of ab = a constant, is the equation of a pair of rectangular hyperbolas (Fig. 3) with a and b as the coordinate axes. One of these hyperbolas, H_1, lies in the field of transpression and the other, H_2, in the field of transtension. If a and b are both negative, and $ab > 1/4$, the instantaneous strain ellipse on the xy plane will lie entirely within the unit circle. In transtensional deformation, if a and b are both positive, and $ab > 1/4$, the instantaneous strain ellipse will lie entirely outside the unit circle. For the type of deformation in which $ab > 1/4$, the strain rate does not become zero in any direction. When a and b are both negative, the hyperbola of equation 57 separates on the ab plane two types of transpressive deformation depending on whether the strain rate vanishes in some directions or remains negative for all directions. Similarly, when a and b are both positive, the hyperbola separates on the ab plane two types of transtensional deformation depending on whether the strain rate vanishes in some directions or remains positive for all directions.

SEPARATION OF DEFORMATION TYPES IN WHICH $R_1 > R_3$ AND $R_1 < R_3$ IN INSTANTANEOUS DEFORMATION

For the type of deformation considered here, the z coordinate axis parallel to the vorticity vector always remains parallel to a principal axis (R_3). The other two principal axes, R_1 and R_2, lie on the xy plane (Fig. 2). For instantaneous deformation the directions of R_1 and R_2 on the xy plane are given by the two values of θ of the following equation:

$$\tan 2\theta = \gamma'/(\varepsilon_x - \varepsilon_y). \tag{58}$$

These directions are obtained from equations 53 and 54 by putting $d\varepsilon/d\theta = 0$. Dividing the numerator and the denominator in the right side of equation 58 by γ', we obtain:

$$\tan 2\theta = (a - b)^{-1}. \tag{59}$$

After determining the expressions for $\sin^2\theta$, $\cos^2\theta$, and $\sin 2\theta$ from equation 59 and substituting these expressions in equation 54, we obtain, after simplification:

$$\varepsilon_{(R_1)}/\gamma' = \tfrac{1}{2}\{[(a - b)^2 + 1]^{1/2} + (a + b)\}. \tag{60}$$

Because $\varepsilon_z/\gamma' = -(a + b)$,

$$\varepsilon_{(R_1)}/\varepsilon_{(R_3)} = -\tfrac{1}{2}(\{[(a - b)^2 + 1]^{1/2}/(a + b)\} + 1). \tag{61}$$

$\varepsilon_{(R_1)}$ and $\varepsilon_{(R_3)}$ are the strain rates along the instantaneous R_1 and R_3 axes.

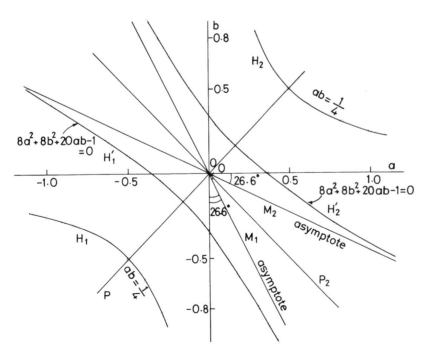

Figure 3. Representation of hyperbolas H_1 and H_2 and H'_1 and H'_2, their M_1 and M_2 asymptotes, and principal axes on ab plane.

The strain rates along the R_1 and the R_3 directions are equal when

$$-\tfrac{1}{2}(\{[(a-b)^2 + 1]^{1/2}/(a+b)\} + 1) = 1, \quad (62)$$

or

$$8a^2 + 8b^2 + 20ab - 1 = 0. \quad (63)$$

The equation is of the form

$$Ax^2 + By^2 + 2hxy + C = 0, \quad (64)$$

of a central conic section and, because $h^2 > AB$, it is the equation of a pair of hyperbolas (see Fig. 3 for hyperbolas H'_1 and H'_2). The equation of the hyperbolas becomes much simpler if we rotate the coordinate axes through 45° to coincide with the principal axes of the hyperbolas. In the new coordinate axes a', b', equation 63 becomes

$$9a' - b' = 1. \quad (65)$$

The asymptotes of the hyperbolas make an angle of

$$\tan \theta' = \pm 3 \quad (66)$$

with respect to the a' axis. With respect to the original axes, the asymptotes (M_2 and M_1 in Fig. 3) make angles of

$$\tan\theta = b/a = -1/2 \text{ and } -2, \quad (67)$$

$$\theta = -26.6°, \quad (68)$$

and

$$\theta = -63.4°. \quad (69)$$

On the ab plane the hyperbolas represented by equation 63 separate types of deformation in which instantaneous $R_1 > R_3$ and $R_1 < R_3$.

TYPES OF TRANSPRESSION AND TRANSTENSION

As shown here, the nature of instantaneous deformation depends only on the two parameters a and b, with $a = \varepsilon_x/\gamma'$ and $b = \varepsilon_y/\gamma'$, because, in constant volume deformation, $c = -(a+b)$. The ab plane is traversed by the pair of hyperbolas, H_1 and H_2, as given by equation 57, their asymptotes, the a and the b axes, the 45° line P_1 for $\varepsilon_x = \varepsilon_y$, or $a = b$, i.e., a pure constriction or a pure flattening for the coaxial component of deformation; the hyperbolas H'_1 and H'_2 as given by equation 63, their asymptotes M_1 and M_2 (see equations 67–69), and their principal axes.

As shown by Figure 4, these curves and lines divide the ab plane into 18 areas, 9 for transpression below the a axis and 9 for transtension above it. Each of these areas contain points (a, b), showing a similar type of deformation. For transpression, the areas and the corresponding types of deformation are designated by I, II, ... IX, and for transtension the corresponding areas are designated by TI, TII ... TIX (Fig. 4). For types I, II, and III, and for types TI, TII, and TIII there are subtypes (p) and (q) depending on whether a point (a, b) lies below or above the 45° line of pure constriction and pure flattening (Figs. 3 and 4). In addition, there are special types of transpressional and transtensional deformations for points lying on the following lines and curves: (A) the 45° line P_1 for $\varepsilon_x = \varepsilon_y$ or $a = b$, i.e., for a pure constriction or pure flattening as the coaxial component; (B) the line $a = 0$; (C) the asymptote M_1 of the hyperbola of equation 63; (D) the −45° line P_2, which represents the condition $a + b = 0$, i.e., plane strain; (E) the asymptote M_2 of the hyperbola of equation 63; (F) the hyperbolas H_1 and H_2; and (G) the hyperbolas H'_1 and H'_2.

Deformation of each of these special types A–G also has distinctive characteristics. The point $a = 0$, $b = 0$, lying on the line of plane strain P_2, represents simple shear. The special types B–G delineate regions of different deformation types. The line P_1 representing special type A separates the domains of subtypes (p) and (q).

The first five of the special types coincide with the deformation types A–E of Fossen and Tikoff (1998). The 45° P_1 line for $a = b$, or $\varepsilon_x = \varepsilon_y$ represents the type A of Fossen and Tikoff (1998). The line $a = 0$ represents their type B, and the −45° line P_2 for plane strain represents type D. The two asymptotes of the inner hyperbolas H'_1 and H'_2 on the ab plane in Figure 4 represent types C and E of Fossen and Tikoff (1998). As shown by

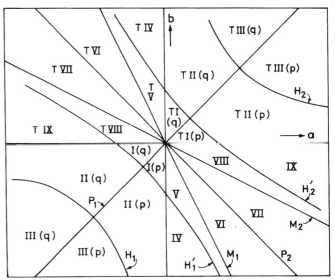

Figure 4. Domains of transpressional deformation types I to IX and of transtensional deformational types TI to TIX on ab plane. For I, II, and III as well as for TI, TII, and TIII, there are subtypes (p and q).

equations 67–69, the asymptotes make angles of $\tan^{-1}(-1/2) = -26.6°$ and $\tan^{-1}(-2) = -63.4°$. The first of these angles represents the condition $b/a = -1/2$ (see also equation 67), or $a + 2b = 0$. Because for constant volume deformation, $a + b + c = 0$, we have $b - c = 0$, or $\varepsilon_y = \varepsilon_z$, the condition of type E of Fossen and Tikoff (1998). The other asymptote with $b/a = -2$ (see also equation 67) represents the case $2a + b = 0$, i.e., $2\varepsilon_x + \varepsilon_y = 0$. Because $\varepsilon_x + \varepsilon_y + \varepsilon_z = 0$, we have $\varepsilon_x = \varepsilon_z$, the condition of type C deformation of Fossen and Tikoff (1998).

The characteristic changes in the relative lengths of R_1, R_2, and R_3 are shown in Figure 5 for the types I to IX of transpressional deformation. Figure 6 shows similar changes for types TI to TIX of transtensional deformation. Some examples of continuous variation of the lengths of these axes are shown in Figures 7 to 10. The changes in the nature of deformation for these types can also be seen by the variation of k′ (Flinn, 1965) with progressive deformation (Figs. 11 and 12). Recall that k′ = [ln(X/Y)/ln(Y/Z)]: k′ = infinity in pure constriction, and

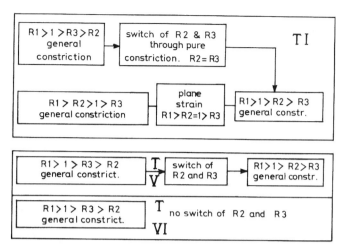

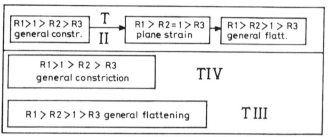

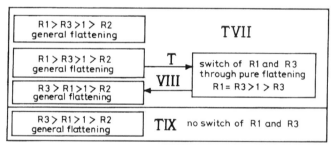

Figure 6. Change in relative lengths of R_1, R_2, and R_3 with progressive deformation in different types of transtension.

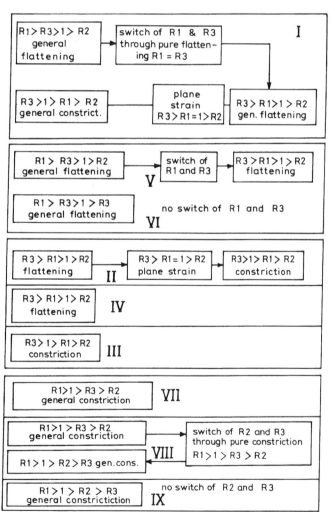

Figure 5. Change in relative lengths of R_1, R_2, and R_3 with progressive deformation in different types of transpression.

lies between 1 and infinity in general constriction; k′ = 1 in plane strain, and lies between 0 and 1 in general flattening; and k′ = 0 in pure flattening. For the special types A–G, the nature of variation of R_1, R_2, and R_3 for instantaneous and progressive deformation in transpression is shown in the following.

Type A depends on which type of deformation I, II, or III is represented by this line. Thus, for the segment of P_1 lying within the region of I, the sequence of change is $(R_1 > R_3 > 1 > R_2)$, $(R_1 = R_3 > 1 > R_2)$, and $(R_3 > R_1 > 1 > R_2)$.

For type B, and for $a = 0$, $(R_3 > R_1 > 1 > R_2)$ all throughout.

For type C, the sequence is $(R_3 > 1 > R_1 = R_2)$, $(R_3 = R_1 > 1 > R_2)$, and $(R_1 > R_3 > 1 > R_2)$.

For type D, plane strain is $(R_1 > R_3 = 1 > R_2)$.

For type E, and for the asymptote M_2, $(R_1 = R_2 > 1 > R_3)$, $(R_1 > R_2 = R_3)$, and $(R_1 > 1 > R_3 > R_2)$.

For type F, and for hyperbola H_1, $ab = 1/4$, the sequence is $(R_3 > 1 > R_1 = R_2)$, and $(R_3 > 1 > R_1 > R_2)$.

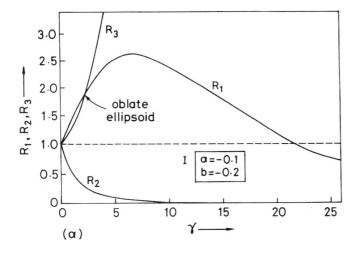
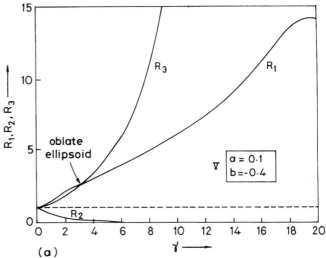
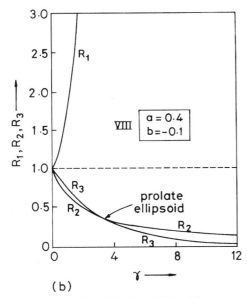
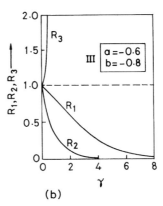

Figure 8. Changes in lengths of R_1, R_2, and R_3 with progressive deformation (γ) for types V and III.

Figure 7. Changes in lengths of R_1, R_2, and R_3 with progressive deformation (γ) for types I and VIII.

For type G, and for hyperbola H′ (separating the domains IV and V), the sequence is ($R_3 > 1 > R_1 = R_2$), and ($R_3 > R_1 > 1 > R_2$).

RATES OF ROTATION OF LINES

The change in angle of a line on the xy plane in simple shear is given by the equation $\cot\theta = \cot\theta_o + \gamma$, where θ_o and θ are the initial and final angles of a line with the x axis (Ramsay, 1967). Differentiating with respect to time, we have $-\theta'\csc^2\theta = \gamma'$, or

$$\theta'_\gamma = -\gamma' \sin^2\theta. \tag{70}$$

For the coaxial part we have

$$x = x_o \exp(\varepsilon_x t), \tag{71}$$

$$y = y_o \exp(\varepsilon_y t), \tag{72}$$

or

$$\cot\theta = \cot\theta_o \exp(\varepsilon_x t - \varepsilon_y t). \tag{73}$$

Differentiating with respect to t and eliminating $\cot\theta_o$ with the help of equation 73, we have, after simplification

$$\theta'_\varepsilon = (\varepsilon_y - \varepsilon_x) \sin\theta \cos\theta. \tag{74}$$

The rate of rotation of a line by simultaneous simple shearing and coaxial straining is obtained by adding θ'_ε and θ'_γ.

$$\theta' = (\varepsilon_y - \varepsilon_x) \sin\theta \cos\theta - \gamma' \sin^2\theta, \tag{75}$$

or

$$\theta'/\gamma' = (b - a) \sin\theta \cos\theta - \sin^2\theta. \tag{76}$$

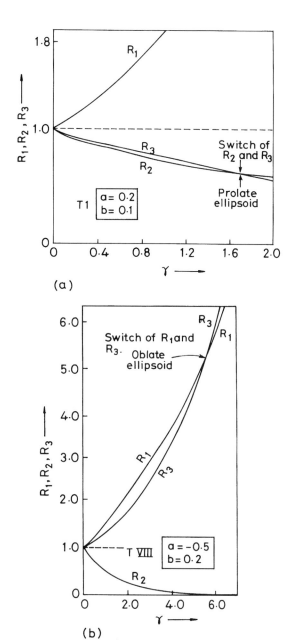

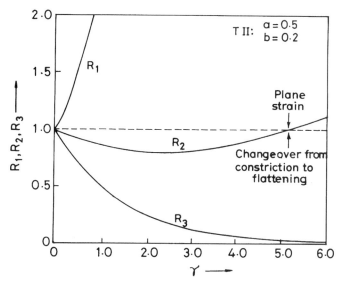

Figure 10. Changes in lengths of R_1, R_2, and R_3 with progressive deformation (γ) for type TII.

Figure 9. Changes in lengths of R_1, R_2, and R_3 with progressive deformation (γ) for types TI and TVIII.

If $\theta'/\gamma' = 0$, then

$$\tan\theta = b - a. \tag{77}$$

This is the same direction as given by equations 32–34 for the apophysis of the particle path.

The absolute value of θ'/γ' is a maximum when

$$\tan 2\theta = b - a. \tag{78}$$

For clockwise or forward rotation of the lines θ'/γ' is negative; for counterclockwise rotation it is positive. Because the dextral simple shear γ has been considered to be positive, a clockwise rotation is a forward rotation and a counterclockwise rotation is a backward rotation. Figure 13 shows the variation of θ'/γ' with θ for different values of $(a - b)$. The figure shows that each of the curves passes through the point $-\theta'/\gamma' = 1$ and $\theta = 90°$. For $a - b = 0$, i.e., when the coaxial part of the deformation is pure constriction or pure flattening, the curve does not cross the line $\theta'/\gamma' = 0$. In other words, apart from the direction of $\theta = 0$, there is no other direction in which the rate of rotation of a line vanishes. When $b \neq a$, the curves cross the line of zero rotation. Among these, there are two types of curves. When $(a - b)$ is negative, the curve intersects the zero rotation line for a positive acute angle θ. When $(a - b)$ is positive, the corresponding point gives a negative acute angle. Thus, when $(a - b) < 0$, there is a range of positive acute angle θ in which a line rotates backward. If $(a - b) > 0$, there is a range of negative acute angle in which the line rotates backward. In the ab plane all points lying below the 45° line (Fig. 4) have $(a - b) > 0$, and all points lying above the line have $(a - b) < 0$.

PARTICLE PATHS AND CRITICAL ORIENTATIONS ON THE xy PLANE

The nature of the particle paths (Figs. 14–19) depends on the orientation of the apophyses and the directions of the vanishing strain rates. Among the three apophyses of the particle paths, two are parallel to the x and z axes. The third, making an angle θ_1 with the x axis, has the orientation $\tan\theta_1 = b - a$.

There are also, in general, two directions in the ab plane in which $\varepsilon = 0$. As obtained from equation 56, these orientations are

$$\tan\theta_2 = (1/2b)[-1 + (1 - 4ab)^{1/2}] \tag{79}$$

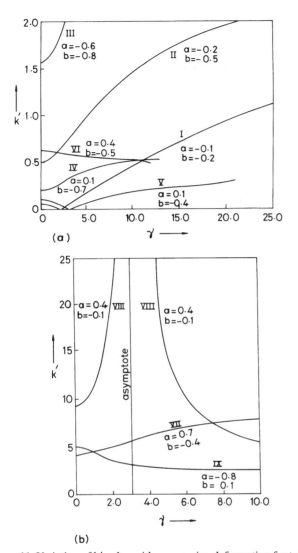

Figure 11. Variation of k' value with progressive deformation for types (A) I to VI and (B) VII to IX.

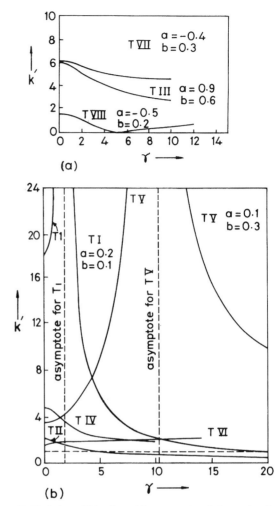

Figure 12. Variation of k' value with progressive deformation for types (A) TIII, TVII, and TVIII, and (B) TI, TII, TIV, TV, and TVI. Note that TI and TV curves rise from either side toward corresponding asymptotes. Asymptotes represent infinite k'.

and

$$\tan\theta_3 = (1/2b)[-1 - (1 - 4ab)^{1/2}]. \qquad (80)$$

The particle paths are orthogonal to θ_2 and θ_3. Along θ_2 the particle paths bend back, the tangent to the particle path being at a right angle to θ_2. Both θ_1 and θ_2 are orientations that separate fields of instantaneous contraction and instantaneous extension of material lines (Fig. 17). Figure 19 summarizes the relevant features of different types of particle paths on the xy plane, and shows that in transpression, the oblique apophysis (making an angle of θ_1 with the x axis) is always a direction of contraction. In transtension it is always a direction of stretching. In both transpression and transtension, if $(a - b)$ is positive, material lines move away from the neighborhood of the oblique apophysis toward the x axis, the direction of simple shear (Figs. 14, A and C, 15A, and 16A). Lines on one side of the oblique apophysis rotate forward and lines on the other side rotate backward, and both sets of lines rotate toward the x axis. The long axis of the strain ellipse on the xy plane also rotate towards the x axis. If $(a - b)$ is negative (Fig. 16, C and D), the lines move toward the oblique apophysis (material line attractor of Passchier, 1997) and away from the x axis (material line repulsor of Passchier, 1997). In this case, the long axis of the strain ellipse does not rotate toward the x axis but rotates toward the oblique apophysis. With very large deformation it becomes nearly parallel to the oblique apophysis. In both transpression and transtension, the apophysis along the z axis is a direction of contraction if $(a + b) > 0$; it is a direction of extension if $(a + b) < 0$.

Particle paths can also be constructed for vertical planes that do not rotate with progressive deformation. There are two such planes; one contains the vertical apophysis along z and the horizontal apophysis along x. The other plane is defined by the vertical z axis and the horizontal oblique apophysis. Figure 15 shows

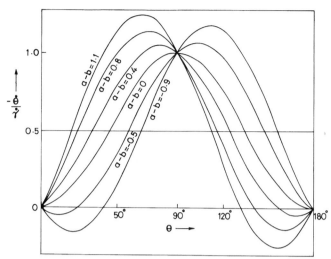

Figure 13. Rates of rotation of lines at different orientations with respect to x axis. Note that negative or backward rotation rates occur at acute angles when $(a-b)$ is negative and at obtuse angles when $(a-b)$ is positive. Points of inflection of each curve lie on line $-\theta'/\gamma' = 0.5$. All curves pass through single point, $-\theta'/\gamma' = 1$, $\theta = 90°$. For curve $(a-b) = 0$, rotation rate vanishes only when $\theta = 0$.

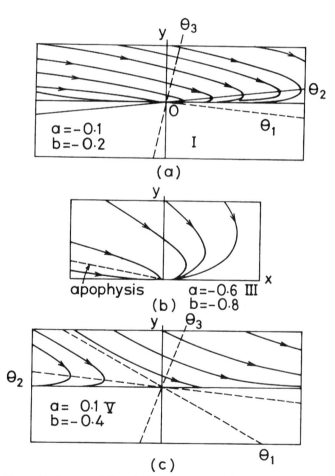

Figure 14. Some particle paths for types I, III, and V. Only paths above x axis are shown.

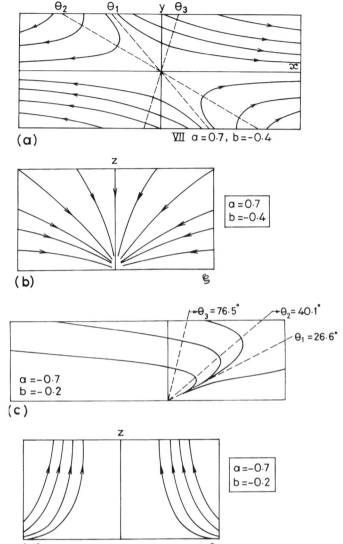

Figure 15. Particle paths on horizontal xy plane and on vertical plane containing z axis and oblique apophysis (ζ). Both these planes are fixed in space and do not rotate with progressive deformation. A: Particle paths and orientations of oblique apophysis θ_1 on xy plane for $a = 0.7$, $b = -0.4$. B: Particle paths on vertical $z\zeta$ plane for same values of a and b as in A. C: Particle paths and orientations of directions of oblique apophysis (θ_1) and of directions of zero strain rates (θ_2 and θ_3) on xy plane for $a = 0.7$, $b = -0.2$. D: Particle paths on vertical $z\zeta$ plane for same values of a and b as in (C).

two examples of such particle paths. The coordinate axis parallel to the horizontal oblique apophysis has been designated ζ.

If $(1 - 4ab) < 0$, the strain rate does not vanish in any direction, and hence the critical orientations of θ_2 and θ_3 do not exist (Fig. 14B). If $ab = 1/4$, there is a single direction in which the strain rate vanishes. Although the particle path is perpendicular to this direction (Fig. 18A), there is no change in the sign of ε while the path crosses this orientation. The strain rate in this

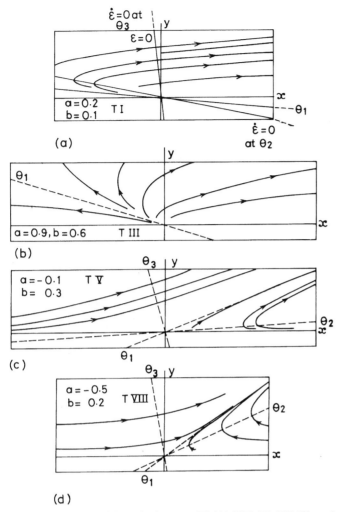

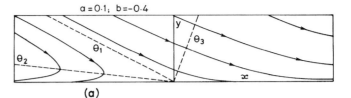

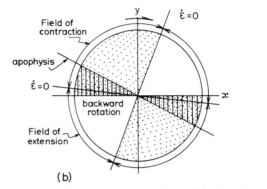

Figure 16. Some particle paths for types TI (A), TIII (B), TV (C), and TVIII (D).

Figure 17. Fields of contraction, extension, and backward rotation as indicated by particle paths.

direction is zero because the instantaneous R_1, parallel to this direction, has a value of 1. The strain rate shows a maximum at $\varepsilon = 0$ (Fig. 18B). The instantaneous strain ellipse lies entirely within the unit circle, but the long axis of the ellipse touches the circle along θ_3. Both forward-rotating and backward-rotating paths move toward the x axis and tend to become parallel to it.

If $a = 0$, the equation reduces to $\tan\theta \, (b \tan \theta + 1) = 0$, which gives $\theta_2 = 0$ and $\tan\theta_3 = -(1/b)$. The equation of the apophysis becomes $\tan\theta = b$. Hence, the direction of θ_3 is perpendicular to the apophysis. The particle paths appear as straight lines parallel to the apophyses that end on the x axis. With further deformation the point does not move from this position of zero strain rate (Fig. 18C).

If θ_1 and θ_2 are both negative but θ_3 is positive (Fig. 14C), the particle path moves from the region of instantaneous shortening to the region of instantaneous extension while crossing θ_2. All particle paths, both for forward-rotating and backward-rotating lines, move toward the x axis.

KINEMATICAL VORTICITY NUMBER AS A FUNCTION OF a AND b

How is the kinematical vorticity number (W_k) related to the different types of deformation? To enquire into this problem let us express W_k as a function of a and b.

For the type of deformation considered here the kinematic vorticity number is

$$W_k = \gamma'/[2(\varepsilon_x^2 + \varepsilon_y^2 + \varepsilon_z^2 + \gamma'^2/2)]^{1/2} \quad (81)$$

(cf. equation 16 of Tikoff and Fossen, 1995), or, because for constant volume deformation $\varepsilon_z = -(\varepsilon_x + \varepsilon_y)$,

$$W_k = \gamma'/(4\varepsilon_x^2 + 4\varepsilon_y^2 + 4\varepsilon_x\varepsilon_y + \gamma'^2)^{1/2}. \quad (82)$$

With $a = \varepsilon_x/\gamma'$ and $b = \varepsilon_y/\gamma'$, we have

$$W_k = 1/(4a^2 + 4b^2 + 4ab + 1)^{1/2}, \quad (83)$$

or

$$4a^2 + 4b^2 + 4ab + [1 - (1/W_k^2)] = 0 \quad (84)$$

With a and b as coordinate axes, this is the equation of a conic section of the general form

$$Ax^2 + By^2 + 2hxy + C = 0. \quad (85)$$

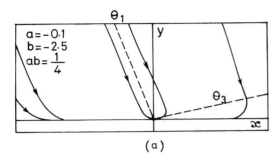

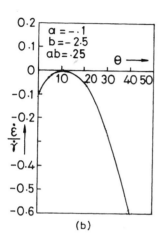

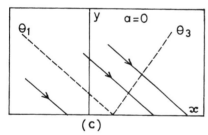

Figure 18. A: Particle paths on xy plane for $ab = 1/4$. In this special case, $\theta_2 = \theta_3$ as given by equations 88–90. Note that although particle paths are curved and direction of θ_3 is orthogonal to particle path, latter does not bend back. Consequently direction of θ_3 (or θ_2) does not separate, as in other cases, fields of instantaneous extension and contraction. All marker lines are shortened progressively as they rotate toward x axis. B: For same values of a and b as herein figure shows changes of strain rate ε/γ' with angle θ between marker line and x axis. This special case of deformation is such that instantaneous strain ellipse lies entirely within unit circle, and major axis touches unit circle. Hence in this diagram strain rates along all directions are negative; only strain rate along θ_3, direction of instantaneous R_1 axis, is zero. C: Straight line particle paths for special case, $a = 0$.

Because $AB - h^2 > 0$, equation 84 represents the equation of a family of ellipses on the ab plane with different values of W_k. The principal axes of the ellipses are oriented at $\pm 45°$ with the a axis, the semimajor and semiminor axes being

$$[(1 - W_k^2)/2W_k^2]^{1/2} \tag{86}$$

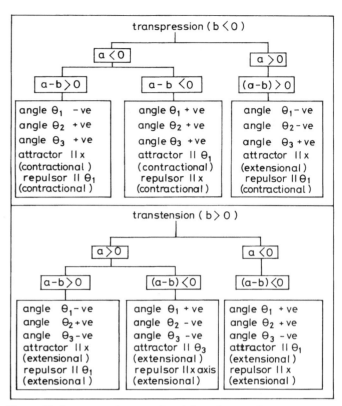

Figure 19. Summary of nature of particle paths and of critical orientations θ_1, θ_2, and θ_3 on xy plane in transpression and transtension. Figure also shows attracting or repulsing nature of oblique apophysis (θ_1) and of apophysis parallel to x axis. If a and b are given, attracting or repulsing nature of vertical axis parallel to z can be easily determined. Vertical apophysis is contractional if $(a + b) > 0$; it is extensional if $(a + b) < 0$.

and

$$[(1 - W_k^2)/6W_k^2]^{1/2}. \tag{87}$$

Figure 20 shows the ellipses for different values of W_k. The figure shows that each contour of equal W_k traverses domains of different deformation types. With increase in W_k the ellipses become smaller. At the origin $a = b = 0$, representing simple shear, $W_k = 1$.

HYPERBOLIC CONTOURS OF $\theta_3 - \theta_2$ ON THE ab PLANE

In order to find the relation of types of deformation to the angular difference between the two directions of vanishing strain rate, we find from equation 56:

$$\theta_3 = \tan^{-1}\{(1/2b)[-1 - (1 - 4ab)^{1/2}]\} \tag{88}$$

$$\theta_2 = \tan^{-1}\{(1/2b)[-1 + (1 - 4ab)^{1/2}]\} \tag{89}$$

$$\tan(\theta_3 - \theta_2) = [-(1 - 4ab)^{1/2} \cdot (a + b)^{-1}]. \tag{90}$$

If $m = \tan(\theta_3 - \theta_2)$, we find, after simplification

$$m^2 a^2 + m^2 b^2 + 2(m^2 + 2)ab - 1 = 0. \quad (91)$$

This is the equation of a family of hyperbolas on the ab plane with principal axes at $\pm 45°$ with the a axis. Figure 21 shows a set of hyperbolas for different values of $(\theta_3 - \theta_2)$. Each hyperbola traverses through several domains of deformation types. Each pair of hyperbola intersects a W_k ellipse at four points. Thus, a unique type of deformation is not obtained from a pair of values of W_k and $(\theta_3 - \theta_2)$.

From equation 90 we find:

$$\begin{aligned} 1/\cos^2(\theta_3 - \theta_2) &= 1 + \tan^2(\theta_3 - \theta_2) \\ &= 1 + [(1 - 4ab)/(a + b)^2] \\ &= [(a - b)^2 + 1]/(a + b)^2 \end{aligned} \quad (92)$$

or

$$\cos(\theta_3 - \theta_2) = \pm(a + b)/[(a - b)^2 + 1]^{1/2}. \quad (93)$$

Equation 61 shows that the ratio of stretching rates along the instantaneous stretching axes R_1 and R_3 is

$$\varepsilon_{R_1}/\varepsilon_{R_3} = -1/2(\{[(a - b)^2 + 1]^{1/2}/(a + b)\} + 1), \quad (94)$$

or

$$-[1 + (2\varepsilon_{R_1}/\varepsilon_{R_3})] = [(a - b)^2 + 1]^{1/2}/(a + b). \quad (95)$$

From equations 93 and 95, it is found that

$$-(1 + 2\varepsilon_{R_1}/\varepsilon_{R_3}) = \pm 1/\cos(\theta_3 - \theta_2), \quad (96)$$

or

$$\cos(\theta_3 - \theta_2) = \pm 1/[1 + (2\varepsilon_{R_1}/\varepsilon_{R_3})]. \quad (97)$$

If $\varepsilon_{R_1}/\varepsilon_{R_3} = 1$, then

$$\cos(\theta_3 - \theta_2) = \pm 1/3, \quad (98)$$

so that

$$1/\cos^2(\theta_3 - \theta_2) = 9 = \tan^2(\theta_3 - \theta_2) + 1, \quad (99)$$

and

$$m^2 = \tan^2(\theta_3 - \theta_2) = 8. \quad (100)$$

Substituting this value of m^2 in equation 91, we find

$$8a^2 + 8b^2 + 20ab - 1 = 0, \quad (101)$$

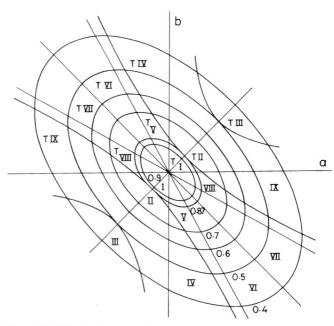

Figure 20. Elliptical contours of equal W_k in relation to domains of different deformation types. For $W_k = 1$ ellipse degenerates to point $a = 0$, $b = 0$, representing simple shear.

which is the same as equation 63 of the pair of hyperbolas H'_1 and H'_2. In other words, for the special case of $\varepsilon_{R_1}/\varepsilon_{R_3} = 1$, equation 91 becomes the same as equation 63. However, even for this special case, the value of $(\theta_3 - \theta_2)$ does not indicate transpression or transtension, whether there is a shortening or stretching along the x axis, or whether the deformation is dominated by simple shear or by coaxial deformation.

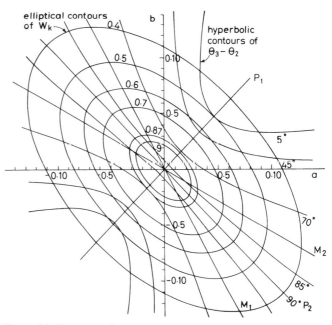

Figure 21. Representation of hyperbolic contours of $\theta_3 - \theta_2$ with respect to elliptical contours of W_k.

SUMMARY OF NUMERICAL RESULTS

The classification of transpressional and transtensional deformation can be done with the help of different pairs of parameters. The classification presented here is based on the two parameters, a and b, with $a = \varepsilon_x/\gamma'$ and $b = \varepsilon_y/\gamma'$. Because it is assumed that the volume remains constant, the ratio of strain rates $c = \varepsilon_z/\gamma'$ is a function of a and b. The ab plane is traversed by two sets of hyperbolas. One of these separates domains in which the strain rate does not vanish in any direction from other domains. The other pair of hyperbolas is obtained from the condition that instantaneous R_1 and R_3 are equal. These two pairs of hyperbolas, their two pairs of asymptotes, and their two sets of principal axes divide the ab plane in nine regions of transpressional and nine regions of transtensional deformation types. This method of representation of the different types of transpression and transtension as discrete domains on the ab plane enables us to compare the degree of similarity of the deformation types. It also enables a geometrical representation on the ab plane of the relation between the kinematical vorticity number (as well as the angular difference between the two directions of zero strain rate) and the different domains of the deformation types. Given any pair of a and b, the type of deformation can be immediately visualized on the ab plane. Each of these types of deformation has a characteristic deformation history. Types I, II, and III are further subdivided into subtypes (p) and (q) depending on whether $(a - b)$ is positive or negative. In addition, there are seven special types of deformation, A, B, C, D, E, F, and G, represented by curves and lines that separate the domains of the nine main types and of the subtypes.

The nature of the particle path on the xy plane, showing the fields of forward and backward rotation of lines and the fields of instantaneous stretching and shortening, depends mostly on the orientations of the apophyses and the directions in which the strain rate becomes zero. The general nature of the particle paths and of the rotation history of material lines depends to a great extent on whether $(a - b)$ is positive or negative. In both transpression and transtension, material lines and the major axis of the strain ellipse on the ab plane rotate toward the direction of simple shear (x axis) if $(a - b)$ is positive. In contrast, if $(a - b)$ is negative, the material lines and the major axis of the strain ellipse rotate toward the oblique apophysis, i.e., the apophysis that is making an angle θ_1 with the x axis on the xy plane. In the latter case, even if the intensity of deformation is very large, a vertical cleavage never becomes essentially parallel to the vertical shear zone.

On the ab plane, contours of equal kinematical vorticity number, W_k, appear as a family of ellipses, the major axes being at an angle of $-45°$ to the a axis. Thus, the kinematical vorticity number alone does not specify a type of deformation. We need an additional piece of information, e.g., the ratio a/b, to fix a type of deformation. The angle β between the two directions of zero strain rates is also a function of a and b. On the ab plane, contours of equal β appear as hyperbolas. A contour of W_k intersects a contour of β at four points, two each in the domains of transpression and transtension. The two points in each of these domains represent two distinct types of deformation.

DISCUSSION

For the deformations in shear zones considered here, the two-fold division of transpression and transtension is based on whether there is shortening or extension across the shear zone. There is, however, another broad two-fold grouping that transects the transpression and transtension division. These two groups of deformation are separated by the criterion of whether $(a - b)$ is positive or negative (Fig. 22). If we consider only the coaxial part of the instantaneous deformation, the major axis of the strain ellipse on the horizontal xy plane is parallel to the shear zone trend if $(a - b)$ is positive; the major axis of the strain ellipse is perpendicular to the shear zone trend if $(a - b)$ is negative. For the component of *instantaneous simple shear alone*, the major axis of the strain ellipse on the xy plane makes an angle of $45°$ with the shear-zone trend. Hence, for combined simple shear and coaxial deformation, the major axis of the instantaneous strain ellipse (R_1) on the xy plane makes an angle of $<45°$ with the shear zone trend if $(a - b)$ is positive (Fig. 22, domains i, j, and k). The instantaneous R_1 makes an angle of $>45°$ if $(a - b)$ is negative (Fig. 22, domains l, m, and n).

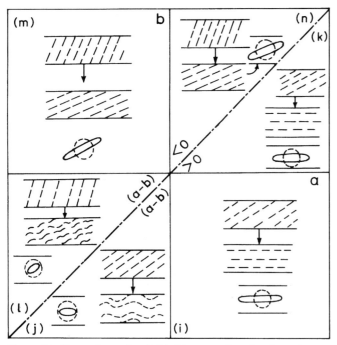

Figure 22. Plane ab may be divided into two domains depending on whether $(a - b)$ is positive (domain below dot-dash line) or negative (domain above dot-dash line). If $(a - b) > 0$, initial cleavage is at angle of $<45°$ with shear-zone strike. If deformation is very large it rotates toward shear-zone trend. If $(a - b)$ is negative initial cleavage may develop at angle of $>45°$ with shear-zone strike. With large deformation it rotates toward oblique apophysis and is stabilized in this position. Note that in domains j and l cleavage becomes folded.

The net effect of these differences is that there may be two types of shear zones depending on whether $(a - b)$ is positive or negative. In one type the vertical cleavage will always initiate at an angle of <45° with the shear zone and will tend to become parallel to the shear zone with large deformation. In the other type the vertical cleavage may initiate at an angle of >45° with the shear zone in both transpression and transtension; it will tend to rotate toward the oblique apophysis with progressive deformation, but will never be essentially parallel to the shear zone, even if the deformation is very large.

In addition to these two situations, a horizontal cleavage may develop in some shear zones when the Z axis of the strain ellipsoid becomes parallel to the vertical axis R_3 (Figs. 5 and 6).

In steady progressive plane strain (with nonpulsating strain ellipsoids of Ramberg, 1975), a line that has once been extended cannot undergo shortening at a later stage. A folded layer may be boudinaged at a later stage, but a boudinaged layer cannot undergo buckle folding in steady progressive deformation. In contrast, the history of transpressional and transtensional deformations may be more complex. The problem may be clarified by taking a specific example of transpressive deformation of type I (p) (Figs. 7A, 14A, and 23). Because the deformation belongs to the domain j of Figure 22, the instantaneous R_1 axis makes an angle of <45° with the x axis. With progressive deformation, the R_1 axis on the xy plane rotates toward the x axis. Figure 23, for example, shows strain ellipses in five stages of progressive deformation. In the initial stage, as Figure 7A shows, the rate of increase of R_1 is greater than that of R_3. With progressive deformation the rate of increase of R_3 becomes greater, and R_3 becomes equal to R_1, giving rise to an oblate strain ellipsoid. The rate of increase of R_1 continues to decrease. However, because R_1 is still in the field of instantaneous extension, it continues to increase (Fig. 23B). At $\gamma = 6.5$ for the case shown in Figures 7A and 23C, R_1 attains a maximum value and at the same time attains the orientation of zero strain rate (θ_2 in Fig. 14A). This is the position of the maximum of the R_1 curve in Figure 7A. With further deformation the orientation of R_1 enters into the field of instantaneous shortening and starts to decrease (Fig. 23D). Finally, with a still larger deformation, the length of R_1 becomes equal to the radius of the unit circle (the point where the R_1 curve in Fig. 7A crosses the $R_1 = 1$ line); after this stage R_1 becomes smaller than the radius of the unit circle (Fig. 23E). Cleavage developing parallel to the vertical $R_1 R_3$ plane will be crenulated when the R_1 axis rotates into the field of instantaneous shortening.

The strain history of a planar structure (such as a preexisting layer or vein) may be more complex. The particle paths (Fig. 14A) for the same type of deformation (i.e., $a = -0.1$, $b = -0.2$) show that if a planar structure is parallel to the z axis and its trace on the xy plane makes, e.g., an initial angle between 90° and 180° with the x axis, it will first undergo a shortening. With progressive deformation, the rate of instantaneous shortening will first increase, then decrease, and then become zero as it reaches the orientation of θ_3 (Fig. 14A). With progressive forward rotation it will undergo instantaneous extension until the line reaches the orientation of zero strain rate of θ_2. Beyond this position it will again undergo instantaneous shortening as it converges toward the direction of simple shearing (x axis). Consequently, the planar structure may show an initial stage of buckle folding, and a set of boudin lines orthogonal to the vertical fold axis, because the z axis (R_3) at this stage is a direction of extension (Fig. 5 for type I). This stage is followed by chocolate tablet boudinage, the boudin lines being parallel and perpendicular to the fold axis. There will be a late stage of buckle folding in the same continuous deformation.

In both transpression and transtension the complexity may be further increased because of the switching of the direction of maximum stretching. In transpression, for example, there is a switching (e.g., Sanderson and Marchini, 1984; Tikoff and Teyssier, 1994; Tikoff and Greene, 1997) of the direction of maximum stretching with progressive deformation in types I and V. For both types, the R_2 axis represents the Z axis of the finite strain ellipsoid throughout the course of deformation. For type I, the cleavage develops along the vertical $R_1 R_3$ plane parallel to the XY plane of the strain ellipsoid (Fig. 24, top). In the early stage of deformation both the X axis (R_1) and the Z axis (R_2) are horizontal, and the Y axis (R_3) is vertical and parallel to the vorticity vector. In the early stage of deformation the Y axis is a direction of stretching. Hence, cleavage-parallel veins and layers may undergo chocolate tablet boudinage (Ghosh, 1988). With the switching of the direction of maximum stretching ($R_3 = X$), the cleavage surface still remains in the field of general flattening. However, with continued deformation, the orientation of the R_1 axis (the current Y axis) crosses the θ_2 direction of zero strain rate and enters into the field of instantaneous shortening. With further deformation the horizontal R_1 axis becomes a direction of finite shortening (Fig. 24, top). It is

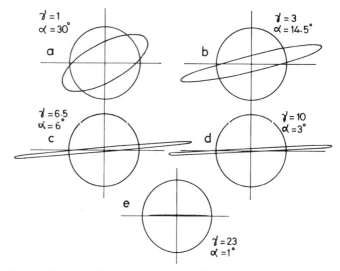

Figure 23. Strain ellipses on xy plane for five stages of deformation in type I transpression. At last stage ellipse (represented by bold line) is almost parallel to x axis.

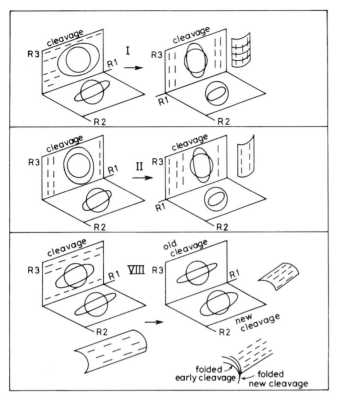

Figure 24. Schematic representation of some complex deformation histories. See text for details.

unlikely that a well-developed cleavage, especially in rocks like mica schists, would be obliterated when the Y axis is still greater than 1 but has become a direction of instantaneous shortening. It is more likely that the cleavage would be folded, the fold axis parallel to the vertical R_3 axis along the vorticity vector. These types of transpressive deformation provide a mechanism by which a cleavage can develop and be folded in the same continuous steady deformation.

The switching of the direction of maximum stretching in types I and V may be associated with a switching of the stretching lineation from the horizontal to the vertical direction in transpression. The occurrence of vertical stretching lineations in vertical shear zones was first interpreted by Hudleston et al. (1988) as a result of transpression. Similar vertical lineations in transpressional shear zones have been reported by others (e.g., Robin and Cruden, 1994; Greene and Schweickert, 1995; Fossen and Tikoff, 1993; Tikoff and Teyssier, 1994). The occurrence of both subhorizontal and subvertical stretching lineations in the Northern Sierra Nevada crest shear zone was explained by Tikoff and Greene (1997) as due to variation of strain along the strike.

What is the effect of switching of the maximum stretching axis on an already existing stretching lineation? Will the early (i.e., horizontal) lineation be completely obliterated when the cleavage becomes a plane of pure flattening ($R_1 = R_3$)? This may not happen in all cases. A stretching lineation marked by well-developed elongate flakes of mica may not be obliterated. Moreover, a stretching lineation may be marked by two different families of elongated minerals, the partial strains being different for the two families and different from the bulk strain (Fig. 1 of Means, 1994). After the switching of the bulk stretching axis, the lineation switch may occur in one family and not in the other. In the event that the early lineation is not entirely destroyed, a later lineation will be superimposed orthogonal to the earlier lineation. The horizontal early lineation and one set of boudin lines will be folded over the vertical folds and a late lineation will be produced parallel to the fold axis. In the late stage of types I and III, the deformation is a general constriction and layers subparallel to the horizontal xy plane may show nonplane noncylindrial folding (Ghosh et al., 1995). The switching of the maximum stretching direction cannot produce a gradual variation in the plunge of the stretching lineation along a strike-parallel strain gradient.

A late folding of the cleavage plane may occur even without a switch of the direction of maximum stretching. In type II, for example (Fig. 24, middle), both $X = (R_3)$ and $Y = (R_1)$ are directions of stretching and the cleavage plane may show a vertical stretching lineation in the initial stage, whereas cleavage-parallel layers or veins may show two orthogonal sets of boudin lines (see Ghosh, 1988). At a later stage, when the horizontal Y axis (R_1) becomes a direction of shortening, the cleavage plane may be folded, the lineation being parallel to the fold axis. In type III, however, the Y axis is a direction of shortening throughout the entire course of deformation. In this case, the development of the cleavage and its folding with a vertical axis should occur synchronously. In type VII, the X axis (R_1) is horizontal and the vertical Y (R_3) is a direction of shortening at all stages (Fig. 5). Hence, the development of a vertical cleavage and its recumbent folding with a horizontal axis may occur synchronously. In type VIII (Fig. 24, bottom), in which there is a switch of the Y and the Z axes, the deformation history can be quite complex. In the early stage, the X axis (R_1) is horizontal and there is a shortening along the vertical Y axis (R_3), so that development of a vertical cleavage and its folding with horizontal axis and axial plane should occur simultaneously. At a later stage, with switching of the Y and Z axes, the Z axis becomes vertical. A new cleavage will form parallel to the horizontal XY plane. Because the Y axis is a direction of shortening, this new cleavage should also be folded, with a horizontal axis and a vertical axial plane (Fig. 24, bottom). Similar complex histories of progressive deformation are also obtained in transtension.

In the model proposed by Sanderson and Marchini (1984), transpressional deformation gives rise to flattening strain and transtensional deformation produces constrictional strain. As shown in Figures 5 and 6 and as pointed out by several workers (e.g., Corsini et al., 1991; Dias and Ribeiro, 1994; Jones et al., 1997; Fossen and Tikoff, 1998), such a generalization cannot be made for all types of transpression and transtension considered here. If we consider only the instantaneous transpressional deformation, flattening is produced in only five (types I, II, IV, V, and VI) of the nine types of deformation. Even among these five, a general constrictional deformation may result in the late stages of progressive deformation in types I and II (Fig. 5). The instantaneous deformation is constriction in types III, VII, VIII, and IX. The deformation remains in the field of general constriction throughout the entire course of progressive deformation. In type VIII, the deformation passes through a stage of pure constriction with the switching of the Y and Z axis (producing a prolate ellipsoid); both the Y and the Z axes remain directions of shortening before and after the switch.

In transtension, the instantaneous deformation is a constriction in types TI, TII, TIV, TV, and TVI. Among these, TI and TII may give rise to a flattening type of deformation in the late stage of progressive deformation. In the other types, i.e., TIII, TVII, TVIII, and TIX, the instantaneous deformation causes a flattening, and the deformation remains in the flattening field in all stages of progressive deformation. These conclusions are consistent with the field observations of Dias and Ribeiro (1994). According to Dias and Ribeiro, constrictional deformation is associated with transpression in a large part of the Centro-Iberian autochthon.

For the model of vertical tabular zone considered here, the vorticity vector of simple shearing is vertical and the displacement vector of one zone-boundary block relative to the other lies in a fixed horizontal plane. However, the model is equally applicable to inclined shear zones with a true thrusting or normal sense of movement. For such inclined zones, however, the vorticity vector will be horizontal and the displacement vector of the zone-boundary block will lie in a vertical plane that is perpendicular to the vorticity vector. It is likely that certain types of transpressional or transtensional deformations are more likely to be found in inclined shear zones of thrust or normal type than in vertical shear zones with strike-slip movement. Where the vorticity vector is not perpendicular to the plane containing the zone-boundary displacement vector, the displacement will be oblique, and the orientation of the vorticity vector may change during progressive deformation (Jones and Holdsworth, 1998).

The model presented here is idealized, especially in its assumption of homogeneous strain and constant ratios of strain rates (ε_x/γ' and ε_y/γ'). The analysis may be modified to include heterogeneous strain (resulting from no free slip at the boundary walls) by taking a series of boundary-parallel deforming elements, as suggested by Fossen and Tikoff (1998, their Fig. 9). In most cases the strain rates or the ratios a and b are also likely to change in time. The exact nature of change will depend upon specific boundary conditions, especially on how one zone boundary approaches the other. For certain simple boundary conditions that clearly prescribe how the strain rates or their ratios vary in time, the finite deformation can be modeled by a succession of incremental deformations. In general, however, it is difficult to determine the history of progressive deformation without a knowledge of how the strain rates vary in time. For certain boundary conditions it may be possible to suggest the deformation type (or a group of closely related deformation types on the *ab* plane in Fig. 4) that is most likely to occur (cf. Fossen and Tikoff, 1998).

ACKNOWLEDGMENTS

I thank the Indian National Science Academy and the Council of Scientific and Industrial Research for financial support for this work. I am grateful to R.R. Jones and K.M. Cruikshank for their suggestions for improvement of the paper. I also thank Suspa Sinha and Sreemati DasGupta for their help during preparation of the manuscript.

REFERENCES CITED

Corsini, M., Vouchez, A., Archanjo, C., and Sa, E., 1991, Strain transfer at continental scale from a transcurrent shear zone to transpression fold belt: The Patos-Serido system, northeastern Brazil: Geology, v. 19, p. 586–589.

Dias, R., and Ribeiro, A., 1994, Constriction in a transpressive regime: An example in the Iberian branch of the Ibero Armorican arc: Journal of Structural Geology, v. 16, p. 1543–1554.

Flinn, D., 1965, Deformation in metamorphism, *in* Pitcher, W.S., and Flinn, G.W., eds., Control of metamorphism: Geological Journal, special issue, v. 1, p. 48–72.

Fossen, H., and Tikoff, B., 1993, The deformation matrix for simultaneous simple shearing, pure shearing and volume change, and its application to transpression-transtension: Journal of Structural Geology, v. 15, p. 413–422.

Fossen, H., and Tikoff, B., 1998, Extended models of transpression and transtension, and application to tectonic settings, *in* Holdsworth, R.E., et al., eds., Continental transpressional and transtensional tectonics: Geological Society [London] Special Publication 135, p. 15–33.

Ghosh, S.K., 1988, Theory of chocolate tablet boudinage: Journal of Structural Geology, v. 10, p. 541–553.

Ghosh, S.K., 1993, Structural geology: Fundamentals and modern developments: Oxford, Pergamon Press, 598 p.

Ghosh, S.K., Khan, D., and Sengupta, S., 1995, Interfering folds in constrictional deformation: Journal of Structural Geology, v. 17, p. 1361–1373.

Greene, D.C., and Schweickert, R.A., 1995, The Gem Lake shear zone: Cretaceous dextral transpression in the northern Ritter Range pendant, eastern Sierra Nevada, California: Tectonics v. 14, p. 945–961.

Harland, W.B., 1971, Tectonic transpression in Caledonian Spitsbergen: Geological Magazine, v. 108, p. 27–42.

Hudleston, P.J., Schultz-Ela, D., and Southwick, D.L., 1988, Transpression in Archean Greenstone belt, Minnesota: Canadian Journal of Earth Sciences, v. 25, p. 1060–1068.

Jaeger, J.C., 1964, Elasticity, fracture and flow: London, Methuen, 212 p.

Jones, R.R., Holdsworth, R.E., and Bailey, W., 1997, Lateral extrusion in transpression zones: The importance of boundary conditions: Journal of Structural Geology, v. 19, p. 1201–1217.

Jones, R.R., and Holdsworth, R.E., 1998, Oblique simple shear in transpression zones, *in* Holdsworth, R.E., et al., eds., Continental transpressional and transtensional tectonics: Geological Society [London] Special Publication 135, p. 35–40.

Means, W.D., 1994, Rotational quantities in homogeneous flow and the development of small-scale structure: Journal of Structural Geology, v. 16, p. 437–445.

Means, W.D., Hobbs, B.E., Lister, G.S., and Williams, P.F., 1980, Vorticity and non-coaxiality in progressive deformations: Journal of Structural Geology, v. 2, p. 371–378.

Passchier, C.W., 1986, Flow in natural shear zones—The consequences of spinning flow regimes: Earth and Planetary Science Letters, v. 77, p. 70–80.

Passschier, C.W., 1991, The classification of dialatant flow types: Journal of Structural Geology, v. 13, p. 101–104.

Passchier, C.W., 1997, The fabric attractor: Journal of Structural Geology, v. 19, p. 113–127.

Ramberg, H., 1975, Particle paths, displacement and progressive strain applicable to rocks: Tectonophysics, v. 28, p. 1–37.

Ramsay, J.G., 1967, Folding and fracturing of rocks: New York, McGraw Hill, 568 p.

Robin, P.Y., and Cruden, A.R., 1994, Strain and vorticity patterns in ideally ductile transpressional zones: Journal of Structural Geology, v. 16, p. 447–466.

Sanderson, D.J., and Marchini, R.D., 1984, Transpression: Journal of Structural Geology, v. 6, p. 449–458.

Simpson, C., and De Paor, D.G., 1997, Practical analysis of general shear zones using porphyroclast hyperbolic distribution method: An example from the Scandinavian Caledonides, *in* Sengupta, S., ed., Evolution of geologic structures in micro to macro scales: London, Chapman and Hall, p. 169–184.

Tikoff, B., and Fossen, H., 1993, Simultaneous pure and simple shear: The unified deformation matrix: Tectonophysics, v. 217, p. 267–283.

Tikoff, B., and Fossen, H., 1995, The limitations of three-dimensional kinematic vorticity analysis: Journal of Structural Geology, v. 17, p. 1771–1784.

Tikoff, B., and Greene, D., 1997, Stretching lineations in transpressional shear zones: An example from the Sierra Nevada batholiths, California: Journal of Structural Geology, v. 19, p. 29–39.

Tikoff, B., and Peterson, K., 1998, Physical experiments on transpressional folding: Journal of Structural Geology, v. 6, p. 661–672.

Tikoff, B., and Teyssier, C., 1994, Strain modeling of displacement field partitioning in transpressional orogens: Journal of Structural Geology, v. 16, p. 1575–1588.

MANUSCRIPT ACCEPTED BY THE SOCIETY APRIL 12, 2000

Experimental modeling of strike-slip faults and the self-similar behavior

Martin P.J. Schöpfer* and Hans Peter Steyrer
Institut für Geologie und Paläontologie, Universität Salzburg, Hellbrunner Strasse 34, A-5020 Salzburg, Austria

ABSTRACT

Analogue modeling of strike-slip faults in sandbox experiments gave new insights into the nature of the relationship between size of structures and the thickness of the granular material. Experiments have been conducted with different materials and varying thicknesses. The results emphasize the self-similar behavior of structures modeled in sandbox experiments. Shearing of the sand cake led to a change of the sand properties, such as the frictional angle and the packing density within the shear-zone. Dilatancy is an undesirable effect and a scaling problem. Supressing dilatancy by the use of coarse-grained granular material consisting of well-rounded grains with smooth surfaces led to quasiductile behavior of the shear-zone. Layered models with quasiductile granular material and typical model sand led to results similar to those observed in prestressed sand cakes.

INTRODUCTION

The results of analogue modeling studies have provided many insights into the progressive evolution of fault systems (e.g., Cloos, 1928; Riedel, 1929; Hubbert, 1951). In particular, strike-slip fault experiments have been carried out, with results that vary according to the type of material being used (Sylvester, 1988). However, few studies have addressed the progressive evolution of strike-slip faults in sandbox experiments and the relationships between size, shape, and orientation of structures and the thickness of the sand cake (Emmons, 1969; Naylor et al., 1986; Richard and Krantz, 1991; Dauteuil and Mart, 1998). Few shear-zone experiments have been conducted with granular materials that differ in their physical properties (Richard et al., 1995). Questions remain from such studies concerning the use of layered sand cakes built with sand types differing in grain size, grain shape, and grain-size distribution.

The interpretation of experimentally produced shear-bands as scaled-down tectonic faults has been justified largely by reference to ideal plastic Mohr-Coulomb theory (cf. Mandl, 1988). While the latter is a simplification, it is usually believed to capture the essence of a typical sandbox model of tectonic faulting. However, the actual constitutive response of sand and its dependence on grain shape, grain sorting, and initial packing density are clearly of concern in these experiments. Dilatant behavior is undesirable for a typical model sand, e.g., a grain-size range of 150–300 μm, as in Naylor et al. (1986), or 125–250 μm, as in this work.

The aim of the strike-slip experiments was to determine the dependence (if any) of the initially produced structures on the thickness and properties of the sand layer. Experiments were carried out with three types of sand, the physical properties of which were characterized in detail. The shear zones developing in these three different sand types were then compared. Experiments with varying sand-cake thicknesses, i.e., wedge-shaped cakes and layered sand cakes, were also conducted.

This study does not deal with natural geologic settings, but should be a critical view on the problematics of scaled models.

---

*E-mail: martin.schoepfer@sbg.ac.at

Schöpfer, M.P.J., and Steyrer, H.P., 2001, Experimental modeling of strike-slip faults and the self-similar behavior, *in* Koyi, H.A., and Mancktelow, N.S., eds., Tectonic Modeling: A Volume in Honor of Hans Ramberg: Boulder, Colorado, Geological Society of America Memoir 193, p. 21–27.

EXPERIMENTAL TECHNIQUE

Experimental apparatus

The shear box used for our experiments was similar to those used by Cloos (1928), Riedel (1929), and Wilcox et al. (1973). The base of the experimental apparatus consists of two plates. One of them can be slid past the other, driven by a geared motor, thus simulating a straight basement wrench fault, inducing shear in the overburden, either dextral or sinistral. The basal plates form a square with edge lengths of 65 cm effective area for the experiments; e.g., the area that is not affected by boundary effects is dependent on sand thickness and the offset between the two plates. The effective area decreases with increasing thickness and offset. For typical experiments (sand thickness 3 cm) the effective area is ~40 × 40 cm.

Analogue materials and their physical properties

The following three types of sand were used. (1) The first sand type was sieved (125–250 μm) dry natural quartz sand with a slope angle of 35°. The grains have rough surfaces and are of low sphericity, because the grains were industrially broken. (2) The second type is the same sand as sand type 1, except that it was washed after sieving with a 125 μm mesh-size sieve, thus removing the clay mineral fraction. (3) The third type of sand is so-called Granucol® (produced by Dorfner, Hirschau, Germany). This is a course-grained, industrially produced granular material (finishing coat) consisting of pigmented quartz grains with smooth grain surfaces and a high sphericity.

The grain-size distributions are shown in Figure 1.

Density measurements were carried out with three different shallow containers of known volume and three different preparation techniques: (1) spreading with a mechanical hopper from a height of 25 cm, (2) pressing in the container by hand, and (3) pouring from a low height with a small shovel. For each container and spreading method, measurements were taken five times. The values are listed in Figure 2.

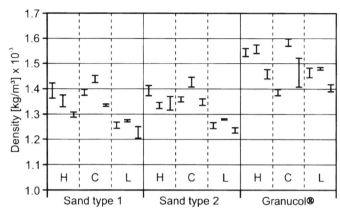

Figure 2. Densities of three different types of sand used. Bars represent calculated standard deviations for five measurements. Columns H, C, and L include density measurements for sands spread with three different handling techniques described in text; i.e., mechanical hopper (H), compacted (C), and loose (L). Three bars in each column were calculated from measurements of three shallow containers with differing volumes, 0.263 L, 1,577 L, and 0.749 L, respectively.

The angles of internal friction (ϕ) were estimated with normal fault-scarp distances (Krantz, 1991). An unconfined sand pile was built against a 4-cm-high moveable wall. The pile was wide enough to neglect side effects. The pile was then leveled by careful scraping to a uniform height. The partition was then displaced until a clear fault scarp appeared on the surface. The distance from the edge of the partition to the fault scarp was measured. For the three different sand types and spreading methods described here, measurements were taken five times. The obtained data and the application for shear-zone experiments are summarized in Figure 3. To assist the interpretation, some experiments were made with a sand cake including marker layers. For the black marker layers sand type 1 was colored with indian ink, dried, and sieved again. After each run the sand cake was covered with more sand so as to conserve the topography of the free surface and then made moist with water. The sand cake was then sectioned perpendicular to the partition (Fig. 3B). Friction along the moving partition and on the base plate led to edge effects and the values for the friction angles in Figure 3 are an approximation only. However, this method is a fast way to determine ϕ, and the values fit well with the $\phi/2$ angles, i.e., the angle between the Riedel shears and the principal displacement zone, measured in homogeneous sand cakes with uniform height.

Experimental procedure

Experiments were conducted with sand type 1—sand cakes of varying thicknesses (0.5–4 cm) and marker horizons positioned every 0.5 cm. The sand was spread over the plates with a mechanical hopper, which guaranteed homogeneous and smooth sand layers, which needed only minor finishing with a metal ruler. Alternative spreading methods, such as flattening a sand cone with a glass plate, led to density inhomogeneities in the

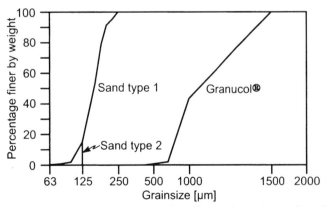

Figure 1. Grain-size distribution of sand type 1, sand type 2, and Granucol.

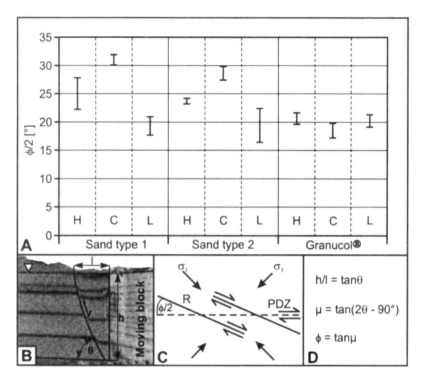

Figure 3. $\phi/2$ values determined with fault-scarp measurement technique described in text. A: Graph showing $\phi/2$ values. Bars represent calculated standard deviations for five measurements. Columns H, C, and L include calculated $\phi/2$ values of sands spread with three different handlings. B: Cross section of experiment. Wooden block was moved toward right. Section is cut obliquely (45°) to strike of partition in order to amplify internal structures. Friction along wooden block led to side effects such as drag. Triangle marks surface after filling graben with black sand, prior to covering with further sand. C: $\phi/2$ in context of wrench-fault pattern. Theoretical angle between R-shears and principal displacement zone (PDZ) is $\phi/2$. D: Trigonometric relationships of ϕ, μ, and θ.

sand cake, and when such methods were used structures first appeared in regions of least density. The mechanical hopper ensured reproducible experiments, which were carried out in dextral strike-slip mode with a rate of displacement of 5 cm/h (which ought to have a negligible effect on the results). For the marker layers and the marker grid on top of the sand cake, sand type 1 was colored with indian ink, dried, and sieved again.

Experiments were also conducted with sand type 2 and Granucol, using the same procedure as described previously. A few experiments were made with a layer of Granucol on the base and a layer of sand type 1 on top in order to study the effects of combining sands with differing physical properties.

Experiments with a sand wedge were conducted for studying the effects of sand thickness variations. This was an economical way to study the dependence of structures on the thickness of the sand cake and on the displacement in a simple experimental procedure. The dip of the wedge surfaces were low and initial differential stresses due to the inclined free surface were neglected.

A number of sand type 1 models were sprinkled with water and sectioned perpendicular to the plate boundaries in order to reconstruct the three-dimensional geometry of the internal structures (Fig. 4A). This technique was not suitable for Granucol, due to the fact that this material does not produce sufficient capillary cohesion. Another procedure was to remove sand at the toe of the natural slope of a dry, sheared sand cake with a ruler perpendicular to the principal displacement zone (PDZ) (Fig. 4, B and C). This technique is suitable for qualitative visualization of the area affected by shearing. Photographs and time-lapse videos were produced, the latter for reconstructing the incremental displacement on the individual faults. A constant grid-line spacing of 1 cm made length measurements straightforward. A low-angle spotlight intensified morphological features.

SCALING PROBLEMS

A detailed presentation of the theory of scale models was given by Mandel (1963), according to whom sand can be regarded in a first approximation as an ideal plastic Coulomb material, the elastic strains remaining small in comparison with the plastic strains (in Mandel's notion of extended similitude). Faults in dry granular material with low cohesion develop as shear bands at low confining pressures. In the absence of grain-size reduction, plastic shearing of dry, densely packed, granular materials typically leads to positive dilatancy (Jaeger and Nagel, 1992). The experimentally produced dilatancy is clearly not to scale and depends on the initial packing density, the grain-size distribution, the sphericity, and the surface condition of the grains (Koopman et al., 1987).

The experimental setup used in this study simulates basement-induced wrenching (type C, Fig. 1.2-72, p. 75 in Mandl, 1988). Close to the surface of the sand cake the material may be regarded as undergoing simple shear induced by the moving base plates. During the onset of simple shearing of an isotropic, nonprestressed elastic material, the maximum compressive stress is at an angle of 45° to the simple shear direction. Under simple elastic shear this direction will be maintained and Coulomb slips will form oriented $\phi/2$ to the strike of the PDZ. Movement along these Riedels causes the material to be compressed in the region of overlapping, and the maximum compressive stress rotates toward the strike of the PDZ (Mandl, 1988). Thus low-angle R-shears and P-shears will

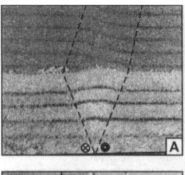
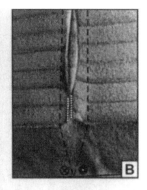

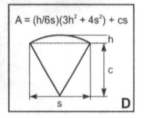

Figure 4. Dilatancy in experiments. A: Sand type 1, dextral shearing experiment. After 0.5 cm offset, before R-shear formation, experiment was sprinkled with water and sectioned perpendicular to principal displacement zone (PDZ). Dashed line marks wedge-shaped area affected by shearing. Marker layers and lines are at 1 cm intervals. B: Sand type 2 model, 2 cm thick, after 2.5 cm dextral displacement, nonwetted. Marker lines at 1 cm intervals. Sand was removed by scraping at toe of slope with ruler perpendicular to PDZ. Wedge-shaped area affected by shearing is marked with dashed line. Note change of natural slope angle within the shear zone; smallest value is in area of shear bands (dotted line). Shear zone is marginally wider than area cut by shear bands. C: Sand cake with 2 cm Granucol on base plates and 1 cm sand type 1 on top, 3.5 cm dextral displacement, nonwetted. Natural slope of sand cake as in B. Dashed line marks probable area affected by shearing. Note that shear zone is much wider than area cut by shear bands. D: Circle sector calculations for approximation of volume increase due to shearing.

form. However, elastic strains are taken to be negligible, and strain is taken up by plastic Coulomb behavior. Thus the theoretical orientation of R-shears is only correct when the amount of plastic deformation is low. The stress field prior to the onset of formation of discontinuities predicts the orientation of the initial shear bands (Dresen, 1991).

EXPERIMENTAL RESULTS

Homogeneous sand layer of uniform height

Experiments were conducted with three different types of sand. The following observations were made with sand type 1 and sand type 2. After about 3 mm offset, deformed marker lines and a central ridge were clearly visible (Fig. 4A). After about 7 mm en echelon arranged Riedel (R) shears, which made an angle of 23° (20° for sand type 2) with the PDZ appeared. We could not observe a conjugated set of Riedels (R′), but a second generation of Riedel shears was at an angle of ~10° with the PDZ. At the tips of the R-shears short-lived splays (S) formed. Parallelogram-shaped lenses, confined by a pair of R-shears and the central ridge boundaries, were popped up and slightly rotated with increasing offset. The shear-zone width and the shear lenses in sand type 2 experiments are smaller due to the lower R-shear angle. The width of the shear zone increased slightly due to the rotation of the shear lenses. P-shears connected the individual lenses and finally Y-shears cut through the lenses and finished the master fault, the orientation of which coincides with the PDZ. The result is a central ridge consisting of popped up and rotated shear lenses with a sharp, deep cutting valley that is made up by R-, P-, and Y-shears.

Circle sector calculations for the initial central ridge give a value for volume increase of >5% within the shear zone. The dip of the planes separating the zone affected by shearing is about 70° and they meet at the PDZ on the base plate (Fig. 4A). The zone affected by dilatancy is thus V shaped and the increase in volume is proportional to the circle sector above the shear zone (Fig. 4D). The ϕ changed within this zone (Fig. 4B).

Numerous experiments with sand type 1 were conducted in order to deduce the relationship between shear-lens geometry and the sand-cake thickness. The shear-lens widths and lengths were measured before the lenses were cut by P- and Y-shears. The ratio of shear-lens width and shear-lens length is independent of sand thickness (Fig. 5).

The progressive development of shear-band patterns in our sand types 1 and 2 obviously correspond to those summarized by Sylvester (1988). The theoretical explanation and mechanical behavior of shear bands was summarized by Mandl (1988, p. 75–80), who referred to Skempton (1966), Morgenstern and Tschalenko (1967), Tschalenko (1970), and Naylor et al. (1986).

Completely different behavior is observed with the analogue material Granucol (Fig. 7). This material was initially chosen in order to suppress dilatancy and hence several experiments with thicknesses ranging from 2 to 4 cm were conducted with Granucol. However, the deformation effect and shear-band development is completely different from that observed in sand type 1 and 2 experiments. After a 0.5 cm offset, a central ridge, an order of magnitude smaller than those observed in the sand type 1 and 2 experiments, was visible (Fig. 7E). In thin sand cakes (<3 cm) this material deformed without any visible discontinuities throughout the experiment in the manner of a ductile shear zone. Shear-zone width measurements showed that the V-shaped zone is bounded by planes dipping 60° toward the PDZ. No discrete shear bands formed. After a 3 cm offset a straight trench above the PDZ formed. In thick sand cakes (4 cm) short-lived low-angle R-shears oriented ~17° to the PDZ appeared after an offset of ~1 cm (Fig. 7D). These were the only shear bands observed in Granucol. A trench above the PDZ formed after a 3 cm offset.

Homogeneous, wedge-shaped layer

Experiments with a wedge-shape geometry were carried out with sand type 1. After an ~5 mm offset, a central ridge that widened with increasing wedge thickness became visible. The R-shears propagated from the toe of the sand wedge toward the

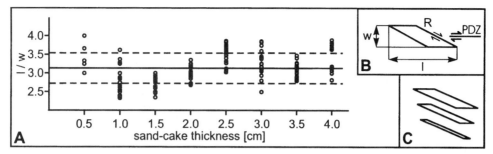

Figure 5. A: Scatter plot of shear-lens length/width ratio vs. sand-cake thickness. Solid line is mean value calculated from all data and is not best fit line. Dashed lines are standard deviations. B: Schematic sketch showing orientation of measured values. C: Theoretical shear-lens geometries for (top to bottom) mean value plus standard deviation, mean and mean minus standard deviation, respectively. In reality, shear lenses are longer, because principal displacement zone–parallel boundaries are curved and only maximum width was measured.

top. Time-lapse videos of the experiments show the delay of R-shear appearance in the thicker part to be very small, e.g., 1 mm offset difference in a 30-cm-long wedge with a dip of ~6°. The shear-zone widens toward the thicker part of the wedge, thus changing the shear-lens geometry (Fig. 7A). Three generations of R-shears were observed in the thicker part of the wedge (>2 cm). P- and Y-shears connect and cut through the shear lenses, as described in the experiments with uniform height. A difference, however, is that the shear bands appear to propagate toward the thick part of the wedge. The ridge is asymmetric, because the cake had no uniform height.

Two-layer sand cake of uniform height

In these experiments 2 cm of Granucol was on the base plates with 1 cm of sand type 1 on top: after a displacement of 0.8 cm, curved marker lines and a low ridge were visible (Fig. 6A). The shear zone had a width of ~2.6 cm. For comparison, the width of the shear zone in a 2-cm-thick Granucol cake without overburden was 2.3 cm. Further displacement led to the formation of low-angle R-shears oriented 10° to the PDZ, and splays that propagated from the R-shear tips (Fig. 6B). After ~2 cm offset, shear lenses confined by R- and P-shears were

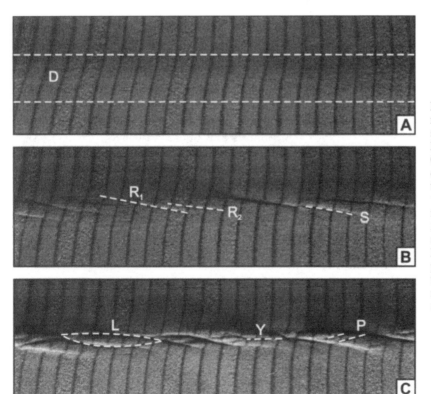

Figure 6. Experiment with 2 cm thickness of Granucol on base plates and 1 cm sand type 1 on top, dextral shearing. Marker lines at 1 cm intervals. A: After 0.8 cm displacement, zone of plastic deformation and ridge, order of magnitude lower than in 3-cm-thick sand type 1 experiments, developed. Dashed lines mark area of shearing, i.e., dilatancy (D). B: After 0.9 cm displacement, Riedel-shears (R) and splays (S) developed. Note that R-shears are oriented ~10° to principal displacement zone (ϕ of sand type 1 ~ 46°). C: After 2.2 cm displacement, shear lenses (L) formed with P-shears (P) and Y-shears (Y) cutting them.

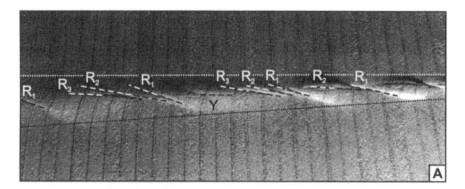
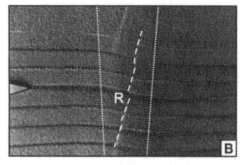
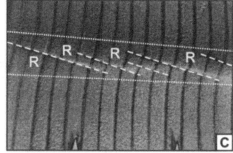
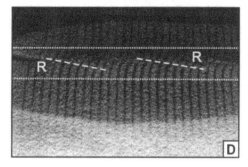

Figure 7. Wedge-shaped experiments and Granucol experiments. Marker lines at 1 cm intervals. All photographs were taken after 1 cm displacement. A: Sand type 1 wedge, 1.5 cm thick at right edge of photograph and 4.5 cm at left edge. Three generations of Riedel shears (R_1, R_2, and R_3) formed. Y-shears started cutting through lenses confined by Riedel shears and shear-zone boundaries (dotted lines). Latter are straight and divergent. B: Model consisting of wedge of Granucol and 1 cm sand type 1 on top. Surface dips by 14° toward bottom of photograph. Arrow on left side marks 4.5 cm total thickness. Small shear zone consisting of low-angle en echelon shear bands (R) in major shear zone (dotted lines) developed in overburden due to Riedel shear in Granucol. C: Same model as in B. Surface dips to right. Arrows at bottom of photograph mark 3.5 cm and 3.0 cm total thickness, respectively. Note that Riedel shear angle decreases toward thicker part of wedge. D: Granucol experiment with uniform height of 4 cm. Riedel shears (R) formed in 4.2-cm-wide shear zone (dotted lines). E: Granucol experiment with uniform height of 3 cm. Shear zone was 2.7 cm wide (dotted lines) and no shear bands appeared.

formed. Further displacement led to throughcutting shearbands, such as P- and Y-shears (Fig. 6C). After the complete obliteration of the shear lenses, a valley, made by R-, P-, and Y-shears, completed the formation of shear bands.

Two-layer, wedge-shaped sand cake

Experiments with a wedge of Granucol—striking 90° to the PDZ with a dip of 14°—and 1 cm sand type 1 as overburden were conducted. At the onset of the experiment, a low ridge with divergent boundaries and widths of 3 cm at 1.5 cm total thickness and 4 cm at 4.5 cm total thickness formed. After a 0.5 cm offset, the first R-shears appeared in the thinner part of the cake, i.e., <1.5 cm. After a 1 cm offset, R-shears were formed over the total length of the shear zone (Fig. 7, B and C). The R-angle decreases toward the thicker part, e.g., 23° at 1.5 cm thickness and 14° at 4.5 cm. The R-shear at 4.5 cm total thickness is made of en echelon arranged shear bands (Fig. 7B). Splays formed at the tips of the R-shears. Further displacement led to the formation of a second generation of R-shears in the thicker part of the cake (>3 cm), whereas in the thin part, P-shears and shear lenses appeared. After a 2 cm offset, shear lenses were formed over the total length of the shear zone. The widths of the lenses decrease and their lengths increase toward the thick part of the cake, because the R-shear angles differ with thickness. Further displacement led to the formation of a trench made by R-, P-, and Y-shears. The morphology is asymmetric due to the wedge shape.

Conclusions

The sandbox experiments with varying thicknesses and different granular materials provide new insights into the progressive development of shear zones and the analogue modeling technique.

1. For the interpretation of analogue models one has to know the physical properties of the materials used. Density measurements and determination of the angle of internal friction and the grain-size distributions are essential. Experiments with homogeneous sand cakes ensure reproducible experiments.

2. In sand cakes with uniform height using typical analogue modeling sand, the shear-band pattern and development are independent of the sand-cake thickness; only the sizes of the struc-

tures change. The first appearance in thicker sand cakes is delayed. The shear zone is V shaped, and the dip of the planes is dependent on the material being sheared. After the formation of the R-shears the zone between overlapping R-shears widens due to compression, leading to additional dilatancy. However, the geometry and orientation of shear bands and lenses are different in wedge-shaped sand cakes due to the difference of shear band lengths and the distances.

3. Changes of the thickness of the sand cake lead to geometrically similar structures, i.e., self-similar behavior. As a result the overburden thickness is the scaling factor. Experiments showed that not only the size of the structures and the width of the shearzone are scaled by the thickness, but also the displacement (Fig. 7A). Structures in the thin part of the wedge are like those that will be formed after further displacement in the thick part.

4. An undesirable effect is dilatancy, i.e., an increase of volume due to shearing. The amount of dilatancy depends on the initial packing density, the grain-size distribution, the sphericity, and the surface condition of the grains. It was possible to minimize dilatancy by using an analogue material in which the grain-size range is small and the grains have smooth surfaces, thus avoiding strong interlocking and friction. However, by using such materials the shear-band development is completely different from that observed in the classical sandbox experiments. In thin sand cakes the material behaves quasiductile throughout the experiment and the formation of Coulomb-slip planes is only observable in thick sand cakes (>4 cm thick for a grain-size range of about 700–1500 μm). This material (Granucol) is thus simulating a quasiductile shear. Although dilatancy can be suppressed by the use of a proper material, other desired effects, such as the development of well-defined shear bands, are suppressed.

5. By using a thick layer of the quasiductile material mentioned herein and a thin overburden of typical model sand, it was possible to simulate a shear zone within a major shear zone (Fig. 7B). An R-shear formed in Granucol and induced wrench faulting in the sand layer on top. The shear zone was oriented 10° to the PDZ, because this angle corresponds to the R-shear in the Granucol in 1 cm depth.

6. The stress-field development in shear-zones is complex. In sandbox experiments the direction of the principal stresses can be estimated from the shear-band orientation. The onset of shear-band formation and the orientation depend on the material properties and the initial stress field. The maximum principal stress is oriented 45° to the strike of the wrench fault and rotates toward it. Strain is taken up by discrete Coulomb slip and quasiductile behavior; the latter fixes the amount of the stress-field rotation before Coulomb slip planes are formed.

7. The use of a quasiductile material on the base and typical model sand on top made it possible to influence the orientation of the shear bands by varying the thickness of the quasiductile layer. The R-angle observed on the free surface decreases with increasing thickness of the quasiductile material, because a bigger amount of strain is taken up by quasiductile behavior. Initial low-angle R-shears were formed without prestressing the sand cake.

ACKNOWLEDGMENTS

Financial support received from the Senat der Universität Salzburg is gratefully acknowledged. The experimental apparatus was built by Gerhard Aigner. We acknowledge the critical reviews of Robert Krantz and an anonymous reader. We thank Alex Densmore, Trinity College, Dublin, and Florian Lehner and Hans Genser, Salzburg University, for useful discussion and comments.

REFERENCES CITED

Cloos, H., 1928, Experimente zur inneren Tektonik: Zentralblatt für Mineralogie, Geologie und Paläontologie, v. 1928B, p. 609–621.

Dauteuil, O., and Mart, Y., 1998, Analogue modeling of faulting pattern, ductile deformation, and vertical movement in strike-slip fault zones: Tectonics, v. 17, p. 303–310.

Dresen, G., 1991, Stress distribution and the orientation of Riedel shears: Tectonophysics, v. 188, p. 239–247.

Emmons, R.C., 1969, Strike-slip rupture patterns in sand models: Tectonophysics, v. 21, p. 93–134.

Hubbert, M.K., 1951, Mechanical basis for certain familiar geologic structures: Geological Society of America Bulletin, v. 62, p. 355–372.

Jaeger, H.M., and Nagel, S.R., 1992, Physics of the granular state: Science, v. 255, p. 1523–1531.

Johnson, A.M., 1995, Orientations of faults determined by premonitory shear-zones: Tectonophysics, v. 247, p. 161–238.

Koopman, A., Speksnijder, A., and Horsfield, W.T., 1987, Sandbox model studies of inversion tectonics: Tectonophysics, v. 137, p. 379–388.

Krantz, R.W., 1991, Measurements of friction coefficients and cohesion for faulting and fault reactivation in laboratory models using sand and sand mixtures: Tectonophysics, v. 188, p. 203–207.

Mandel, 1963, Tests on reduced scale models in soil and rock mechanics, a study of the conditions of similitude: International Journal of Rock Mechanics and Mining Sciences, v. 1, p. 31–42.

Mandl, G., 1988, Mechanics of tectonic faulting: Elsevier Science Publishers B.V., p. 407.

Morgenstern, N.R., and Tschalenko, J.S., 1967, Microscopic structures in kaolin subjected to direct shear: Geotechnique, v. 17, p. 309–328.

Naylor, M.A., Mandl, G., and Sijpesteijn, C.H.K., 1986, Fault geometries in basement-induced wrench faulting under different initial stress states: Journal of Structural Geology, v. 8, p. 737–752.

Richard, P., and Krantz, R.W., 1991, Experiments on fault reactivation in strike-slip mode: Tectonophysics, v. 188, p. 117–131.

Richard, P.D., Naylor, M.A., and Koopman, A., 1995, Experimental models of strike-slip tectonics: Petroleum Geoscience, v. 1, p. 71–80.

Riedel, W., 1929, Zur Mechanik geologischer Brucherscheinungen: Zentralblatt für Mineralogie, Geologie und Paläontologie, v. 1929B, p. 354–368.

Skempton, A.W., 1966, Some observations on tectonic shear zones: First International Congress on Rock Mechanics, Proceedings, Lisbon, v. 1, p. 329–335.

Sylvester, A.G., 1988, Strike-slip faults: Geological Society of America Bulletin, v. 100, p. 1666–1703.

Tschalenko, J.S., 1970, Similarities between shear zones of different magnitudes: Geological Society of America Bulletin, v. 81, p. 1625–1640.

Wilcox, R.E., Harding, T.P., and Seely, D.R., 1973, Basic wrench tectonics: American Association of Petroleum Geologists Bulletin, v. 57, p. 74–96.

MANUSCRIPT ACCEPTED BY THE SOCIETY APRIL 12, 2000

Printed in the U.S.A.

Finite-element approach to deformation with diffusive mass transfer

Brian Bayly
Earth and Environmental Sciences Department, Rensselaer Polytechnic Institute, Troy, New York 12180, USA
Donald H. Minkel
Adirondack Community College, Queensbury, New York 12804, USA

ABSTRACT

In the common form of finite-element calculation for viscous materials, one neglects diffusive mass transfer and solves for a velocity field. By contrast, if diffusive mass transfer is allowed for, one has to solve first for the stress field and derive velocities, if needed, by a secondary calculation.

In estimating the rate of dissipation of energy by diffusive processes, one may postulate that diffusion is driven by a gradient in just the mean stress, or alternatively, that diffusion is affected by all three principal stress magnitudes and their gradients.

The first (and only) two-dimensional problem we have attacked so far concerns a highly viscous long cylindrical inclusion in an infinite less viscous host material. Results gained using the second postulate agree semiquantitatively with results by Finley, who used the first postulate. Results are also compatible with the known behavior of ideal nondiffusing materials.

The underpinnings for the second postulate were provided by Hans Ramberg 40 years ago. We are glad to extend a line of thought on groundwork that Hans provided.

INTRODUCTION

The long-range objective is to improve understanding of material transfer by intergranular processes. Differentiated cleavage zones are a prime example because of the smooth gradation from source region to sink region. Other examples are the pockets of segregated material between boudins or at the sides of augen. Systems of complementary tension gashes and stylolites, of cracks and anticracks, are related phenomena wherein questions of scale considered and diffusion distance are clearly displayed.

In parallel with study of such features seen in outcrop, there have been theoretical treatments where the main geological application regards separation and migration of magma (McKenzie, 1984; Spiegelman, 1993a, 1993b) or compaction of accumulated sediments (Bethke, 1985; Person et al., 1996). The same principles underlie all these fields of application, but different assumptions and objectives lead to different styles of treatment: for example, in magma studies, gravity is important and interfaces between two deforming materials with different properties are not. For these reasons the fields have developed separately. However, there are problems, such as behavior of a fluid-filled aggregate near a rigid cylinder, where the different approaches could overlap and illuminate each other.

Danielewski and co-workers described a finite-element approach to diffusion problems in general terms (Danielewski and Filipek, 1996). The layered assemblies in microelectronic devices have also prompted studies such as those of Greer (1995, theoretical), and Opposits et al. (1998, experimental). In this field of application, the direction normal to the layering is of prime concern; thus a one-dimensional approach is emphasized.

Within the field of structural geology, landmark papers are those of Dieterich and Onat (1969) and Fletcher (1982). Of these, the first drew attention to the finite-element method and its suitability for studying deformation of heterogeneous assemblies of viscous materials; the second showed how to combine the effects of diffusive mass transfer with viscous deformation, but took an analytical approach. The objective here is to incorporate Fletcher's concept in a finite-element scheme; at the same time we explore the use of a proposal by Ramberg (1959, 1963), which suggests a variant on Fletcher's formulation. Specifically, we make three points.

Bayly, B., and Minkel, D.H., 2001, Finite-element approach to deformation with diffusive mass transfer, *in* Koyi, H.A., and Mancktelow, N.S., eds., Tectonic Modeling: A Volume in Honor of Hans Ramberg: Boulder, Colorado, Geological Society of America Memoir 193, p. 29–38.

1. In finite-element models of deforming viscous materials, it is customary to exclude diffusive mass transfer and to solve for the nodal velocities; but if one includes diffusive mass transfer, one has to solve for the nodal stresses instead.
2. There are two ways of setting up the diffusive terms.

    (1) The first is to assume that diffusion is driven by gradients in a field of magnitudes of some variable that has a single magnitude at every point in space. The concentration of a solute in a solvent that fills the space envisaged would be an example of such a single-valued variable.

    Along this line, if diffusion is occurring in a permeable deformable host where at every point there is a nonhydrostatic stress state, one option is to note the mean stress at each point and to take that as the single-valued quantity that controls diffusion.

    (2) In contrast to the preceding, for a field of nonhydrostatic stress states, a second option is to assume that all three of the principal stress magnitudes affect the diffusive behavior. That is, one might assume that the effect of stress on diffusion is not fully represented by just the mean stress, nor by any other variable that has a single value at each point.

    Both of these options seem worth developing; the first is simpler, but may in some circumstances be an oversimplification.
3. Pursuing the second option, one finds that although Hans Ramberg never did any finite-element modeling, a proposal he made 40 years ago bears directly on the procedure.

FINITE ELEMENT APPROACH WITH DIFFUSIVE MASS TRANSFER

In a Newtonian material of viscosity N, a strain rate e_{ij} is given by

$$e_{ij} = \sigma'_{ij}/2N. \tag{1}$$

Here the corresponding stress component is σ_{ij}, the mean stress is σ_0 and the deviatoric stress component $\sigma'_{ij} = \sigma_{ij} - \sigma_0$. Then a rate of dissipation of energy can be written in three ways, $e_{ij}\sigma'_{ij}$, or $e_{ij}e_{ij} \cdot 2N$, or $\sigma'_{ij}\sigma'_{ij}/2N$. In the commonest form of finite-element calculation, the second is used. The strain rate e_{ij} is expressed in terms of nodal velocities and the total dissipation is expressed in quadratic terms of velocity. Then, by virtue of an extremum principle, differentiation with respect to each unknown velocity gives the set of equations needed for a solution.

When diffusive mass transfer is postulated, however, e_{ij} is formed from two terms. In addition to $\sigma'_{ij}/2N$, there is a term of the general type $K \cdot d^2(\text{stress})/dx^2$, where x is a spatial coordinate and K is a material property, a diffusion or transmissivity coefficient (Fletcher, 1982). There are two ways of forming this second term; the simpler is to assume that the mean stress σ_0 is the only influence on diffusion. Then

$$e_{ij} = \frac{\sigma'_{ij}}{2N} - K \cdot \nabla^2 \sigma_0. \tag{2}$$

Even this simple form shows that it is no longer possible to express σ'_{ij} as $e_{ij} \cdot 2N$ and form quadratic terms in velocities. One is forced to seek quadratic terms in stresses instead, as follows.

In two dimensions where axes are x and y and stress components are σ_{xx}, σ_{yy} and τ_{xy}, dissipation by viscous processes is at a rate

$$\frac{(\sigma_{xx} - \sigma_{yy})^2}{4N} + \frac{\tau_{xy}^2}{N}. \tag{3}$$

If diffusion also follows a linear form

$$flux = -K \cdot grad\, \sigma_0, \tag{4}$$

then dissipation by diffusive processes is at a rate $K \cdot (grad\, \sigma_0)^2$. Let two nodes be designated by suffixes 1 and 2; then $grad\, \sigma_0$ is built from terms such as $(\sigma_{01} - \sigma_{02})/(x_1 - x_2)$ or $(\sigma_{xx1} + \sigma_{yy1} - \sigma_{xx2} - \sigma_{yy2})/2(x_1 - x_2)$, so that $(grad\, \sigma_0)^2$ is clearly also quadratic in σ_{xx} and σ_{yy}. Let the total dissipation be the sum of the viscous dissipation and the diffusive dissipation: then in parallel with the usual treatment of velocities, an extremum principle allows us to differentiate with respect to each unknown stress component and so form a set of equations sufficient for solving for the unknown stresses.

Further details of the implementation are given in Appendix 1. An unwelcome feature comes into view when evolution through time is considered. An elegant aspect of the usual treatment is that as soon as one solves for a velocity field, this can be used immediately to advance the evolution of the model through the next time step; an entire kinematic evolution can be followed step by step without any stress magnitude ever being calculated. In the present formulation, after solving for a stress field, it is necessary to calculate a full set of strain rates in order to calculate a velocity field and advance the geometry, before the next stress field can be evaluated.

DIRECTION-DEPENDENT DIFFUSIVE LOSS

The possibility in view here is that when a volume element of material diminishes in volume by diffusive loss, the shrinkage may be greater along some directions than along others; a sphere might become an ellipsoid by anisotropic loss of material as well as by ordinary viscous change of shape. There are two reasons for pursuing this possibility, the first being that it is suggested by outcrops. Many metamorphic rocks of low or moderate grade contain seams poor in quartz and rich in less soluble minerals. It is common to refer to these as insoluble *residues*, from the idea that quartz and other grains have been preferentially dissolved and their constituents have migrated elsewhere. Such seams typically have strong preferred orientation; they give the impression that as quartz dissolved, the rock contracted in dimension mostly normal to the seams, with only much smaller changes in dimension in the plane of the seams. This kind of anisotropic behavior is not allowed for in Fletcher's formulation (1982, p. 279); hence we seek a modification more compatible with what the seams seem to show.

The second reason for proposing anisotropic diffusive loss is much more abstract, as follows.

In an ideal Newtonian continuum, the strain rate along some direction **n** is given by

$$e_{nn} \quad \text{or} \quad \frac{\sigma'_{nn}}{2N} = -\frac{1}{12N}\left(\frac{\partial^2 \sigma_{nn}}{\partial \theta_1^2} + \frac{\partial^2 \sigma_{nn}}{\partial \theta_2^2}\right) \quad (5)$$

(Bayly, 1992, p. 58–62). Here $d\theta_1$ and $d\theta_2$ are small angular deviations from direction **n** in two orthogonal planes containing **n** (see Fig. 1). To aid the imagination, we can introduce a small distance R and write

$$e_{nn} = -\frac{R^2}{12N}\left(\frac{\partial^2 \sigma_{nn}}{\partial (R\theta_1)^2} + \frac{\partial^2 \sigma_{nn}}{\partial (R\theta_2)^2}\right), \quad (6)$$

$$= -\frac{R^2}{12N}\left(\frac{\partial^2 \sigma_{nn}}{\partial s_1^2} + \frac{\partial^2 \sigma_{nn}}{\partial s_2^2}\right). \quad (7)$$

Here ds_1 and ds_2 are small distances along arcuate paths of radius R. These manipulations give a description of viscous behavior that closely parallels the standard form in diffusion,

$$\frac{dV}{V \cdot dt} = -K\left(\frac{\partial^2 \phi}{\partial x^2} + \frac{\partial^2 \phi}{\partial y^2} + \frac{\partial^2 \phi}{\partial z^2}\right), \quad (8)$$

where ϕ is some kind of potential for which the volume flux $= -K \, grad \, \phi$. The parallel is closer still if we consider a line element parallel to **n** of length l so that $e_{nn} = dl/l \cdot dt$.

But if equation 7 is a diffusion equation, what is diffusing and where is the diffusion path? To get value from this equation, it seems necessary to use the concept of infinitesimal "wafers," as follows.

Take a small line element of length l; form a cube with edge length l, with four edges parallel to **n**; after a time-interval dt, remove a wafer normal to **n** and of thickness dl from the cube. If a larger volume element is formed from many such cubes packed in parallel and the removal operation is repeated on all of them at regular intervals dt, the result is a macroscopic homogeneous shortening rate $e_{nn} = dl/l \cdot dt$.

Now consider that each wafer is not removed to a remote site, but is added back to the cube in a new orientation parallel to a side face (i.e., a face parallel to **n**). The gross effect is now a plane strain at constant volume, with a shortening rate along **n** and an equal elongation rate in a direction normal to **n**.

Finally, consider that each wafer is not removed and replaced, but instead changes from its original to its final orientation by sliding around a cylindrical surface, as in Figure 2. If the agent driving the wafers is the compressive normal stress across this surface, with a gradient from a high value σ_{nn} to some smaller value σ_{mm} (**m** normal to **n**), then it is easy to propose that

$$\text{flux of wafers} = -K^* \cdot \frac{\partial(\text{normal stress})}{\partial s}, \quad (9)$$

(where K^* is a material property, some suitable diffusion parameter) from which equations 6 and 7 would follow, except that we have yet to specify the relation among N, K^*, and R.

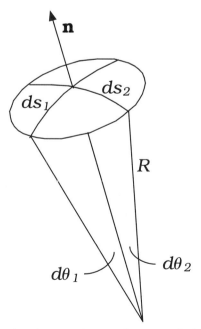

Figure 1. Directions in space that are close to particular direction **n**. Angular deviation from **n** can be expressed either directly, as angle $d\theta$, or indirectly by specifying arbitrary radius R and length of arc ds.

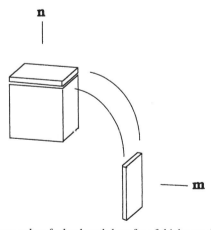

Figure 2. From cube of edge length l, wafer of thickness dl is detached and made to slide in its own plane through 90° around cylindrical surface. Suppose l is small and homogeneous sample of material is composed of compact mass of such cubes; if every cube yields wafer that behaves as in figure, sample as whole will undergo plane strain with principal values dl/l and $-dl/l$. Directions **n** and **m** are normals to wafer in original and final position.

To summarize, homogeneous straining of a continuum in a homogeneous stress field can be seen as a diffusion process; it can be correctly described as a diffusion process by equations 6 and 7. But for a mental picture to supplement the equations, we need to envisage the infinitesimal migrators as wafers and not spheres, the driving agent as the normal stress across the wafer and not the mean stress at the wafer's location, and the diffusion path of any wafer as an arc with a small but finite radius R.[1]

We now transfer the wafer concept to a situation where the stress field is not homogeneous. Consider two small regions $P1$ and $P2$, each wider than the dimension R and separated by a distance two or three times R or more; let the maximum principal stress be σ_1 within $P1$ and σ_2 within $P2$, with $\sigma_2 < \sigma_1$. Then a wafer subject to σ_1 can escape to a new location where the normal stress is lower, either around a short cylindrical path to a new orientation within region $P1$ or along a longer path to a new location within $P2$. We need to imagine a rather free flow of wafers; however, for book-keeping purposes, we can evaluate separately a flux due to change of orientation within a fixed location and a flux due to change of location at a fixed orientation. The point to be emphasized is that if equation 9 applies at all, it applies to both kinds of path. If we could straighten the curved path or twist the longer path while keeping d(normal stress)/ds unchanged at all points, the flux along the path would not be affected; it is axiomatic that where a gradient along some path drives a flux along that path, the relation of flux to gradient is not affected by the shape the path takes in external geographical space.

The wafers described are not real, but are just an aid to the imagination. By their use we write equations that describe a continuum that deforms as if by flow of wafers. These equations describe shortenings by diffusive loss that are anisotropic, and thus possibly bring us closer to a theory for the swarms of seams of insoluble residues and other comparable features in real materials.

As regards energy dissipation by diffusion in the finite-element setting, the result is that we sum terms involving, e.g., $d\sigma_{xx}/dx$, $d\sigma_{xx}/dy$ in one group, and $d\sigma_{yy}/dx$, $d\sigma_{yy}/dy$ in another group. Fluxes driven by these gradients were discussed by Bayly (1992, 1996), but an aspect not covered before is a gradient in shear stress such as $d\sigma_{xy}/dx$. Dissipation that would result from a gradient of this type is considered in Appendix 2. A useful check is that in a correct formulation, the total dissipation rate at a point per unit volume should be invariant under change of coordinate system.

Results

Our purpose is to pursue one of the many lines of research prompted by Hans Ramberg's seminal thinking, but at present we can give only an early progress report. Figure 3 shows preliminary results from our first exploration of a two-dimensional problem.

The problem centers on a highly viscous inclusion in an infinite less viscous host material (see Fig. 3A). The host material is subjected to a strain-rate field that is a homogeneous plane strain except insofar as the inclusion causes deviations. The included cylinder is infinitely long normal to the plane of strain, and we consider a circular section in the plane of strain. For coordinates, let the origin be at the center of the circle and take the principal directions of elongation and shortening as the x and y directions, respectively; then in the host material there is a region of high compression just outside the margin of the inclusion along the y axis, and a region of low compression, a stress shadow, just outside the margin along the x axis. One consequence is a set of diffusive fluxes around the quadrants from high-compression regions to low.

For a specific example, we take the inclusion to be both more viscous and less diffusive than the host by a factor of 4. Assumed magnitudes are:

Model 1: $N_{inc} = 1.00$ $K_{inc} = 0.0256$
          $N_{host} = 0.25$ $K_{host} = 0.1024$
Model 2: same $K_{inc} = 0.10$
                              $K_{host} = 0.40$

Profiles are shown in Figure 3B for the variation of σ_{yy} along the x axis. Also shown is the corresponding profile in absence of diffusion (Muskhelishvili, 1963; Jaeger and Cook, 1984).

The no-diffusion profile has a step at the inclusion's margin and a deep narrow valley just outside the margin. If diffusion occurs, material migrates into the valley; the step discontinuity is replaced by just a discontinuity in slope, and the stress drop outside the margin becomes less severe. In these respects the finite-element output agrees very well with semiquantitative expectations.

A disadvantage of the chosen problem is that, in the presence of diffusion, there is no benchmark analytical solution that our results should match. The first analytical solution admitting diffusion in the manner of Fletcher (1982) is that by Finley (1994, 1996), where two differences stand in the way of an exact comparison: first Finley assumed the mean stress field to be the control on diffusion (no wafer concept), and second she assumed implicitly a certain amount of anisotropy in the materials (as necessary, to permit an analytical solution in tractable form). The general shape of the profiles she shows agrees with that in Figure 3B, but detailed comparison is not justified. A more recent partial solution (Bayly, in press) is also unsuitable for comparison. For a quantitative test of our current approach a different problem will be needed.

HANS RAMBERG'S CONTRIBUTION

The essence of the preceding section is that σ_{xx} is one agent whose gradient might drive diffusion and σ_{yy} is a different agent. Or, admitting that the agent is in fact a spatial varia-

---
[1] Magnitude of R. If equation 8 is made specific by inserting the mean stress σ_0 in place of ϕ, then for equation 6 and 7 to be compatible with 8, one needs $R^2 = 4NK$. Values of K for real materials are not well known, but estimates of R range from a few nanometers in glass to more than a micrometer in olivine, in circumstances where creep is the dominant behavior (Bayly, 1992, p. 120; there the length in question is designated B).

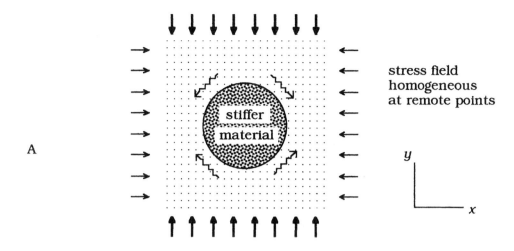

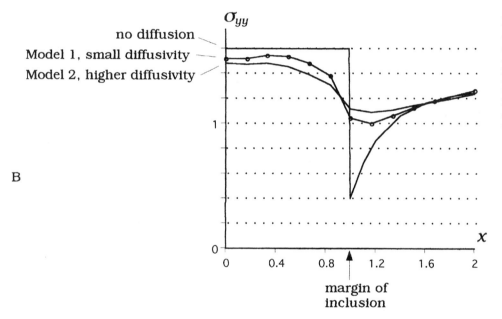

Figure 3. Test problem and preliminary results. A: Long cylindrical inclusion embedded in extensive less viscous host material (viewed in cross section). Straight arrows show uniform conditions imposed far from inclusion and crooked arrows show one part of diffusive response. B: Stress profiles along line outward from center of inclusion along x-direction. Profile with step and deep sharp valley is for materials that permit no diffusion and two less angular profiles are for two diffusive assemblies specified in text. These less angular profiles agree with expectation in semi-quantitative way but have not yet been tested quantitatively for validity.

tion in a single tensor field, for purposes of evaluation we treat σ_{xx} and σ_{yy} separately. At a single point, the agent driving diffusion is a tensor, or a multivalued scalar, or an infinite set of scalar magnitudes, not a single valued scalar. This was recognized and put forward by Hans 40 years ago (Ramberg, 1959; for the same proposal in a more accessible journal, see Ramberg, 1963). In this volume, it is appropriate to look at the context of his proposal.

Gibbs (1878) treated a cube of a solid material bounded by three pairs of fluid cells at pressures P_x, P_y, and P_z; he proposed that at equilibrium the chemical potential of the material of the solid in the fluids would be $\mu_0 + P_x V$, $\mu_0 + P_y V$, and $\mu_0 + P_z V$ (μ_0 = a reference potential, V = molar volume of the material of the solid). He gave no statement about the chemical potential of the material of the solid in the solid phase.

Ramberg proposed that at any point in the solid phase, three potentials could be identified: "The quantities $P_x V$, $P_y V$ and $P_z V$ may . . . be recognized as the potential spatial energy of the solid in the directions of the principal axes of stress" (1959, p. 4; 1963, p. 243). Ramberg did not comment on directions other than principal directions, and avoided explicitly stating that the "chemical potential" had three values at a single point, but he treated the interior of the solid phase and indicated three values coexisting at the same point.

Kamb (1959) enlarged on Gibbs' work in a different direction. He continued to avoid the interior of the solid phase and retained the idea that at a single point, the chemical potential had a single value; he proposed that at an interface where the normal stress across the interface was σ_n, the chemical potential of a component would be $\mu_0 + \sigma_n V$ regardless of whether σ_n was a principal stress.

Bowen (1967) in some sense combined the proposals of Kamb and Ramberg. He proposed that at a point in the interior of the solid phase, the chemical potential would be $\mu_0 + \sigma_n V$ for a surface on which the normal stress component was σ_n and at the same point would be $\mu_0 + \sigma_m V$ for a surface on which the normal stress component was σ_m; the potential varied with orientation, as with Kamb, and took several values at a single point, as with Ramberg. He proposed a chemical potential tensor **M**, where

$$\mathbf{M} = \mu_0 \mathbf{I} + \boldsymbol{\sigma} V \qquad (10)$$

(Bowen, 1976, equation 1.6.22, p. 31); here $\mathbf{I}$ = the unit tensor. A strong endorsement of this proposal was given by Grinfeld (1991, p. 2, 125–132).

Aside from the term in μ_0, the principal values of Bowen's **M** are the three terms named by Ramberg. As far as we know, Hans was the first person to propose a suite of magnitudes with the character of potentials coexisting at a single point. In that respect, his work supports the middle section of this chapter; we are glad to follow his footsteps, or to be under his aegis.

Hans had deep insight regarding both the chemical and the mechanical behavior of rocks; his ideas are the starting point for many aspects of our work. It is for this reason that we give the present report, despite the project being at only a preliminary stage; we hope eventually to develop a useful outgrowth of his work.

ACKNOWLEDGMENTS

We thank Robert Oakberg, Sharon Finley, and Ray Fletcher for unstinted help over many years, and Mark Taylor and Bertram Schott for helpful comments.

The list of our debts would not be complete without naming Mrs. Ramberg. The pleasure of associating with Hans was constantly augmented by Lillemör's presence: her humor, zest, and unique point of view raised working with Hans from the mainly academic to the agreeably convivial. We are glad of the opportunity to acknowledge this long-standing debt.

APPENDIX 1. FINITE ELEMENT IMPLEMENTATION

The five aspects noted in this appendix are the viscous dissipation, the diffusive dissipation, the boundary work rate, conservation of momentum, and conservation of mass. Because of parallels with the more familiar process of solving for velocities, the diffusive dissipation is treated first. The general principles given in the main text can be implemented in many ways. The following paragraphs note the procedures we are using in the present exploratory stage of the work. It is likely that they will be modified in future, as experience grows and other types of physical problem are attacked. Our main guide has been the exposition by Hughes (1987).

We work in just two Cartesian dimensions, x and y. In our present formulation, each element has four nodes. Each node is assigned a stress state specified by values of σ_{xx}, σ_{yy}, and σ_{xy}, and to reduce the number of suffixes on the page, in this appendix these scalar magnitudes are renamed p, q, and t. If we focus attention on the p values, an element will have four of these, p_{1-4}, at its four nodes. These imply a gradient in p across the element, and our first objective is to assess a rate of dissipation of energy of the form $K\,(\text{grad}\,p)^2$. Reverting to the more common procedure, where the nodal unknowns are velocities, e.g., u in the x direction, we remark that the strain rate there is the gradient in u and the dissipation rate has the form $J\,(\text{grad}\,u)^2$. As now defined, p and u are both scalar fields, and the methods for interpolating and integrating across an element apply equally well to either.

We interpolate using a bilinear scheme. This is illustrated in Figure A1.1 for a square element with sides of length one unit, and transforms algebraically for nonsquare elements. By this scheme, variables such as dp/dy and $(dp/dy)^2$ can be evaluated at any point within the element. In principle we evaluate the latter at all points and do a double integration across the element's area. But because of the simple form of the interpolated surface, we can instead evaluate the variable at just a finite number of Gauss points (Hughes, 1987, p. 137), and currently use four such points.

For the viscous dissipation, the scalar fields for t^2 and $(p - q)^2$ are treated in a similar way.

Boundary work

Considering an assembly of many elements, the objective is to minimize the difference between the rate of doing work around the boundary of the assembly and the rate of dissipation within the interior. We currently set up the problem of interest by specifying both velocity components at all boundary nodes. These would ideally be at an infinite distance from the inclusion of interest, where the strain rate field is homogeneous and $u = ax$, $v = -ay$ (a = strain rate). In practice, we apply these two velocity conditions at all boundary nodes and evaluate the boundary work rate (using linear interpolation if necessary), with the boundary not infinitely remote, but as remote as can be conveniently arranged with our current mesh-generating software.

Conservation of momentum

The objective is to solve for the set of nodal magnitudes of p, q, and t that give the minimum difference in energy rates as just described, but all subject to the condition that momentum be conserved throughout the assembly. In the coordinate system used, this requires that

$$dp/dx + dt/dy = 0$$

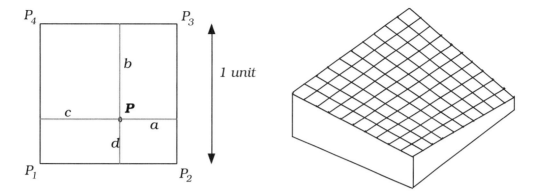

$$P = abP_1 + bcP_2 + cdP_3 + daP_4$$

Figure A1.1. Bilinear interpolation scheme used. It is applied here over unit square, but transforms linearly to apply over any quadrilateral element. P_{1-4} are nodal values; P is value to be interpolated at interior point using distances a,b,c,d.

and

$$dq/dy + dt/dx = 0$$

at all points. In these equations each left side can be considered a single variable, a single-valued function of position; e.g., let $f(x,y) = dp/dx + dt/dy$. Then by the interpolation scheme already described, f at any point in the element can be expressed in terms of p_{1-4}, t_{1-4} and the element dimensions; then f can be integrated over the element via the Gauss points, to yield Int f, an expression that is still linear in p and t. We introduce a Lagrangian multiplier λ_j for every element $j = 1 \ldots M$ (M = number of elements) to form the sum of products $\lambda_j(\text{Int } f)_j$ and a second multiplier to handle the second condition $dq/dy + dt/dx = 0$. The sums thus formed are treated as work-rate quantities and conceptually added to the total dissipation rate for the assembly exactly as in conventional problems without diffusion.

As regards data storage, we find it convenient to assign each element a fifth node, a dummy node with no specific position in the geometry of the element. Then the two Lagrangian multipliers can be treated as unknowns associated with this node, in the same way that a trio of unknowns viz. p, q, and t is associated with each of the four true nodes, for a total of 14 unknowns per element.

Conservation of mass

The work we have done so far has been limited to solving for an instantaneous stress field. However, a long-range aim is to follow the evolution of a structure through time, and for this purpose it will be necessary to calculate a velocity field. At this point, two difficulties will appear, both arising from the fact that the finite-element method only approximates a real situation. First, if the stress results are used to evaluate strain rates point by point for many points within one element, the calculated strain rates may not form a compatible set; second, at the boundary between one element and the next, the postulated fluxes to the boundary on one side and away from it on the other side may not match. Both difficulties can be regarded as failures to conserve mass; they have exact parallels in the more common solutions for nodal velocities, which normally fail to conserve momentum.

As long as one uses finite elements and linear equations, there is no way to eliminate these defects. However, a suitable choice of mesh makes them small, and they can be further reduced by smoothing. Failures to conserve momentum have proved manageable in standard finite-element work, and we expect to develop comparable procedures, but conservation of mass will certainly require attention in future extensions of the present work.

The results shown in Figure 3B are strictly preliminary. We have yet to begin the process of validating, assessing magnitudes of error, and testing alternative solution procedures. We propose only that the results are plausible in a semiquantitative way.

APPENDIX 2. CONSTITUTIVE RELATIONS

In a heterogeneous stress field, at some point P_1 there will be three principal stresses and at another point P_2 three principal stresses again. The essence of the wafer proposal is that diffusive mass transfer from P_1 to P_2 will be affected by all six of these principal stress magnitudes and not by just the two magnitudes of mean stress. For practical calculations, we make the further supposition that with Cartesian coordinates x, y, and z, the stress components on coordinate planes, e.g., σ_{xx} and σ_{yy}, can be used to predict diffusion effects. Thus if stress component σ_{yy} has a gradient along x, we suppose that there will be a flux along direction x of wafers normal to y. Using F for flux, we write

$$F_{yyx} = -K\, d\sigma_{yy}/dx. \qquad (2.1)$$

Here F is measured in m^3 per m^2-s or ms^{-1}, so that K, the diffusivity or transmissivity, is in m^{2}Pa^{-1}s^{-1}. This relation defines K in any instantaneous circumstances; for present purposes we assume that K is constant, giving a linear relation between flux and gradient.

In parallel with K, two more coefficients J and H are introduced. The basis for J is shown in Figure A2.1: if K describes behavior as in part A, a different coefficient is needed for behavior as in part B and J is used; thus,

$$F_{yyy} = -J\, d\sigma_{yy}/dy. \qquad (2.2)$$

However, if $d\sigma_{yy}/dy$ is nonzero and body forces and inertia are absent, there must also be a gradient or gradients in shear stress; in two dimensions $d\sigma_{xy}/dx = -d\sigma_{yy}/dy$. In parallel with the relations for J and K, we might expect a relation

$$F_{xyx} = -H\, d\sigma_{xy}/dx. \qquad (2.3)$$

To link this relation to a physical picture, we need Figure A2.2.

In the figure, the general stress state σ_{xx}, σ_{yy}, σ_{xy} is replaced by two superimposed states, each of which is a pure shear; numerically, $s_1 = \sigma_{xy1}$ and $s_2 = \sigma_{xy2}$. Suppose that s_1 is greater than s_2: then we see a gradient $(s_1 - s_2)/(x_1 - x_2)$ tending to drive a flux from left to right, and a second gradient $(-s_1 + s_2)/(x_1 - x_2)$ driving a flux from right to left. If equation 2.3 describes one of these, then a second equation,

$$F_{yxx} = H\, d\sigma_{yx}/dx, \qquad (2.4)$$

is needed to describe the other.

For a physical picture that is still more specific, consider steady uniform flow between two parallel plates with a parabolic velocity profile, as in Figure A2.3.

In the classical picture, the flow is laminar and the directions one naturally uses as axes are parallel and perpendicular to the flow direction, as shown by x and y in the figure. With respect to these axes, the deformation of any fluid element is a simple shearing. However, if instead one pays attention to lines at 45° to the direction of flow, one sees shortening along d_1 and e_1 and elongation along d_2 and e_2. If we also refer the stress state to these 45° directions, we find compression along d_1 and tension along d_2, or at least a gradient from a larger compression at d_1 to a smaller compression at d_2.

The stress state in Figure A2.3 is very like the stress state in Figure *A*2.2 as regards shear stresses on xy planes and compressive stresses on planes at 45°. Now suppose the gradient in compressive stress from left to right across the fluid in Figure A2.3 drove a flux of wafers of material with 45° orientation across the flow, as suggested in the diagram in the central region d. Consequences would be loss of wafers and shortening along d_1 and gain of wafers and elongation along d_2; the flux of these hypothetical wafers would add to and accentuate the deformation that was already occurring by the classical viscous flow. The same is true of a flux of hypothetical wafers of the other 45° orientation in the opposite direction, affecting dimensions e_1 and e_2 (shortening e_1 and elongating e_2). That is, the fluxes for which coefficient H is proposed would augment the deformation already occurring in the classical manner; they would deepen or steepen the parabolic profile (and distort it slightly so as to be no longer exactly parabolic).

The add-on character of the material transfers proposed can be used as an argument for their being real. One discovers this by starting the argument from the other end. We first postulate a nondiffusing viscous fluid and note the pattern of strain rates that develops between parallel plates. We then grant the fluid a second mechanism for dissipating energy, namely mass transfer or self diffusion. Then the fluid must dissipate more energy per unit time; in such simple circumstances, to do what it was doing before, but faster, seems a natural accommodation.

The idea of wafers of material retaining a definite orientation while migrating down a stress gradient is imaginative, and

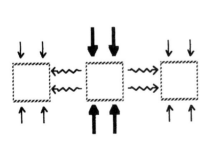

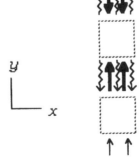

Figure A2.1. Geometrical bases for coefficients K and J. A: Nonuniform compression σ_{yy} (straight arrows) and resulting flux along x (crooked arrows) are linked through coefficient K. B: Nonuniform compression σ_{yy} and resulting flux along y are linked through coefficient J.

A   B

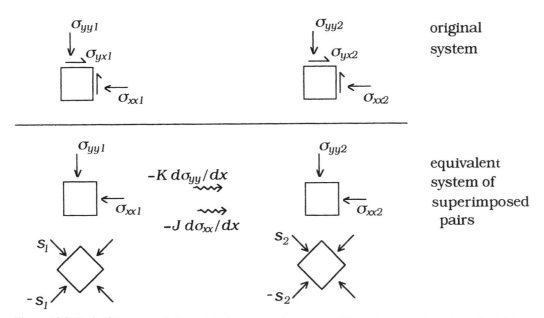

Figure A2.2. Each of two general stress states is expressed as superposition of two pure-shear states. Straight arrows and symbols σ and s show stress components. Crooked arrows show resulting fluxes along x. Flux magnitudes are governed by coefficients K and J as in Fig. A2.1.

to imagine two such fluxes interpenetrating without interfering is highly artificial. But equation 5, viz.

$$e_{nn} \quad \text{or} \quad \frac{\sigma'_{nn}}{2N} = -\frac{1}{12N}\left(\frac{\partial^2 \sigma_{nn}}{\partial \theta_1^2} + \frac{\partial^2 \sigma_{nn}}{\partial \theta_2^2}\right)$$

is fundamental, and it supports the idea that materials behave as if wafers were present with behavior as described. By considering two sets of axes at 45° to each other, one can infer that $H = (J + K)/2$, but in the present exploratory stage, to keep H as a separate parameter seems helpful.

Use of the constitutive equations

The uses are, first, to write expressions for the rates of dissipation of energy by diffusion; these are added into the total dissipation rate to which the extremum principle is applied, as a means to find the stress field; and second, after the stress field has been found, the fluxes are used again in estimating the element strain rates and so constructing a velocity field. The velocity field is needed to advance the model through a time step to a new geometrical configuration.

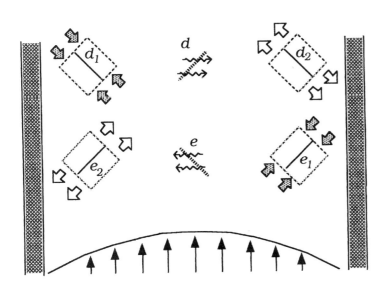

Figure A2.3. Laminar flow between parallel plates. By classical viscous flow, dimension d_1 and e_1 will diminish and d_2 and e_2 will increase. Crooked arrows show hypothetical diffusive fluxes through axial regions d and e; these fluxes would augment classical flow if, in effect, they transported microscopic wafers with specific orientations down stress gradients that run across flow. In practice, diffusive effects are negligible except where fluid's microstructure and wall separation are on comparable length scales, but they are needed for completeness in comprehensive theory.

The result is likely to be that, within an element, the shortening rate by diffusive loss is greater in one direction than in another. One purpose of the constitutive relations proposed here is to permit this outcome, and so give a description that resembles what we seem to see in outcrops. The entire approach has a tentative or exploratory character.

REFERENCES CITED

Bayly, B., 1992, Chemical change in deforming materials: New York, Oxford University Press, 224 p.

Bayly, B., 1996, Stress and diffusion in a one-component viscous continuum, *in* Beke, D.L., and Szabo, I.A., eds., Diffusion and stresses: Defect and Diffusion Forum, v. 129–130, p. 281–288.

Bayly, B., 2000, Deformation with diffusion: The growth of augen, *in* Jessell, M.W., and Urai, J.L., eds., Stress, strain, and structure: A volume in honour of W.D. Means: Journal of the Virtual Explorer, v. 2, ch. 1.

Bethke, C.M., 1985, A numerical model of compaction-driven groundwater flow and heat transfer and its application to the paleohydrology of intracratonic sedimentary basins: Journal of Geophysical Research, v. 90, p. 6817–6828.

Bowen, R.M., 1967, Toward a thermodynamics and mechanics of mixtures: Archive for Rational Mechanics and Analysis, v. 24, p. 370–403.

Bowen, R.M., 1976, Theory of mixtures, *in* Eringen, A. C., ed., Continuum physics, Volume 3: New York, Academic Press, p. 1–127.

Danielewski, M., and Filipek, R., 1996, Generalized solution of interdiffusion problem: Optimal approach for multicomponent bounded systems: Journal of Computational Chemistry, v. 17, p. 1497–1507.

Dieterich, J.H., and Onat, E.T., 1969, Slow finite deformation of viscous solids: Journal of Geophysical Research, v. 74, p. 2081–2088.

Finley, S.G., 1994, Diffusion-enhanced deformation of a circular inclusion, [Ph. D. thesis]: Troy, New York, Rensselaer Polytechnic Institute, 92 p.

Finley, S.G., 1996, Diffusion-enhanced deformation of a circular inclusion, *in* Beke, D. L., and Szabo, I. A., eds., Diffusion and stresses: Defect and Diffusion Forum, v. 129–130, p. 289.

Fletcher, R.C., 1982, Coupling of diffusional mass transport and deformation in a tight rock: Tectonophysics, v. 83, p. 275–291.

Gibbs, J.W., 1878, On the equilibrium of heterogeneous substances: Connecticut Academy Transactions, v. 3, p. 343–524.

Greer, A.L., 1995, Diffusion and reaction in thin films: Applied Surface Science, v. 86, p. 329–337.

Grinfeld, M., ed., 1991, Thermodynamic methods in the theory of heterogeneous systems: Harlow, UK, Longman Scientific and Technical Publishers, 399 p.

Hughes, T.J.R., 1987, The finite element method: Englewood Cliffs, New Jersey, Prentice-Hall Inc., 803 p.

Jaeger, J.C., and Cook, N.G.W., 1984, Fundamentals of rock mechanics: New York, Chapman and Hall, 593 p.

Kamb, W.B., 1959, Theory of preferred crystal orientation developed by crystallization under stress: Journal of Geology, v. 67, p. 153–170.

McKenzie, D., 1984, The generation and compaction of partially molten rocks: Journal of Petrology, v. 25, p. 713–765.

Muskhelishvili, N.I., 1963, Some basic problems of the mathematical theory of elasticity: Groningen, P. Noordhoff, 718 p.

Opposits, G., Szabo, S., Beke, D.L., Guba, Z., and Szabo, I.A., 1998, Diffusion-induced bending of Cu-Ni thin sheet diffusion couples: Scripta Materialia, v. 39, p. 977–983.

Person, M., Raffensperger, J.P., Ge, S.M., and Garven, G., 1996, Basin-scale hydrogeologic modeling: Reviews of Geophysics, v. 34, p. 61–87.

Ramberg, H., 1959, The Gibbs' free energy of crystals under anisotropic stress, a possible cause for preferred mineral orientation: Anais da Escola de Minas de Ouro Preto, v. 32, supplement, p. 1–12.

Ramberg, H., 1963, Chemical thermodynamics in mineral studies: Physics and Chemistry of the Earth, v. 5, p. 225–252.

Spiegelman, M., 1993a, Flow in deformable porous media. Part 1, Simple analysis: Journal of Fluid Mechanics, v. 247, p. 17–38.

Spiegelman, M., 1993b, Flow in deformable porous media. Part 2, Numerical analysis—The relationship between shock waves and solitary waves: Journal of Fluid Mechanics, v. 247, p. 39–63.

MANUSCRIPT ACCEPTED BY THE SOCIETY APRIL 12, 2000

Modeling of anisotropic grain growth in minerals

Paul D. Bons*
Institut für Geowissenschaften—Tektonophysik, Universität Mainz, Becherweg 21, 55099 Mainz, Germany
Mark. W. Jessell, Lynn Evans, Terence Barr
Epsilon Laboratory, Department of Earth Sciences, Monash University, Clayton (Melbourne), VIC 3168, Australia
Kurt Stüwe
Institut für Geologie, Universität Graz, Heinrichstraße 26, A-8010 Graz, Austria

ABSTRACT

The role of surface energy anisotropy during grain growth is investigated using both physical experiments on octachloropropane and numerical experiments using the Elle microstructural modeling system. In particular the effects of anisotropy on growth rates, grain shapes, and lattice-preferred orientations are analyzed. Anisotropic growth in thin polycrystalline sheets of octachloropropane is found to systematically remove certain *c*-axis orientations from the sample, without obviously modifying the grain shapes. By comparison with equivalent numerical experiments, we can explain these observations with a simple boundary-layer model that treats each side of each grain-boundary as an independent system. The observed growth exponent in the physical experiment of 0.35 is significantly below the theoretical level of 1, and although the numerical experiments have growth rates as low as 0.68, apparently some other factor that suppresses the growth rate in the octachloropropane was also active. The anisotropy of surface energies also leads to an increase in the grain-size distribution, and when it occurs in rocks with a strong preexisting crystallographic preferred orientation, may even lead to the development of a shape-preferred orientation that is completely unrelated to the shape of the finite strain ellipsoid.

INTRODUCTION

The driving force for grain-boundary migration in rocks may be a product of deformation, the result of chemical potential gradients, or the inherent surface energy of the boundaries (Urai et al., 1986). Grain boundaries in a grain aggregate have a surface energy (γ) typically in the order of 0.1–1 J/m^2 (Urai et al., 1986, and refs therein). If we take a two-dimensional view of the problem, grain boundaries can migrate to achieve a local minimum surface energy configuration, given by straight boundaries and equal 120° angle intersections. Because the two cannot be achieved simultaneously, smoothly curved grain boundaries develop that meet at ~120°. This balance is very similar to the growth of bubbles in a foam and the result is a grain aggregate that resembles a foam and is therefore often termed a foam texture. Larger than average, concave outward, and many-sided grains on average grow, while smaller, convex, and fewer sided grains shrink and eventually disappear. Grain growth results, as the average grain-size must increase as the total volume of grains remains the same, but their number decreases. For reviews on the topic, see Smith (1964), Weaire and Rivier (1984), and Anderson (1986).

Grain growth is an important process in metamorphic rocks. Grain growth can, for example, modify or in some cases completely obliterate an existing deformation microstructure (Fig. 1) (Bons and Urai, 1992). Although most investigations have focused on static grain growth, it can also occur during deformation (Jessell, 1986; Gleason et al., 1993). It can balance grain-size reduction mechanisms such as dynamic recrystallization or cataclasis,

*E-mail: bons@mail.uni-mainz.de

Figure 1. Example of unusually coarse foam texture in quartzite. Traces of isoclinal folds in hand specimens of quartzite indicate that rocks were once intensely deformed and probably strongly recrystallized and fine grained. Nothing of that deformation microstructure has remained. Hidden Valley, Arkaroola Property, South Australia. Crossed polars.

leading to stable grain sizes (Tullis and Yund, 1982; Olgaard and Evans, 1986; Urai et al., 1986; Ree, 1991; de Bresser et al., 1998).

The average grain-size (S) increases as a function of time (t), which can be described by

$$S = kt^G, \quad (1)$$

where k is a material department parameter and G the growth exponent (Anderson, 1986). If the grain-size is defined by the average grain radius or diameter, the growth exponent G_R is 0.5 for the ideal case where the surface energy (γ) and the grain-boundary mobility are constants. If one takes the average cross-sectional area (A) as a measure for grain size, one gets a growth exponent $G_A = 2 \cdot G_R = 1$ for the ideal case. Growth exponents of $G_A=1$ are found in numerical models and some experiments (Karato, 1989), but often lower values are observed, ranging from 0.2 to 1 (Haessner and Hofmann, 1978; Tullis and Yund, 1982; Anderson et al., 1984; Grest et al., 1985). Slowing down of grain-boundary migration rates by pinning or dragging of second phase particles or fluid inclusions can be a cause for a lower than unity growth exponent (e.g., Olgaard and Evans, 1986; Randle et al., 1986). Srolovitz et al. (1984), however, argued that pinning does not affect the growth exponent noticeably, until an abrupt transition occurs and grain growth stops, although the exact behavior probably depends on both the size and spatial distributions of the pins. Another cause for a reduction in the growth exponent can be that the grain-boundary energy is not a constant, but a function of the relative orientations of the grain boundary and the adjacent crystal lattices (Grest et al., 1985). Anisotropy of grain boundary energy leads to euhedral grain shapes for many minerals during metamorphic or igneous growth, but little is known of its role in single phase grain growth in rocks. Grest et al. (1985) simulated anisotropic grain growth and found that it can cause as much as 50% reduction in the growth exponent.

In this study we investigated anisotropic grain growth to gain more understanding on its role and effects in rocks. The primary aims are to determine the effect of anisotropy on the growth exponent and to find the microstructural characteristics of anisotropic grain growth, so it may be recognized in thin sections of rocks. Following this, we may also gain insight to the behavior and migration mechanisms of grain boundaries of rock-forming minerals at depth. In the spirit of Hans Ramberg, we used an analogue material to experimentally study grain growth. We first describe an experiment with octachloropropane, which exhibits anisotropic growth. With physical experiments, however, one is constrained by the limited range of properties and behavior of the available materials. This problem can be overcome with numerical models, where one can freely vary all parameters. The advent of numerical modeling has not made experiments with analogue materials obsolete, but instead has increased their importance. Such experiments are still very useful to physically simulate geological processes. They have also gained a new role as well-defined test cases to validate numerical modeling. In the second part of this chapter, we present grain growth simulations with the numerical model Elle and used the octachloropropane experiment as such a test case.

EXPERIMENTAL GRAIN GROWTH IN OCTACHLOROPROPANE

Octachloropropane (abbreviated to OCP) has been used in several studies on grain growth (McCrone, 1949; Bons and Urai, 1992; Ree and Park, 1997) and dynamic recrystallization (Jessell, 1986; Means, 1983, 1989; Means and Ree, 1988; Ree, 1990, 1991, 1994). OCP is a crystalline organic compound (C_3Cl_8) with a melting temperature of 160 °C that deforms and recrystallizes readily, even at room temperature. It is therefore well suited for in situ experiments with a thin sample between glass plates. The sample can then be studied with a microscope during an experiment (Tamman and Dreyer, 1923; Means, 1989). Under static conditions, OCP develops a foam texture with equidimensional grains in minutes to hours, depending on temperature. A possible role of anisotropy on the grain growth process has not been studied in any detail, although McCrone (1949) suggested that it is unimportant in materials such as OCP. Ree and Park (1997) studied the effect of postdeformational recrystallization on an existing lattice-preferred orientation (LPO) in OCP and found that it did not alter the LPO noticeably, but they did not follow growth over an extended period of time. They also did not specifically consider anisotropic grain growth. Considering that OCP is an anisotropic crystalline material with a hexagonal crystal symmetry, one would expect the anisotropy to have some effect on grain growth. To investigate this we carried out one static recrystallization experiment with OCP.

Sample preparation

We made a thin sample of OCP by melting a small quantity on one (thin section) glass plate and then quenching the sample by pressing a cold second glass plate on top of the molten sample. The quenching produces a fine grain-size (smaller than the thickness of 50–100 µm of the sample). Recrystallization and grain growth commences immediately, first driven by both internal strain energy from the quenching and pressing, and by grain-boundary energy reduction. Internal energy is released from the crystal lattice when a grain boundary sweeps through it. After a 10-fold increase in the mean grain radius, 99.9% of the material is swept at least once by grain boundaries and one can regard any ongoing recrystallization as purely driven by the reduction of grain-boundary energy. We estimated that this stage was well reached when the mean grain area was about 0.1 mm², a foam texture was established, and all grain boundaries had oriented perpendicular to the glass plates, because this minimizes the area of each boundary. From that point onward, taken as the start ($t = 0$) of the experiment, gray-scale digital video images (756 × 512 pixels) were taken at 1 h intervals (Fig. 2A), while the sample was left at room temperature.

Results

During the 96 h of observation, the mean grain size (A) increased from 0.1 mm² to 0.45 mm² (Fig. 2B). Growth as a function of time could be approximated (see Appendix 1) with a power law: $A = 0.0046 \times t^{0.35}$. The size frequency distribution can be expressed in terms of the second moment (μ_2), defined by:

$$\mu_2 = \frac{1}{N} \sum \left(\frac{A_i}{A} - 1\right)^2, \quad (2)$$

where N is the number of grains. At $t = 0$, the second moment is 0.7. The c-axis orientations in a selected ~4 mm² area with 72 grains were measured at $t = 0$ (Fig. 2, C–E). The c-axes distribution shows no preferred azimuth (Fig. 2D), but steep plunges are distinctly overrepresented relative to shallow plunges (Fig. 2E). A possibility is that the LPO is a result of the initial pressing of the sample, and Ree and Park (1997) showed that an LPO can be preserved during static recrystallization. However, we do not believe this is the case, because the LPO strengthens toward the end of the experiment. Of 24 grains with a plunge of <45° only one grain survives, while 15 out of 48 of the remaining grains survive to $t = 96$ h. The chance that this is a result of nonpreferential removal of grains with a shallow plunge is <2%. This does not exclude the possibility that grains with a steep plunge were on average relatively large in the initial aggregate and therefore had a more than average chance of survival during grain growth (Y. Park, 1999, personal commun.). We have no measurements on the aggregate immediately after quenching and pressing and therefore have no evidence in favor or against this hypothesis. However, we think it is most likely that the LPO formed as a result of anisotropic static grain growth.

NUMERICAL EXPERIMENTS WITH ELLE

The numerical model Elle is designed to simulate two-dimensional microstructural developments. Although it can simulate the operation of many different concurrent processes, here we use it to simulate static grain growth in isolation. In Elle, grains are represented by polygons. Their boundaries are defined by nodes that are linked by straight boundary segments. A maximum distance (L_{max}) between nodes is defined by the user. The average spacing of nodes is kept roughly constant by adding or removing nodes. Grain-boundary migration is simulated by moving each node in turn over a small distance. The grain-boundary network is checked after every node movement to check for topological inconsistencies and the removal of small grains. To avoid boundary effects, the entire aggregate has periodic boundaries.

Definition of anisotropy type

To simulate anisotropic grain growth, we have to define the function (F) that relates the grain boundary energy (γ) to the angular relationship (defined by five angles α_i) between the orientation of the grain boundary and the lattices of the crystals A and B on either side of the grain boundary:

$$\gamma = F(\alpha_1, \alpha_2, \alpha_3, \alpha_4, \alpha_5). \quad (3)$$

In materials (metals, ceramics) where a grain boundary can be regarded as a planar lattice defect, the function can be complicated, with distinct peaks and troughs at certain orientations (e.g., Gleiter, 1969). The troughs occur where the two lattices fit together relatively well in coincident lattice geometries. Such low-energy coincident lattice grain boundaries should be overrepresented in this type of material. This has not, however, been observed in minerals such as quartz (Geoff Lloyd, 1999, personal commun.), which suggests that a boundary layer model (Urai et al., 1986) for the grain boundary is more appropriate for these materials. Here, the grain boundary is envisaged as a layer with two parallel surfaces that face crystals A and B. Each surface has a surface energy that is a function of the relative orientation of that surface and one of the two neighboring grains. The total grain boundary energy is simply the sum of the two interfacial surface energies, and for each two angles that describe the relative orientation of one lattice relative to the grain boundary surface:

$$\gamma = F(\alpha_{1A}, \alpha_{2A}) + F(\alpha_{1B}, \alpha_{2B}). \quad (4)$$

Not only grain-boundary energy, but also mobility can be lattice orientation controlled. In our modeling, we assumed the mobility to be isotropic. This is justified if the mobility is controlled by the transport rate across the boundary layer. Hence, isotropic (probably diffusional) transport controls the rate of migration for a given driving force, which is determined by the anisotropic grain-boundary energy. This assumption is probably correct for wet mineral aggregates (Tullis and Yund, 1982), although perhaps not when the aggregate is stressed (Karato and Masuda, 1989). Whether this assumption is correct for OCP remains to be determined.

Next, we have to make a choice for the function $F(\alpha_1, \alpha_2)$. For simplicity, we took an axial symmetric crystal model, with the c-axis as the axis of rotation. The angular relationship

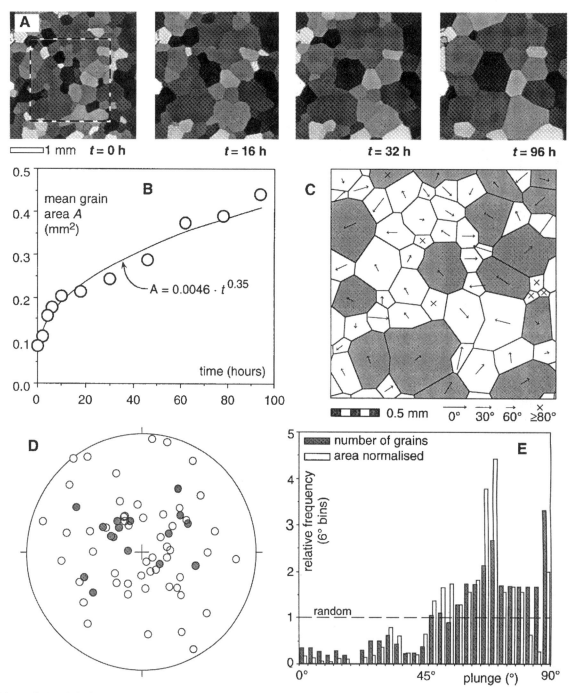

Figure 2. A: Digital video images of octachloropropane sample that show more than four-fold increase in mean grain area over 96 h. Cross-polarized light. Rectangle in image for time, $t = 0$ is outline of area where c-axes were measured. B: Mean grain area versus time with least squares best fit for power-law type growth function, giving low growth exponent $G_A = 0.35$. C: Drawing of area where c-axes were measured at $t = 0$ h. Orientations of c-axes are indicated by arrows (long arrows for shallow plunges; short arrows for steep plunges; cross for plunge >80°). Grains that are shaded gray survived until end of experiment. D: Equal-area stereo projection of c-axes in selected area, showing no preferred azimuth. Gray circles are grains that survived until end of experiment. E: Plot of c-axes plunge distribution at $t = 0$ h, normalized to frequency expected for random c-axis distribution (dashed line), as function of plunge. Dark bars give frequency distribution in number of grains per 6° bin and white bars in normalized area per 6° bin (both calculated every 3°). Distribution is clearly not random: steep plunges are overrepresented. Grains with steep plunges are also on average larger than grains with shallow plunges.

between lattice and grain boundary can now be defined by a single angle (α) between the c-axis and the grain boundary surface (Fig. 3A). The function $F(\alpha)$ is determined by a geometrical construction for a cylinder. The surface energy for a given α is taken as proportional to the distance from the center of a cylinder, to the surface of the cylinder along a straight line normal to the grain boundary (Fig. 3B). The c-axis runs up the center of the cylinder, which is defined by the diameter:height ratio of $D{:}h$. This procedure is identical to the Wulff construction (Wulff, 1901; Kim et al., 1994) for the crystal facets (here cylinder surfaces), but differs for generally oriented surfaces, and is known as the γ^{-1} polyhedron (Frank, 1963). The Wulff construction predicts convex-upward sections in the γ versus α graph, as general surface orientations are assumed to be stepped or faceted. Our construction produces concave-upward sections, and thus avoids this faceting (Cabrera and Coleman, 1963). Different functions $F(\alpha)$ can be chosen, depending on which crystal habit, defined by the ratio $D{:}h$, one wants to model (Fig. 3C). All curves consist of two concave-upward sections, and hence all possible surface orientations are stable and do not fall apart in stepped surfaces. The grain-boundary energy anisotropy of OCP, which has a flat hexagonal natural habit (Bons, 1993), is approximated with a $D{:}h$ ratio larger than one.

Node movement routine

The local grain boundary energy state (E) for a node, at position p, and its surroundings is described by:

$$E = L_1(\gamma_{\alpha 1} + \gamma_{\beta 1}) + L_2(\gamma_{\alpha 2} + \gamma_{\beta 2}), \quad (5)$$

where the distances between the node and its two neighbors are given by L_1 and L_2, and the four angles α_1 to β_2 are the different angles between the grain-boundary and the c-axes of the two grains (A and B) for each of the two grain-boundary segments (Fig. 3D). For a triple junction between grains A, B, and D (Fig. 3E), the equation is given by:

$$E = L_1(\gamma_{\alpha 1} + \gamma_{\beta 1}) + L_2(\gamma_{\beta 2} + \gamma_{\delta 2}) + L_3(\gamma_{\alpha 3} + \gamma_{\delta 3}), \quad (6)$$

We effectively model a thin polycrystalline sheet in which the c-axes have three-dimensional orientations and the grain boundaries are assumed to be perpendicular to the modeling plane.

A small displacement of the node changes the total energy state by ΔE. ΔE is calculated for a small trial displacement vector (Δp) of distance $L_{max}/100$ in a random direction. This procedure is repeated for three more displacements, each time rotating the trial displacement vector by 90°. The displacement vector that gives the largest reduction of the local energy state is then taken as the movement direction of the node. Taking ΔE positive for a reduction in energy, we calculate the actual displacement vector of the node (v) with

$$v = \frac{\Delta p}{|\Delta p|} \cdot \Delta E \cdot M, \quad (7)$$

where M is the mobility of the grain boundary.

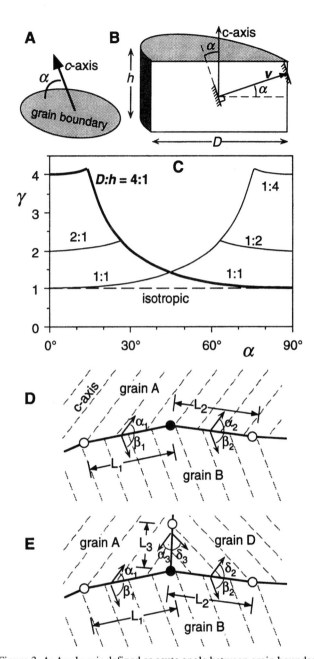

Figure 3. A: Angle α is defined as acute angle between grain boundary surface and c-axis. B: Surface energy for grain boundary is taken as absolute length of normal to that boundary (vector v) from center to surface of cylinder, which has its c-axis as axis of rotation. Cylinder dimensions are defined by diameter D and height h. C: Surface energy γ as function of α for different crystal models, ranging from flat tablets ($D > h$) to rods ($D < h$). Minimum surface energy in all cases is one. Concave-upward curves have one peak each and overlap part of range. Whole curve for $D{:}h = 4{:}1$ is shown in bold. γ is independent of α in case of isotropic growth (dashed line). D: Total energy state around node that links two boundary segments (equation 5) is given by lengths of these segments and four angles between boundary and c-axes of adjacent grains. E: Three boundary segments and six angles must be taken into account to derive equation 6 for triple junction.

As only four displacements are tried each time, the incremental displacements are not exactly in the direction of maximum energy reduction. However, because only very small displacements are made each time, the average movement direction over many steps is in the direction that gives the maximum energy reduction. The validity of the routine can be tested by simulating the shrinking of a circular grain embedded in an infinite grain (Anderson, 1986; Bons, 1993). The area of such a grain should decrease linearly with time, if the grain-boundary energy is isotropic, and the shrinkage rate should be proportional with the grain-boundary mobility for a given isotropic grain boundary energy. Figure 4 shows that our routine gives the desired linear relationships.

Results of Elle simulations

Simulations with 25 000 time steps were performed with a grain aggregate of initially 2227 grains in a square box of dimensions 1 × 1. The starting aggregate had an inclined shape-preferred orientation (SPO), which is an artefact from the creation of the initial grain aggregate (Fig. 5A). This SPO disappears in a few thousand steps, as observed in analogue experiments (Ree and Park, 1997) and numerical experiments (Bons, 1993). To avoid errors due to the SPO, we only used the last 15 000 steps for our analyses. The maximum node distance was set at 0.005 and the mobility at $M = 10^4$.

Isotropic growth produced a classical foam texture (Fig. 5A). The average grain area increased linearly (within error, $G_A = 0.99$, see equation A.1, Appendix 1) over the entire experiment (Fig. 5B). For ideal isotropic grain growth, the average area and average area change for N-sided grains should be a linear function of N, according to Von Neumann's law (Von Neumann, 1952) and Lewis's hypothesis (e.g., Weaire and Rivier, 1984; Glazier et al., 1987). These relationships are found in our model (Fig. 5, C and D).

We used five different crystal models, ranging from tablets to rods (Fig. 6), to investigate the effect of anisotropy on grain growth. All settings and the starting grain aggregate were the same as for the previously described isotropic growth test. Each grain in the starting aggregate was assigned a random c-axis orientation. The crystal model has a distinct effect on the resulting topology of the grain aggregate (Fig. 6A). Tablet models ($D > h$) give on average equidimensional grains, while the rod models ($D < h$) produce mainly elongate grains. All crystal models produce a distinct LPO. Tablet models give mostly steep plunges, whereas the rod models give mostly shallow plunges (Fig. 6B). The $D = h$ crystal model gives an intermediate result, with square grains and both shallow and steep plunges surviving, although there is still a maximum for steep plunges. The growth function can be approximated with a power law for all crystal models (Fig. 6C). In all cases the growth exponent G_A is <1; the lowest value is for the $D:h = 2:1$ tablet model, which gives a growth exponent of 0.68.

Grest et al. (1985) observed in their modeling that the growth exponent was reduced by anisotropy and that the grain-size distribution increased. Our anisotropic simulations also show an increase in grain-size distribution, with μ_2 increasing from the starting value of 0.4 to 0.8 (Fig. 7A). Isotropic growth and the $D:h = 1:1$ model, which have the highest growth exponents, have the lowest μ_2 values. Average μ_2 values do not correlate well with the growth exponent (Fig. 7B), but the rate of μ_2 increase shows some correlation (Fig. 7C). This indicates that the low growth increase is wholly or partly an effect of the change in microstructure. This may be a transient phenomenon, the most anisotropic models showing a stabilization of μ_2 toward the end of the simulation.

DISCUSSION

We have shown that OCP exhibits an effect of crystallographic anisotropy on grain growth. With our numerical model, Elle, we have been able to simulate its topological behavior during grain growth. The grain shape, lattice-preferred orientation and grain-size distribution can be duplicated with Elle when OCP is modeled as having a flat tablet crystal habit (Fig. 8, A and B). A lower than unity growth exponent is predicted by Elle, but not as low as found in the experiment. Drag by submicroscopic impurities, or more likely by the glass-sample interface, can be suggested as causes for the low growth exponent of only 0.35 in the OCP.

The OCP experiment served as a test for the Elle model. Figure 8 (C and D) shows that the simulation for the rod model ($D:h = 1:4$) closely resembles the topology in statically grown Al_2O_3 (Rödel and Glaeser, 1990). This gives us confidence that our model can adequately simulate anisotropic grain growth. It can be applied to different crystal models, and we can then use the observations from the simulations, where making such observations in experiments or nature are difficult or impossible.

In accordance with Grest et al. (1985), we found that anisotropy reduces the growth exponent in our simulations to 30%, and that it increases the grain-size distribution, raising μ_2 to 0.8. We found the reduction of the growth exponent not

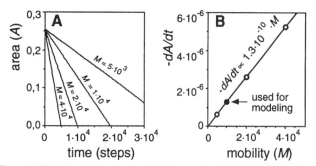

Figure 4. Test of node-movement routine as described in text. A: Graph of area (A) of shrinking circular grain that is surrounded by infinite grain, as function of time (in number of steps) for different grain-boundary mobilities (M). B: Graph of shrinkage rate (dA/dt) as function of mobility. Routine is validated by linear relationships in these graphs. Model deteriorates if M is too large for given distance between nodes; $M = 1·10^4$ was used in models because it is well inside linear zone.

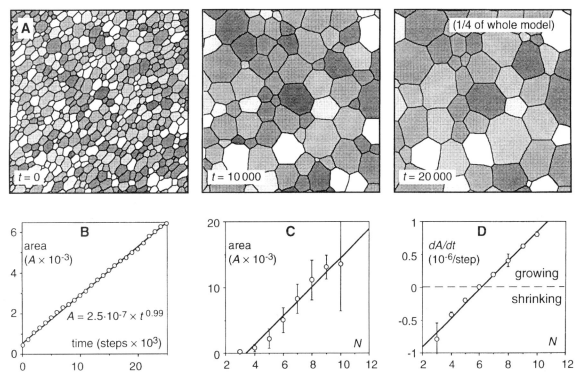

Figure 5. A: Three stages of isotropic growth simulation. Only one quarter of entire model is shown. B: Graph showing that, within error, mean grain area (A) increases linearly with time (t). C: In accordance with Lewis's hypothesis, mean grain area of N-sided grains is proportional to N. D: In accordance with Von Neumann's law, mean growth rate (dA/dt) of N-sided grains is proportional to number of sides. Grains with ≤6 sides shrink on average, while 7 or more sided grains grow. Graphs C and D are for $t = 10\,000$ when 192 of original 2227 grains were left.

directly related to μ_2, but rather to an increase in μ_2. Anderson et al. (1985) showed that grain-size distributions resulting from grain growth are virtually identical in two and three dimensions. This means that the results of the two-dimensional Elle simulations can be applied to real rocks. Measuring the grain-size distribution in statically recrystallized rocks may give insight to the significance and type of anisotropic growth.

Anisotropic grain growth may have implications for paleopiezometry. Derby and Ashby (1987) and de Bresser et al. (1998) proposed that a stable grain size at a certain stress is due to a balance between grain-size reduction processes and grain growth. Anisotropy, by affecting the growth function, plays a role in determining at what grain size a balance is reached.

Another observation that can be made from the OCP experiment and our simulations, and also described by Grest et al. (1985) and Hong and Messing (1998), is that the grain aggregate evolves toward the highest proportion of lowest energy grain boundaries. This may not be immediately evident, because the boundaries we see in the OCP and the $D \geq h$ simulations are mostly the high-energy boundaries. However, we must not forget that both experiment and simulations were thin-sheet cases, where the largest grain-boundary area is parallel to the sheet or plane of view. An evolution toward low-energy grain-boundary dominance is expected, because reduction of grain-boundary energy is driving the process.

From the preceding, one would predict that anisotropic grain growth in a grain aggregate with a lattice-preferred orientation (LPO) would not lead to a destruction of that LPO (cf. Ree and Park, 1997), but instead to the formation of a shape preferred orientation (SPO), which is determined by the LPO and the crystal anisotropy. This SPO may or may not be similar to the finite strain ellipsoid. A well-known case is that of a mica foliation. Shearing normally leads to an alignment of the basal planes of micas, which would only be strengthened by grain growth, because this is also the minimum surface energy configuration (Etheridge et al., 1974; Ishii, 1988).

It is perhaps more interesting to speculate on the effect on the mineral quartz. It is likely that quartz displays a hexagonal prismatic γ^{-1} polyhedron (Heavans and Ashbee, 1975), that is similar to the cylindrical system applied here. We have carried out an anisotropic grain growth experiment in Elle with a starting material having both a strong single maximum c-axis LPO parallel to the vertical boundary of the model (Fig. 9A) and a strong SPO oriented at about 50° clockwise from the LPO maximum orientation (Fig. 9B). We used the same $D{:}h = 1{:}2$ γ^{-1} polyhedron as that used in Figure 6, and carried out an anisotropic grain growth experiment with $M = 10^3$; that lasted 100 000 time steps, leading to an ~20 times increase in grain area (Fig. 9C). At the end of the experiment we were able to obtain a significant reorientation of the SPO so that it

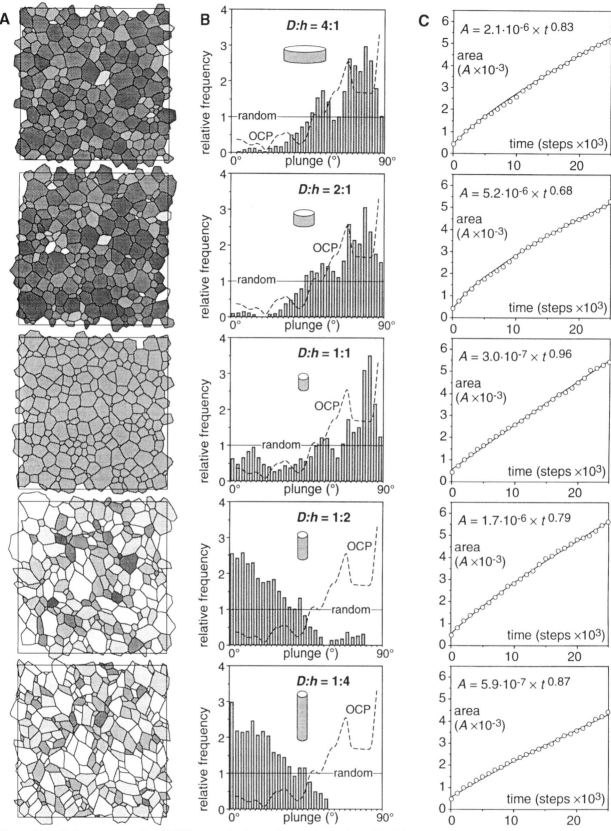

Figure 6. A: Grain aggregate after 20 000 steps of anisotropic growth, starting with 2227 grains as shown in Figure 5A. Shading of grains is proportional to c-axis plunge, dark grains having steepest plunge. B: Frequency of c-axes plunges, normalized to expected frequency for random distribution. Dashed line shows c-axes distribution of octachloropropane experiment. C: Graphs of mean area (A) versus time (t) in steps, with least-squares best fit power-law growth function drawn in. Rows are for different crystal models, ranging from $D:h = 4:1$ at top to 1:4 at bottom.

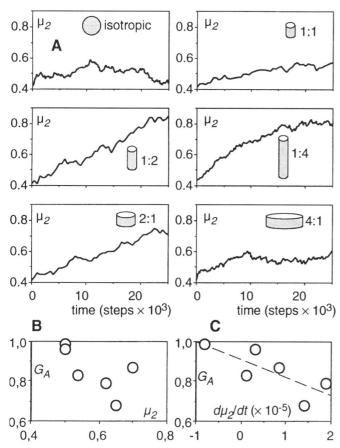

Figure 7. A: Evolution of second moment (μ_2) of grain-size distribution as function of time for different crystal models. B: Growth exponent (G_A) plotted against average value μ_2 over period $t = 10\,000$ to $25\,000$ steps, showing no obvious correlation. C: G_A plotted against average increase in μ_2 over period time, $t = 10\,000$ to $25\,000$ steps, showing vague negative correlation (dashed line).

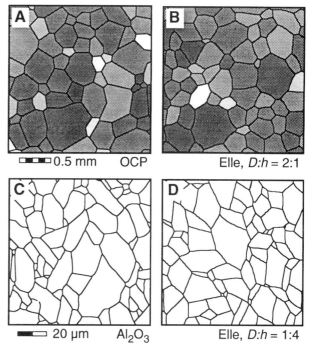

Figure 8. Comparisons between numerical simulations and experiments. A: Octachloropropane (OCP) aggregate at time, $t = 0$ (Fig. 2) with grain shaded according to c-axis plunge. B: One quarter of grain aggregate for $D{:}h = 2{:}1$ tablet model simulation at $t = 20\,000$ steps. C: Statically grown Al_2O_3, after Figure 2c of Rödel and Glaeser (1990). Al_2O_3 has rod-like crystal habit. D: One quarter of grain aggregate for $D{:}h = 1{:}4$ rod model simulation at $t = 20\,000$ steps.

became much closer in orientation to the strengthened c-axis LPO point maximum (Fig. 9D).

In shear zones quartz usually develops LPOs that range from oblique girdles at a high angle to the shear plane to point maxima normal to the shear plane and shear direction or close to the shear direction (Schmid and Casey, 1986). The SPO due to the deformation (stretching lineation) is typically an oblate ellipsoid parallel or at a small angle to the shear direction. Postdynamic grain growth can quickly destroy this SPO (Bons and Urai, 1992; Ree and Park, 1997) and our results here show that it could then develop a new SPO. In case of a girdle LPO, this would lead to a foliation at a high angle to the shear plane, and in case of a point maximum normal to the stretching lineation, it may lead to a mineral lineation perpendicular to the shear direction. Similarly, anisotropic grain growth during shearing could produce oblique grain-shape foliation at angles $>45°$ to the shear plane in type II S-C fabrics of Lister and Snoke (1984), which show quartz geometries similar to those found in our simulations, which in turn have an average grain-shape aspect ratio of 1.9:1.

CONCLUSIONS

In conclusion, we have accomplished the following.

1. We have developed a numerical routine that can simulate anisotropic grain growth and opens the way for further numerical modeling of processes that involve or interact with grain growth.
2. We have shown that anisotropic grain growth reduces the growth exponent, which has possible implications for the determination of the balance between grain growth and grain-size reduction processes, and which therefore should be taken into account in palaeopiezometers.
3. We have shown that anisotropic grain growth increases the grain-size distribution, which may provide a useful indicator to determine its importance in real rocks.
4. We have found indications that, if an LPO exists, anisotropic grain growth can produce an SPO that is determined only by the LPO and crystal anisotropy and may be completely unrelated to the shape of the finite strain ellipsoid.

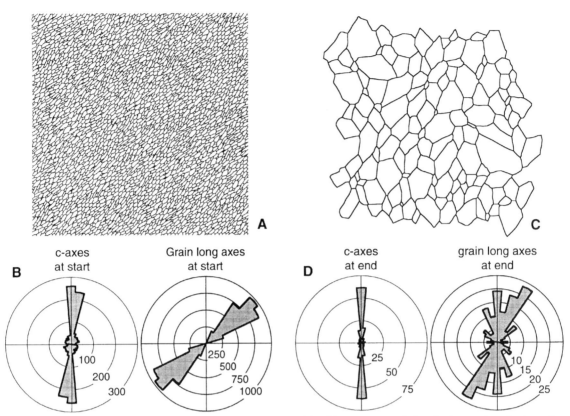

Figure 9. Modification of shape-preferred orientation in material with strong preexisting lattice-preferred orientation (LPO). A: Starting aggregate of 3600 grains. B: Rose diagrams show strong alignment of c-axes (LPO) parallel to vertical axis of model (left) and strong alignment of long axes of grains (shape-preferred orientation, SPO) at about 50° to vertical (right). C: Grain-boundary configuration after 100 000 time steps, using $D{:}h = 1{:}2$ model and $M = 1000$. D: Rose diagrams showing preservation of strong vertical LPO (left) and reorientation of SPO (right) after 100 000 growth steps. SPO data were calculated using best fit ellipsoids to grain shapes using NIH-Image image processing package, equivalent to PAROR technique of Panozzo (1993).

ACKNOWLEDGMENTS

Funding through an Australian Research Council Large Grant and Monash University SMURF grant and Logan Research fellowship are gratefully acknowledged. We thank Win Means and Young-Do Park for their kind reviews.

APPENDIX 1: DETERMINATION OF THE GROWTH FUNCTION

The expected function to describe the area (A) as a function of time (t) is of the form (e.g., Anderson, 1986):

$$A = k \cdot t^G \quad \text{(A.1)}$$

The constant k and G can be determined with a least square linear fit in $\log(A) - \log(t)$ space, using:

$$\log(A) = \log(k) + G \cdot \log(t). \quad \text{(A.2)}$$

In experiments and simulations, the starting grain area is nonzero. If t is the time from the start of the experiment, then equation (A.1) becomes:

$$\log(A) = \log(k) + G \cdot \log(t + \Delta t), \quad \text{(A.3)}$$

where Δt is the unknown time interval it would have taken to reach the starting mean grain area. We iterated over Δt until a linear least-squares best fit was found in log-log space.

Other growth laws than equation (A.1) are used in the literature (e.g., Tullis and Yund, 1982; Rödel and Glaeser, 1990), e.g.,

$$A - A_0 = k \cdot t^n \quad \text{(A.4)}$$

We prefer not to use such a growth law, because it contains an experiment-dependent parameter A_0, which is the grain-size at the start of the experiment. As such, the growth law is not a universal growth law for a material, but one that is restricted to a single experiment.

REFERENCES CITED

Anderson, M.P., 1986, Simulation of grain growth in two and three dimensions, in Hanse, N., et al., eds., Annealing processes—Recovery, recrystallization and grain growth: Roskilde, Risø National Laboratory, p. 15–34.

Anderson, M.P., Srolovitz, D.J., Grest, G.S., and Sahni, P.S., 1984, Computer simulation of grain growth—I. Kinetics: Acta Metallurgica, v. 32, p. 783–791.

Anderson, M.P., Grest, G.S., and Srolovitz, D.J., 1985, Grain growth in three dimensions: A lattice model: Scripta Metallurgica, v. 19, p. 225–230.

Bons, P.D., 1993, Experimental deformation of polyphase rock analogues: Geologica Ultraiectina, v. 110, 207 p.

Bons, P.D., and Urai, J.L., 1992, Syndeformational grain growth: Microstructures and kinetics: Journal of Structural Geology, v. 14, p. 1101–1109.

Cabrera, N., and Coleman, R.V., 1963, Theory of crystal growth from the vapour, in Gilman, J.J., ed., The art and science of growing crystals: New York, John Wiley & Sons, p. 3–28.

de Bresser, J.H.P., Peach, C.J., Reijs, J.P.J., and Spiers, C.J., 1998, On dynamic recrystallization during solid state flow: Effects of stress and temperature: Geophysical Research Letters, v. 25, p. 3457–3460.

Derby, B., and Ashby, M.F., 1987, On dynamic recrystallization: Scripta Metallurgica, v. 21, 879–884.

Etheridge, M.A., Paterson, M.S., and Hobbs, B.E., 1974, Experimentally produced preferred orientation in synthetic mica aggregates: Contributions to Mineralogy and Petrology, v. 44, p. 275–294.

Frank, F.C., 1963, Metal surfaces: Structure, energetics and kinetics. Papers presented at a joint seminar of the American Society for metals and the Metallurgical Society of AIME, October 27 and 28, 1962: Metals Park, Ohio, American Society for Metals, 408 p.

Glazier, J.A., Gross, S.P., and Stavans, J., 1987, Dynamics of two-dimensional soap froths: Physical Review A, v. 36, p. 306–312.

Gleason, G.C., Tullis, J., and Heidelbach, F., 1993, The role of dynamic recrystallization in the development of lattice preferred orientations in experiemntally deformed quartz aggregates: Journal of Structural Geology, v. 15, p. 1145–1168.

Gleiter, H., 1969, Theory of grain-boundary migration rate: Acta Metallurgica, v. 17, p. 853–862.

Grest, G.S., Srolovitz, D.J., and Anderson, M.P., 1985, Computer simulation of grain growth—IV. Anisotropic grain-boundary energies: Acta Metallurgica, v. 33, p. 509–520.

Haessner, F., and Hofmann, S., 1978, Migration of high angle grain boundaries, in Haessner, F., ed., Recrystallization of metallic materials: Stuttgart, Riederer Verlag, p. 63–96.

Heavans, J.W., and Ashbee, K.H.G., 1975, On the relative fracture surface energies of the major and minor rhombohedra in low-quartz: Journal of Materials Science, v. 10, p. 1938–1942.

Hong, S.-H., and Messing, G.L., 1998, Anisotropic grain growth in diphasic gel-derived titania-doped mullite: American Ceramical Society Journal, v. 81, p. 1269–1277.

Ishii, K., 1988, Grain growth and re-orientation of phyllo silicate minerals during the development of slaty cleavage in the South Kitakami Mountains, NE Japan: Journal of Structural Geology, v. 10, p. 145–154.

Jessell, M.W., 1986, Grain boundary migration and fabric development in experimentally deformed octachloropropane: Journal of Structural Geology, v. 8, p. 527–542.

Karato, S.I., 1989, Grain growth kinetics in olivine aggregates: Tectonophysics, v. 168, p. 255–273.

Karato, S.I., and Masuda, T., 1989, Anisotropic grain growth in quartz aggregates under stress and its implication for foliation development: Geology, v. 17, p. 695–698.

Kim, D.Y., Wiederhorn, S.M., Hockey, B.J., Handwerker, C.A., and Blendell, J.E., 1994, Stability and surface energies of wetted grain boundaries in aluminium oxide: American Ceramical Society Journal, v. 77, p. 444–453.

Lister, G.S., and Snoke, A.W., 1984, S-C mylonites: Journal of Structural Geology, v. 6, p. 617–638.

McCrone, W.C., 1949, Boundary migration and grain growth: Discussions of the Faraday Society, v. 5, p. 158–166.

Means, W.D., 1983, Microstructure and micromotion in recrystallization flow of octachloropropane: A first look: Geologische Rundschau, v. 72, p. 511–528.

Means, W.D., 1989, Synkinematic microscopy of transparent polycrystals: Journal of Structural Geology, v. 11, p. 163–174.

Means, W.D., and Ree, J.H., 1988, Seven types of subgrain boundaries in octachloropropane: Journal of Structural Geology, v. 10, p. 765–770.

Olgaard, D.L., and Evans, B., 1986, Effect of second phase particles on grain growth in calcite: American Ceramical Society Journal, v. 69C, p. 272–277.

Panozzo, R., 1993, Two-dimensional analysis of shape-fabric using projections of digitized lines in a plane: Tectonophysics, v. 95, p. 279–294.

Randle, V., Ralph, B., and Hansen, N., 1986, Grain growth in crystalline materials, in Hansen, N., et al., eds., Annealing processes—Recovery, recrystallization and grain growth: Roskilde, Risø National Laboratory, p. 123–142.

Ree, J.H., 1990, High temperature deformation of octachloropropane: Dynamic grain growth and lattice reorientation, in Knipe, R.J., and Rutter, E.H., eds., Deformation mechanisms, rheology and tectonics: Geological Society [London] Special Publications 54, p. 363–368.

Ree, J.H., 1991, An experimental steady-state foliation: Journal of Structural Geology, v. 13, p. 1001–1011.

Ree, J.H., 1994, Grain boundary sliding and development of grain-boundary openings in experimentally deformed octachloropropane: Journal of Structural Geology, v. 16, p. 403–418.

Ree, J.H., and Park, Y., 1997, Static recovery and recrystallization in sheared octachloropropane: Journal of Structural Geology, v. 19, p. 1521–1526.

Rödel, J., and Glaeser, A.M., 1990, Anisotropy of grain growth in alumina: American Ceramical Society Journal, v. 73, p. 3292–3301.

Schmid, S.M., and Casey, M., 1986, Complete texture analysis of commonly observed quartz c-axis patterns, in Hobbs, B.E., and Heard, H.C., eds., Mineral and rock deformation: Laboratory Studies—The Paterson Volume: American Geophysical Union Geophysical Monograph 36, p. 263–286.

Smith, C.S., 1964, Some elementary principles of polycrystalline microstructure: Metallurgical Reviews, v. 9, p. 1–48.

Srolovitz, D.J., Anderson, M.P., Grest, G.S., and Shani, P.S., 1984, Computer simulation of grain growth—III. Influence of particle dispersion: Acta Metallurgica, v. 32, p. 1429–1438.

Tamman, G. Von, and Dreyer, K.L., 1923, Die Rekristallisation leicht schmelzender Stoffe und die des Eises: Zeitschrift für anorganische und allgemeine Chemie, v. 182, p. 289–313.

Tullis, J., and Yund, R.A., 1982, Grain growth kinetics of quartz and calcite aggregates: Journal of Geology, v. 90, p. 301–318.

Urai, J.L., Means, W.D., and Lister, G.S., 1986, Dynamic recrystallization of minerals, in Hobbs, B.E., and Heard, H.C., eds., Mineral and rock deformation: Laboratory studics—The Paterson Volume: American Geophysical Union Geophysical Monograph 36, p. 161–199.

Von Neumann, J., 1952, Written discussion on grain topology and the relationships to growth kinetics, in Metallic materials: Metals Park, Ohio, American Society for Metals, p. 108–113.

Weaire, D., and Rivier, N., 1984, Soap, cells and statistics—Random patterns in two dimensions: Contemporary Physics, v. 25, p. 59–99.

Wulff, G., 1901, Zur Frage der Geschwindigkeit des Wachstums und der Auflösung der Krystallflächen: Zeitschrift für Kristallographie, v. 34, p. 449–530.

MANUSCRIPT ACCEPTED BY THE SOCIETY APRIL 12, 2000

Printed in the U.S.A.

Modifications of early lineations during later folding in simple shear

Sudipta Sengupta*
Department of Geological Sciences, Jadavpur University, Calcutta 700032, India
Hemin A. Koyi*
Hans Ramberg Tectonic Laboratory, Institute of Earth Sciences, Uppsala University, Villavagen 16, SE 75236 Uppsala, Sweden

ABSTRACT

A series of experiments was carried out with analogue models to study deformation of early lineations on layers undergoing simultaneous buckling and flattening under a bulk simple shear. The results show that when buckling is accompanied by layer-parallel strain, the angle between the lineation and the fold axis changes differently in different parts of the fold. A U-shaped pattern of deformed lineation on the unrolled fold surface develops in bulk simple shear when (1) the initial orientation of the layer is in the field of shortening, (2) the layer is inclined to a plane that is perpendicular to the vorticity vector of simple shear (i.e., the plane perpendicular to the Y-axis of bulk strain), and (3) the early lineation is initially perpendicular to the later fold axis. With progressive simple shear, an initially symmetric U pattern of lineation may become asymmetric. The nature of the U pattern is modified over parasitic folds that grow on limbs of larger folds. When the initial angle between the early lineation and the later fold axis is moderate, a sigmoidal pattern of deformed lineation develops, the angle remaining a minimum at the hinge and progressively increasing on the limbs. If the curvature of the hinge region is large in a relatively thick layer, the pattern of deformed lineation is considerably modified by tangential longitudinal strain.

INTRODUCTION

When an early lineation is deformed by flexural slip folding, the initial angle between the lineation and the fold axis remains the same in all parts of the fold (small circle pattern in stereographic projection). As a result the lineation straightens out when the folded surface is unrolled. If the lineated surface is deformed by shear folding, the angle between the fold axis and the lineation varies from place to place at different parts of the fold, but all the deformed lineations are parallel to a plane (great circle pattern in stereographic projection) defined by the initial orientation of the lineation and the shear direction (Weiss, 1955, 1959; Ramsay, 1960). In many areas, however, the deformation of early lineations deviates from these two ideal situations; the lineations can neither be unrolled as they do in case of buckling or external rotation, nor do they lie on a plane as expected in shear folding. Although deformed lineations occurring in a great circle pattern do not straighten out when the folded surface is unrolled, the term "nonunrollable lineation pattern" has been used here only for those folded lineations that are neither unrollable nor show a great circle pattern. If the folding of the early lineation is by a single phase of deformation, the occurrence of such nonunrollable lineation patterns would indicate that they have formed by both external rotation (buckling) and broadly synchronous layer-parallel strain (Ramsay, 1967; Ghosh and Chatterjee, 1985).

The geometry of the folded lineation can be conveniently described by the patterns produced when the folded surface is unrolled about its axis. The pattern clearly indicates the manner in which the angle between an early lineation (e.g., L_1) and the later fold axis (e.g., F_2) varies over the fold. Evidently, the patterns can be best obtained when the surface of the fold is exposed over both the limbs and the hinges. The angle between the lineation and the fold axis at the hinge zone is often of crucial importance. The lineation pattern can also be obtained from

*E-mails: Sengupta, jugeossg@iacs.res.in
Koyi, Hemin.Koyi@geo.uu.se

a stereographic projection of the folded surface and lineations from different parts of a fold. Along with the variation in the angle between the lineation and the fold axis, it also gives the attitudes of the lineation and of the folded surface at different points of the fold. It does not, however, show the continuous variation of the angle between the lineation and the fold axis. The stereographic projection also cannot distinguish between a symmetric and an asymmetric pattern of the curved lineation. For the present purpose and especially for folds on the mesoscopic scale, for known values of fold tightness, drawing the lineation pattern directly on a transparent sheet placed over the fold is much more convenient and useful.

With a slight simplification of the classification of Ghosh and Chatterjee (1985), these diverse patterns on the unrolled surface of the fold can be grouped into the following types (Fig. 1): (1) type 1, a constant angle between the early lineation (L_1) and the later fold axis (F_2); (2) type 2, the angle between L_1 and F_2 having a minimum value over hinge lines and increasing symmetrically or asymmetrically on either side; (3) type 3, the angle between L_1 and F_2 continuously increasing from one limb to the other across the hinge line. If the angle changes through more than 90° the pattern is an asymmetrical U shape, and the turning point of U (the point of maximum curvature) is on one side of the hinge; (4) type 4, a U-shaped pattern, the turning point of U being at the hinge line of the fold; (5) type 5, the angle between L_1 and F_2 is a maximum at the hinge and progressively decreases symmetrically or asymmetrically on either side; and (6) type 6, the angle between L_1 and F_2 varies along the hinge line. If we look along the direction of fold axis, the type 3 and type 4 lineation patterns are in the shape of a U or an inverted U (Ghosh et al., 1999). In type 4 the point of maximum curvature is located on the fold hinge (Fig. 1), and the lineation is at a right angle to it. In type 3, the maximum curvature occurs on one side of the hinge line (Fig. 1). The asymmetry of the U pattern may result because the point of maximum curvature is shifted toward one side of the hinge line or because the two arms of the U are of unequal lengths and make unequal angles with the hinge line. These two types are described as type A and type B asymmetry. Both types of asymmetry are often associated.

The final pattern of the deformed lineation depends mainly upon the following factors (Ghosh, 1993, p. 374): (1) the relative competence of the folded layer, (2) the initial angle between the early lineation and the later fold axis, (3) the initial geometry of the fold and of the lineation over it with respect to the bulk strain, and (4) the nature of bulk deformation.

Lineation patterns that remain curved on unrolled fold surfaces have often been described from areas of superposed deformations (e.g., Ramsay, 1958, 1960; Hudleston, 1973; Naha and Halyburton, 1977; Ghosh and Chatterjee, 1985). In recent years such lineation patterns have been frequently reported from ductile shear zones. Among them the U pattern of lineation is the most important (Ghosh and Sengupta, 1987, 1990; Sengupta and Ghosh, 1997; Ghosh et al., 1999). A central theme of this series of papers on ductile shear zones is the deciphering of the deformation history of folds from patterns of deformed lineations over unrolled form surfaces. Field studies indicate that U patterns of lineations develop in shear zones when the initial angle between the early lineation and the later fold axis was large and the fold hinge line rotated through a large angle. Figure 2 shows some hand specimens with deformed lineation patterns of folded mylonites from different shear zones. A convenient method of representing the lineation patterns is to place a transparent polythene sheet over the exposed surface of the fold and to trace out the lineation pattern. When the polythene sheet is straightened out we get the lineation pattern on the unrolled surface. Some of these lineation patterns are shown in Figure 3.

Deformation of early lineation in buckling and flattening folds was studied by Ghosh (1974) in experiments with bulk coaxial deformation. In the mathematical model for the deformation of early lineations over buckling and flattening folds, Ghosh and Chatterjee (1985) assumed that the bulk deformation was coaxial and the fold was symmetrical. Different types of lineation patterns, including the U pattern, were obtained from this model. This model, however, does not explain the development of these patterns by noncoaxial deformation in ductile shear zones (Figs. 2 and 3; see Figs. 5 and 8 of Ghosh and Sengupta, 1987). The aim of the present work is to study the development

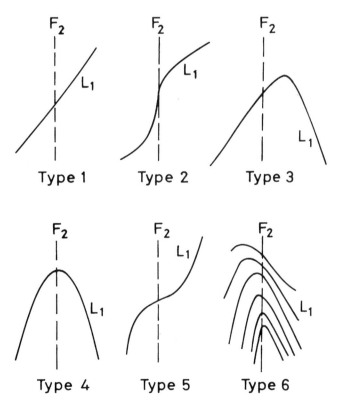

Figure 1. Six types of deformed lineation patterns on unrolled fold surfaces (after Ghosh and Chatterjee, 1985). L_1—early lineation, F_2—hinge line of later fold.

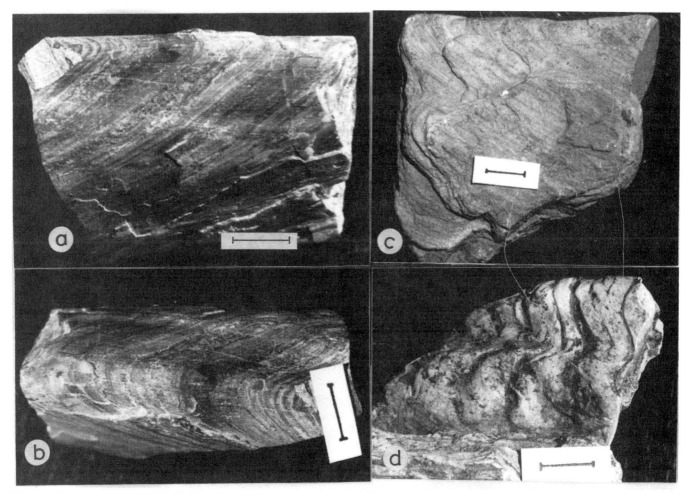

Figure 2. Some examples of deformed lineation patterns over mesoscopic folds of different shear zones. A, B: Side view and top view of specimen of folded mylonite with early lineation deformed to U pattern. Pattern of lineation on unrolled surface is shown in Figure 3B. Lineation makes lower angles with fold axis on limbs but is at right angle to it at hinge. Sample from Phulad shear zone, India. C: Isoclinally folded mylonite with early lineation making variable angles with fold axis on limbs and at hinge. Lineation pattern on unrolled fold surface is shown in Figure 3C. D: Early lineation showing curved pattern over minor folds on limb of larger fold. Pattern on unrolled fold surface is shown in Figure 3D. Samples C and D are from Singhbhum shear zone, India. Scale bar 1 cm.

of deformed lineation patterns under different situations with the help of analogue models in bulk simple shear.

MODEL PREPARATION AND MATERIAL PROPERTIES

The experiments were carried out in a simple shear box with silicone putty and a mixture of silicone putty with modeling clay. The competent layers were made up of a mixture of modeling clay (plastilina) and Dow Corning silicone putty. The matrix is a mixture of Dow Corning silicone putty and magnetite powder with some acid oil. Both the materials are non-Newtonian. The material properties of the layer and the matrix are shown in Figure 4. For these experiments, the effective viscosity ratio between the layer and the matrix was ~15:1.

Marker lines representing lineations were printed at different angles on the competent layer. The layer was then placed on the upper surface of a slab of silicone putty mixture and the model was deformed in a simple shear box. In one of the experiments the folds were first generated in the layer outside the shear box, and the model was then deformed by simple shear in the shear box. In the simple shear apparatus used for these experiments, a model could be deformed to a maximum value of simple shear of gamma (γ) = 1.3. In a few models, a larger value of simple shear was obtained by a two-stage deformation. For this, the model was taken out after a simple shear of $\gamma = 1.3$. The model was then trimmed and placed again in the shear box for a further deformation of $\gamma = 1.3$.

The simple shear deformation is described as follows with reference to the coordinate axes x, y, and z, x being parallel to the shear direction and y being the axis of shear (i.e., parallel to the vorticity vector), so that a point (x, y, z) changes to (x', y',

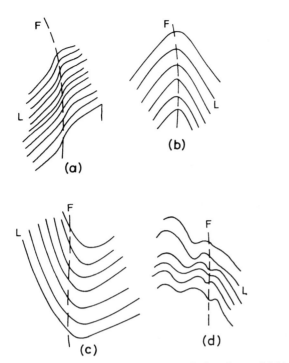

Figure 3. Deformed lineation patterns on unrolled surfaces of folds. L—early lineation, F—hinge line of later fold. A: Type 2 pattern over synformal fold hinge of quartzite mylonite. Specimen from Olden Window, Scandinavian Caledonides. B: Symmetric U-shaped pattern (type 4) of specimen shown in Figure 2, A and B. C: U-shaped (type 3) lineation pattern of sample shown in Figure 2C. Pattern shows asymmetry of type A. D: Complex pattern of deformed lineation of specimen shown in Figure 2D.

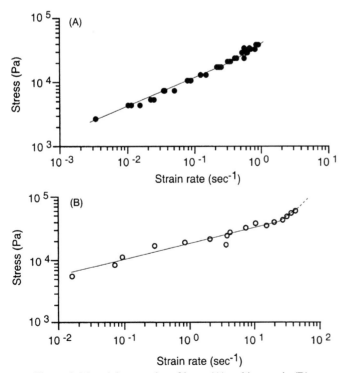

Figure 4. Material properties of layer (A) and its matrix (B).

z') according to the equation $x' = x + \gamma z$; $y' = y$ and $z' = z$, where γ is the amount of simple shear. The major and minor axes of the strain ellipsoid are parallel to the xz-coordinate plane. The intermediate strain axis Y is parallel to the coordinate axis y. The experiments were carried out in three groups. In the first series of experiments, the initial layers were horizontal and parallel to the xz-plane (Fig. 5A). In the second series, the layers were inclined to this plane (Fig. 5B). In the third series, a previously folded layer was deformed by simple shear, the enveloping surface of the folds being parallel to the horizontal xz-plane (Fig. 5C). All the models, except the last one, were deformed by sinistral shear. The initial orientations of the layers and the lineations with respect to the shear direction in different models are shown in Figure 6.

BUCKLE FOLDING BY SIMPLE SHEAR DEFORMATION OF HORIZONTAL LINEATED LAYERS

In this series of experiments a layer was placed parallel to the xz-plane and normal to the y-axis (the vorticity vector of simple shear). Four sets of initial lines were printed at angles of 0°, 45°, 90°, and 135° to the direction of simple shear. These lineations are referred to as L_0, L_{45}, L_{90}, and L_{135}, respectively (Figs. 6A and 7A).

Perceptible folds initiated at a value of simple shear of $\gamma = 0.6$; hinge lines were oriented at an average angle of 40° with the shear direction (Figs. 7B and 8A). The folds were very open; the approximate interlimb angle was 140°. The hinges were slightly noncylindrical and the angle between the lineation and the fold hinges remained more or less the same over the hinge and the limbs of a fold. A polythene sheet was placed carefully on the folded layer and the lineations were traced out. The initial and final angles between the fold axis and the lineations (α) are shown in Table 1.

With continued deformation the folds become tighter (Fig. 7C and D). At the final stage of deformation, with $\gamma = 1.3$, the interlimb angles of the folds decreased to ~90° over the antiforms and 60° over the synforms. As expected in class 1B or 1C folds and as commonly seen in experiments of buckle folding of competent layers with a free upper surface, the folds were often of a cuspate-lobate geometry (Ramsay, 1967). In such folds, the inner arcs of the synforms were significantly less rounded and were tighter than the outer arcs of the antiforms. Because the exposed surface showed the outer arc of antiforms and inner arc of synforms, the tangential longitudinal strain (Ramsay, 1967; Ramsay and Huber, 1987) was extensional over antiforms and contractional over synforms. Because of superimposition of the tangential longitudinal strain, the layer-parallel contractional strain, perpendicular to the fold axis, was reduced over antiformal hinges and accentuated over synformal hinges. Because the bulk simple shear in the experiments was relatively small, the characteristic lineation patterns were much better developed over the synforms than over the antiforms. In the following

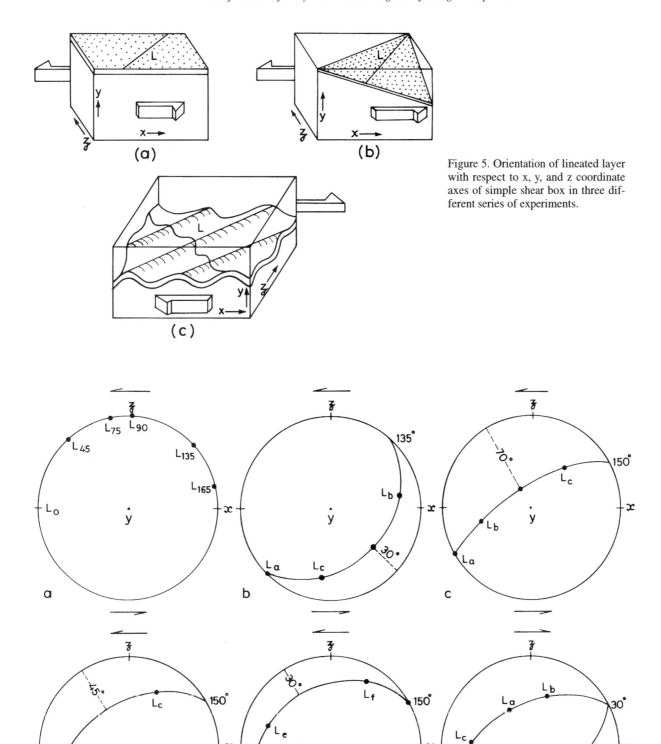

Figure 5. Orientation of lineated layer with respect to x, y, and z coordinate axes of simple shear box in three different series of experiments.

Figure 6. Stereographic plot showing initial orientation of layers and lineations (L) with respect to shear direction in different models; x, y, and z are coordinate axes. Arrows indicate direction of shear.

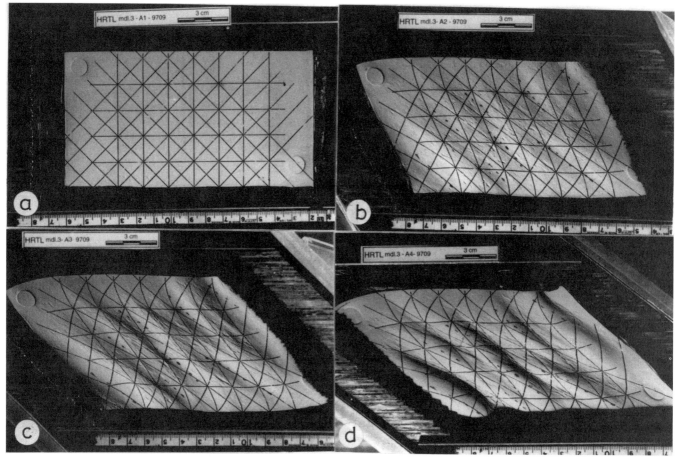

Figure 7. Model with horizontal layer deformed by bulk strain of simple shear. A: Initial stage with lineations at 0°, 45°, 90°, and 135° with shear direction. B: Deformed model after shear strain of $\gamma = 0.6$. Initiation of very open folds. C: Deformed model after shear strain of $\gamma = 0.9$. D: Deformed model after shear strain of $\gamma = 1.3$.

description we have mostly considered the lineation patterns on the upper surface of the synforms. On the unrolled surface the different sets of deformed lineations showed different patterns.

The deformed L_{135} could not be straightened out when the polythene sheet was unrolled, indicating that the deformation of the lineation was caused by the combined effects of buckling and layer-parallel strain. The L_{135} on the unrolled surface showed an asymmetrical sigmoidal pattern, the angle between the lineation and the fold axis being smaller at the fold hinges than at the limbs (Fig. 8B). The angle between the fold axis and the lineation was slightly larger on one set of limbs than on the other. However, the acute angle between the fold axis and the lineation opens in the same sense in the hinge and in both the limbs. The pattern is similar to the type 2 (Fig. 1) pattern of Ghosh and Chatterjee (1985).

At the final stage of deformation ($\gamma = 1.3$), the deformed L_{90} made a very low angle with the fold axis in all parts of the fold, the angle being a minimum at the fold hinges. However, this angular variation is so small in the central part of the model that the lineation is almost unrollable (Fig. 8B). Because the L_{45} becomes virtually parallel to the fold axis in all parts of the model at the final stage of deformation, it is not shown in Figure 8.

The patterns of deformed lineation can be explained with the following simplified theoretical model with a three-stage deformation (Fig. 9). Let us consider a horizontal layer with a lineation occurring at 135° (L_{135}) with the shear direction. After an initial stage of deformation of $\gamma = 0.6$, the fold (F in Fig. 9A) initiates with axis at an angle of 37° with the shear direction, and the L_{135}, deformed by homogeneous strain (plotted as $L_{(h)}$ in Fig. 9A), makes an angle of 112° with the shear direction on the horizontal plane. The hinge surfaces of the folds remain parallel to this plane (i.e., the plane perpendicular to the vorticity vector of simple shear) and the limbs are assumed to be inclined to it on either side at an angle of 30°. Let us also assume that in this initial stage, the $L_{(a)}$ and $L_{(b)}$ on the two limbs remain unrollable. This would imply that the angle between the fold axis and the deformed lineation is 75° in all parts of the fold. The structural elements of the fold are then deformed by a homogeneous strain of $\gamma = 0.7$ and 1.5.

Because the hinge surface and the limbs are differently oriented with respect to the principal axes of the strain ellipsoid, the

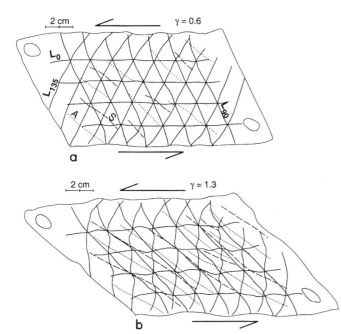

TABLE 1. LINEATION PATTERNS ON HORIZONTAL LAYERS

Lineation	α_1*	α_2 at hinge	α_2 at limbs	Type of lineation pattern
L_0	35–45	15–23	27–33	Type 2
L_{45}	5–10	<2	2–5	Type 1
L_{75}	15–20	5–10	10–15	Type 1
L_{90}	22–27	5–12	12–17	Type 1
L_{135}	80–85	30–40	55–65	Type 2
L_{165}	50–65	18–30	45–55	Type 2

Note: *α_1 and α_2 are the angles (in degrees) between the fold axis and the lineation at $\gamma = 0.6$ and 1.3, respectively.

Figure 8. A: Unrolled surface of folded layer shown in Figure 7B, at $\gamma = 0.6$. Antiforms (A) and synforms (S) are shown in dotted and dashed lines. All lineations are nearly straight lines. B: Unrolled surface of folded layer after simple shear of $\gamma = 1.3$. L_0 and L_{135} lineations show type 2 patterns; L_{90} is nearly straight (type 1).

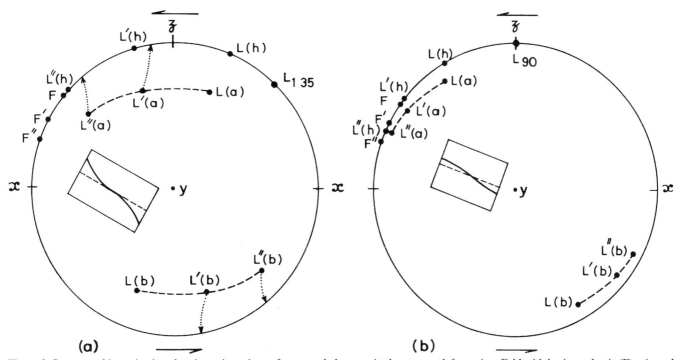

Figure 9. Stereographic projection showing orientations of structural elements in three-stage deformation. Fold with horizontal axis (F) oriented at 37° with shear direction deformed by simple shear of $\gamma = 0.7$ and 1.5 with F' and F" as deformed fold axes for two stages. A: For $\gamma = 0.7$, initial L_{135} lineations make angles of 46° at hinge (L'_h) and 52° on limbs (L'_a or L'_b). For $\gamma = 1.5$ corresponding values are 22° (L''_h) and 30° (L''_a or L''_b). B: For L_{90}, at $\gamma = 0.7$ rotated lineation makes angle of 12° with fold axis at hinge and 14° on limbs. For $\gamma = 1.5$ corresponding angles are 6° and 10°. Deformed lineation patterns are schematically shown in insets.

lineation is reoriented differently in the hinge and in the limbs (Fig. 9A). The angle between the fold axis (F′) and the $L'_{(h)}$ at the hinge is smaller than the corresponding angles of $L'_{(a)}$ or $L'_{(b)}$ with F′ at the limbs. This is the reason for the sigmoidal geometry of the deformed L_{135} in the experiments. For $\gamma = 1.5$, the fold axis is F″ and the lineations at the hinge and the limbs are $L''_{(h)}$, $L''_{(a)}$, and $L''_{(b)}$. The unrolled pattern of the lineation is schematically shown in the inset. Note that in this theoretical model, the passive rotation due to layer-parallel strain is superposed on the buckled layers. In the experiments as well as in natural structures, however, buckling and layer-parallel strains are simultaneous.

For a similar three-stage deformation with the initial lineation at an angle of 90° with the shear direction (L_{90}), the initial and deformed structural elements are shown in Figure 9B. The rotated L_{90} produces a nearly unrollable pattern (inset in Fig. 9B) similar to the patterns developed in the experiments. In other words, on the unrolled form surface of the fold the L_{90} remains almost straight, making a very low angle with the fold axis. In the same manner the L_{45} also becomes virtually parallel to the fold axis in all parts of the fold.

From the theoretical model discussed herein, it is evident that each increment of buckling rotates the limbs away from the enveloping surface parallel to the xz-coordinate plane. The amount of rotation of the lineation is dependent on the aspect ratio of the sectional strain ellipse on the layer surface. This ratio is largest on the horizontal plane containing the X and Z axes of the strain ellipsoid. The aspect ratio of the sectional strain ellipse (calculated from the equations given in Appendix 1) decreases on planes making larger and larger angles to this horizontal plane. Consequently, on the folded surface, the lineations rotated through a smaller angle on the limbs than on the horizontal hinge region.

Although this theoretical model qualitatively agrees with the lineation pattern obtained in the experiments, the actual angle between the lineation and the fold axis at the hinge of the fold in Figures 7D and 8B was distinctly smaller than what was predicted by the theoretical model. The experiments showed that this low angle between the fold axis and the lineations was obtained only over the synformal hinges and not over the antiformal hinges. This difference is explained by the tangential longitudinal contraction at the cores of the synforms. The tangential longitudinal strains will be different at the outer and inner arcs of a fold (cf. Ramsay, 1967, p. 464). We are concerned here with the lineation on the upper surface of the folded layer. Because the wavelength: thickness ratios of the folds are relatively small (often ranging between 3 and 6), the tangential longitudinal strain is expected to be large enough to significantly rotate the lineation. The strain depends essentially on the curvature of the fold as well as on the thickness of the layer. Because the curvature of the cuspate synformal hinges is much greater than that over the lobate antiforms, the rotation of lineation is much more effective over the upper surface of the synformal hinges.

The L_0, which was initially parallel to the shear direction, showed a weakly wavy pattern when the form surface was unrolled. This pattern of deformation of the L_0 was unexpected in the sense that, theoretically, a line parallel to the shear direction on the xz-plane should not show any rotation. In the experiments, however, because of tangential longitudinal strain, the angle between the fold axis and the lineation had consistently decreased over the synformal hinges and had slightly increased over the antiformal hinges. Because the synforms were tighter than the antiforms, the decrease in angle over the upper surface of the synforms was more than the increase in the angle over the upper surface of the antiforms. Although the lower surface of the layer is not exposed, it is expected that the tangential longitudinal strain would be contractional at the antiforms and extensional at the synforms.

In another experiment of the same series, the initial angles of the lineations with the shear direction were 75° and 165° (Figs. 6A and 10A). The folds initiated at $\gamma = 0.6$ (Figs. 10B and 11A). At this stage, the L_{75} and L_{165} rotated to make angles of 55° and 135°, respectively, with the shear direction. The initial and final angles between the lineations and the fold axes (α) are shown in Table 1.

With continued deformation, the folds were tightened and the lineations were further deformed. The L_{75} was rotated to become nearly parallel to the fold axes in most parts of the model (Fig. 11B). The L_{165} showed a type 2 lineation pattern of Ghosh and Chatterjee (1985). However, in some domains, this angle is slightly larger at the hinges than at the limbs (Fig. 11B). This is probably due to hinge migration during progressive deformation. The lineation patterns in these zones deviated from the type 2 pattern.

BUCKLE FOLDING BY SIMPLE SHEAR DEFORMATION OF INCLINED LINEATED LAYERS

The next series of experiments was conducted with models in which an inclined competent layer was embedded within a slab of silicone putty (Fig. 5B). The initial orientations of the layer and of the lineations are shown in Figure 6B. The layer is placed in the shear box in such a manner that its line of intersection with the horizontal plane (strike) is 135° with reference to the x-coordinate axis. The layer has an inclination (dip) of 30°. The L_a, L_b, and L_c had, respectively, a pitch of 0°, 45°, and −45° with the strike (Fig. 6B).

The model was deformed by a sinistral shear of $\gamma = 1.3$ (Fig. 12A). The deformed lineations were traced out on a polythene sheet. Figure 13A shows the lineation patterns on the unrolled form surface at $\gamma = 1.3$. The fold axis was approximately downdip on the enveloping surface. The dip of the enveloping surface reduced from 30° to 23°. There was an extension of 8% along the direction of the fold axis. The L_a on the fold hinge remained parallel to the strike of the rotated enveloping surface. Although the overall trend of the lineation remained perpendicular to the fold axis, the L_a showed a gentle U-shaped curved pattern (when viewed along the fold axis) on the unrolled form

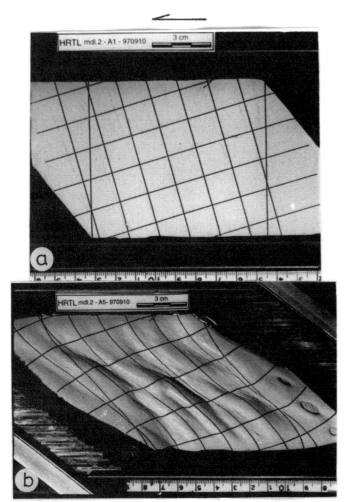

Figure 10. Model containing layer parallel to xz-plane (horizontal) of simple shear. A: Initial stage with lineations at 75° and 165° with shear direction. B: Final stage of deformation at $\gamma = 1.3$. Scale bar 3 cm.

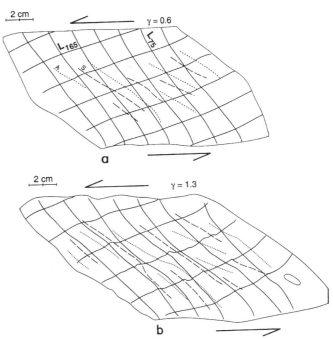

Figure 11. Unrolled surface of model shown in Figure 10. Antiforms (A) and synforms (S) are shown as dotted and dashed lines. A: Initiation of folds at $\gamma = 0.6$. L_{75} and L_{165} lineations make similar angles over hinges and limbs of folds. B: Patterns of deformed lineations at $\gamma = 1.3$. L_{165} show type 2 patterns, whereas L_{75} lineations are nearly unrollable.

surface over synformal hinges (Figs. 13A and 14A). The angles between the fold axes and the lineations are shown in Table 2.

The term "U pattern of lineation" is used here to include both type 3 and type 4 of Ghosh and Chatterjee (1985). For both these types, if one looks along the direction of fold axis on the unrolled fold surface, the curved pattern of lineation appears like U (or inverted U), the point of maximum curvature being on or slightly to one side of the hinge line. The two arms of the U are on one side of a line drawn perpendicular to the hinge line through the point of maximum curvature of the lineation pattern (dashed lines in Fig. 14). The U patterns produced at very large strains in ductile shear zones generally have a small angle between the arms of the U (Fig. 3). Such large strains could not be achieved in the simple shear experiments. The maximum value of simple shear for this apparatus was $\gamma = 1.3$. Consequently, the curving of the lineation in the experiments is very gentle. For lineations with suitable initial orientations, however, the pattern on the unrolled fold surface can be clearly recognized as a U pattern as defined herein. The theoretical studies by Ghosh

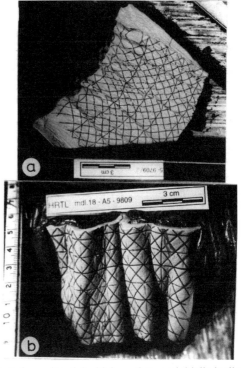

Figure 12. Deformed model with layer that was initially inclined at angle of (A) 30° and (B) 70° to XZ-plane of simple shear at $\gamma = 1.3$.

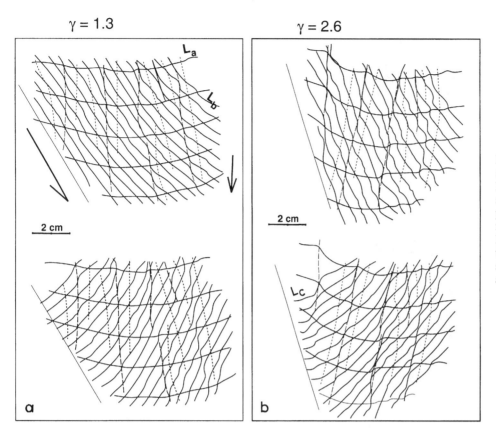

Figure 13. Unrolled surface of model shown in Figure 12A. A: At $\gamma = 1.3$, L_a lineation shows gentle U pattern over synformal hinges. L_b and L_c show type 2 pattern. B: Unrolled surface of same model after $\gamma = 2.6$. U pattern of L_a lineations becomes asymmetric. Lineation patterns are more emphasized at synforms. Closure of U is opposite at antiforms and synforms.

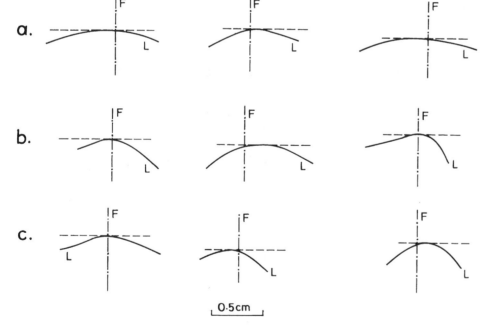

Figure 14. Enlarged view of some of experimental patterns from different domains of models. A: From model shown in Figure 13A. B: From model shown in Figure 13B. C: From model shown in Figure 17. F is hinge line, L is lineation L_a. Dashed line is perpendicular to F. All deformed lineations show U patterns because two arms of curved lineation remain on same side of dashed line. Note that point of maximum curvature may or may not be on fold hinges.

and Chatterjee (1985) and Ghosh et al. (1999) indicate that, once a gentle U pattern is produced in the initial stage, the angle between the two arms continues to decrease with progressive deformation.

At a greater value of simple shear ($\gamma = 2.6$), the U pattern of L_a was maintained but it had become asymmetrical, showing both type A and type B asymmetry, the turning point of U being to one side of the hinge line (Figs. 13B and 14B). The pattern is

TABLE 2. LINEATION PATTERNS ON INCLINED LAYERS

Lineation	α_1^*	α_2 on hinge	α_2 on limbs	α_3 on hinge	α_3 on limbs	Types of lineation pattern
L_a	≅90	≅90	75–80	76–85	66–84	Type 4, Type 3
L_b	≅45	5–10	30–35	0–5	25–30	Type 2
L_c	≅−45	5–10	35–40	0–5	30–38	Type 2
L_e	≅40	5–10	30–35	N.D.	N.D.	Type 2
L_f	≅−60	10–25	45–55	N.D.	N.D.	Type 2

*α_1, α_2 and α_3 are the angles (in degrees) between the fold axis and the lineation at $\gamma = 0.6$, 1.3, and 2.6, respectively.

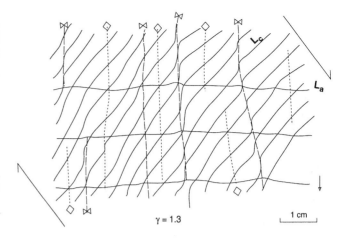

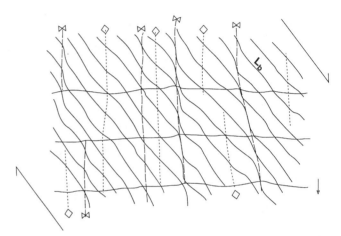

Figure 15. Unrolled surface of deformed model shown in Figure 12B. Lineation patterns are similar to those shown in Figure 13. L_a shows U patterns and L_b and L_c show type 2 pattern.

similar to the type 3 pattern of Ghosh and Chatterjee (1985). On the unrolled form surface, the L_b and L_c show sigmoidal patterns (Fig. 13). Due to tangential longitudinal contraction, this pattern is accentuated over synformal hinges. In another experiment of the same series, the results were similar (Figs. 12B and 15) when the layer had a strike of 150° with the shear direction and an inclination of 70° with the xz-plane (Fig. 6C).

BUCKLE FOLDING BY SIMPLE SHEAR DEFORMATION OF INCLINED LINEATED MULTILAYERS

The next series of experiments was carried out to study lineation patterns in disharmonic folds. Two layers, one thin (3 mm) and one thicker (6 mm), separated by a 3-mm-thick layer of silicone putty, were embedded in a thick slab of silicone putty. The whole unit was then placed in the shear box in such a way that the trace of the layering on the horizontal xz-plane (strike) made an angle of 150° with the shear direction. The layering was inclined to the horizontal xz-plane at an angle (dip) of 45° (Fig. 6D). Lineations L_a, L_b, and L_c, were printed on the upper surface of the thin layer at 0°, 45°, and −45° (angles of pitch) with the strike of the layer (Fig. 16). The whole unit was then sheared for an initial stage of $\gamma = 0.6$. The shearing resulted in folding of the layers with nearly downdip fold axes (Fig. 16A). With continued deformation two orders of folds were produced; the small folds were deformed by contact strain by the larger folds on the thicker layer (Figs. 16B and C). The development of the small folds on the thin layer was restricted to the cores of larger synforms of the thick layer (Figs. 15C and 16). The small folds were absent over the large antiforms presumably because of extension of the thin layer over the outer arc of the large antiform. At the core of a synform of a larger fold, the small folds showed S and Z folds on the limbs and M folds in the hinges. The axes of the small folds were essentially down the dip of the enveloping surface. The lineations of the thin layer were traced at $\gamma = 1.3$ (Fig. 17). The following description is only for the lineation pattern over the small folds of the thin layer.

The L_a, which was nearly perpendicular to the fold axes, showed symmetrical and asymmetrical U-shaped patterns in different parts of a large fold (Figs. 14C and 17). As in all previous experiments, these patterns were more prominent over the synforms than over the antiforms. Over the hinges of the symmetrical synforms the pattern was symmetrically U shaped. Over the asymmetric hinges, the U pattern showed a type B asymmetry. Often the type A and type B asymmetries were closely associated. The sense of asymmetry was opposite on opposite limbs of the larger fold (Figs. 14C, 17, and 18). Thus, in Figure. 17, the topmost L_a shows asymmetric U patterns; the sense of asymmetry over the S_1 and S_2 synforms is opposite to that over S_3 and S_4 synforms. The pattern along the S_5 synform is more or less a symmetrical U, whereas over S_6 the pattern is asymmetrical. The S_2 synform becomes more symmetric along the hinge and the central L_a becomes more or less symmetrical

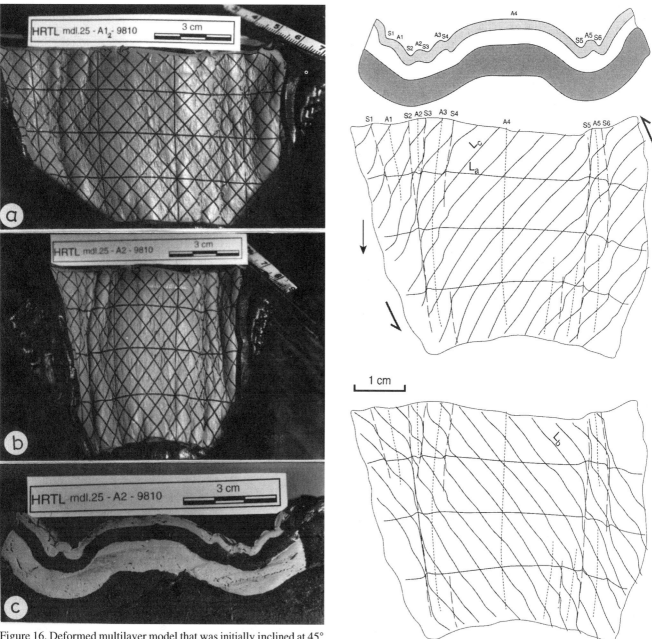

Figure 16. Deformed multilayer model that was initially inclined at 45° with XZ-plane of simple shear. A: Top surface of folded thin layer at $\gamma = 0.6$. Lineations initially made pitches of 0° (L_a), 45° (L_b), and −45° (L_c) with strike of layer. B: Top surface of same layer at $\gamma = 1.3$. C: Profile of same model at $\gamma = 1.3$.

Figure 17. Pattern of deformed lineation on unrolled surface of model shown in Figure 16. Note symmetric and asymmetric U patterns of L_a over symmetric and asymmetric synformal hinges. L_b and L_c lineations show type 2 patterns.

and U shaped over the hinge there. The lowermost L_a shows a symmetrical U pattern over the S_5 synform where the fold is more or less symmetric. In all other parts the folds are asymmetric and the lineation patterns are also asymmetrical.

The L_b and L_c, with a moderate initial angle between the lineation and the fold axis, showed sigmoidal type 2 patterns similar to those in the previous experiments. The lineations made very low angles over the synformal hinges due to greater tangential longitudinal strain at the inner arcs of the synforms (Fig. 17). A similar experiment, but with different orientations of the layers and the lineations (Figs. 6E, 19, and 20), showed results similar to those of the previous experiment. The U pattern of the L_a is particularly well developed in this experiment (Fig. 20, upper).

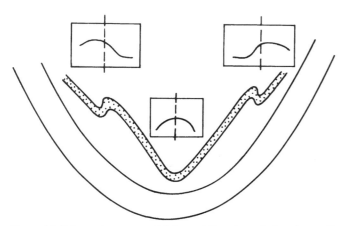

Figure 18. Schematic pattern of deformed lineations (in inset) over S, M, and Z folds.

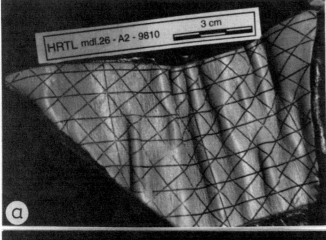

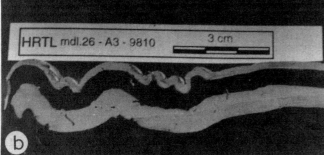

Figure 19. Deformed multilayer that was initially inclined at 30° to XZ-plane of simple shear. Lineations made pitches of 0° (L_a), 50° (L_e), and −30° (L_f) with strike of layer. A: Top surface of layer at $\gamma = 1.3$. B: Profile section of model.

SIMPLE SHEAR DEFORMATION OF LINEATION OF PREVIOUSLY FOLDED LAYERS

In another series of experiments the initial model was prepared with artificially produced sets of cylindrical folds on a competent sheet on a slab of silicone putty. The folding was such that each of the four sets of lineation made a constant angle with the fold axis in all parts of the fold. The lineations were at angles of 90°, 45°, and −45° with the fold axis. They are referred to as L_a, L_b, and L_c, respectively (Fig. 6F). The model was then trimmed and placed in the shear box, the hinge lines making an initial angle of 30° with the shear direction and lying in the field of extension (Figs. 5C and 6F). A thin polythene sheet was stuck over the folded surface (for easy removal of overburden as well as to stop smearing of the marker lines), and an overburden of silicone putty was placed above it. The fold limbs at the inflection points were dipping at an angle of 50° on either side, and the enveloping surface was parallel to the horizontal xz-plane (Fig. 21A). After a dextral simple shear of $\gamma = 1.3$, the overburden of the model was removed to trace the deformed lineation pattern. At this stage (Fig. 21B), the fold axis made an angle of 20° with the shear direction and the limb dips increased to 65°. The deformed model showed an angular hinge migration of 5°.

The deformed lineation pattern on the unrolled surface is shown in Figure 22. The lineations were deformed not only by an external rotation caused by tightening of the preexisting folds and development of small new folds, but were also reoriented by strain. As a result the angle between a lineation and the fold axis changed in the course of progressive deformation. The L_c was reoriented to a maximum extent and made a much smaller angle with the fold axis over the domain of synformal hinges

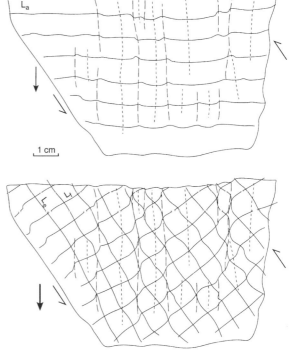

Figure 20. Unrolled surface of model shown in Figure 19. L_a shows symmetric and asymmetric U patterns, while L_e and L_f show type 2 patterns.

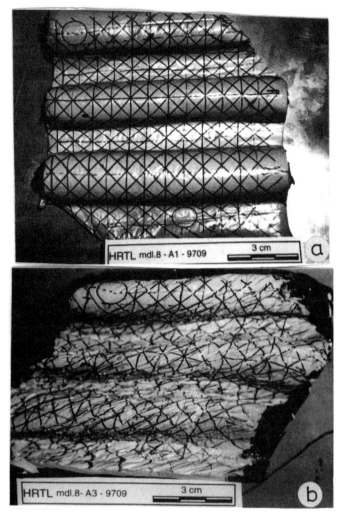

Figure 21. A: Undeformed model with round antiforms and synforms. B: Model after dextral shear of $\gamma = 1.3$. There are small wrinkles at angle with large folds imprinted on layer (see text for details).

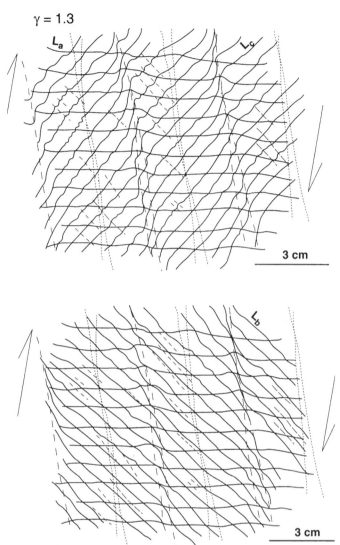

Figure 22. Unrolled surface of folded layer of Figure 21b. L_a lineations show gentle U pattern over large folds. L_c lineations show small U patterns over crenulations.

than anywhere else (Fig. 22). This is because of combined effects of rotation of lineation by layer-parallel strain and tangential longitudinal strain at the inner arc of the synforms.

The L_a was initially at 90° with the fold axis. Because the fold axis was in the extension field of the bulk simple shear, there was a stretching parallel to the fold axis in all parts of the fold. The deformed lineation showed a type B asymmetric U-shaped pattern on the unrolled form surface (Fig. 22). At the initial stages of deformation the open U pattern of the lineation was more or less symmetrical when the fold axis and the general orientation of the lineation were orthogonal. The asymmetry of the U pattern increased with progressive deformation by the combined effects of hinge migration, layer-parallel strain, and tangential longitudinal strain. Because of hinge migration the lineations showed both type A and type B asymmetry.

A new set of small folds developed on the polythene sheet that was stuck over the competent layer. The surface of the layer produced small folds by the effect of these wrinkles. The hinge lines of these folds were at an angle with the axes of the larger folds (Fig. 22). The L_c was reoriented to be nearly perpendicular to the hinge lines of these small folds. When the form surface of the fold was unrolled, the L_c showed small U-shaped curvatures. The U pattern of the lineation was produced by the combined effects of layer-parallel strain and buckling. This pattern was produced only when the small fold axes were nearly at a right angle to the overall orientation of the L_c. The L_b lineations were essentially parallel to the axes of the small folds.

DISCUSSION

The types of lineation patterns obtained in these experiments are similar to those that have been described from the field (e.g., Ramsay, 1958, 1960; Hudleston, 1973; Naha and

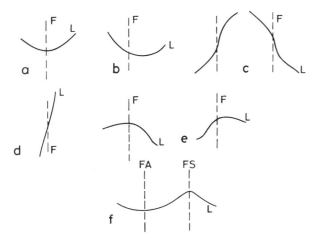

Figure 23. Summary of deformed lineation patterns developed in experiments. A, B: Symmetric (type 4) and asymmetric (type 3) U patterns. C: Sigmoidal type 2 pattern. D: Type 1 (unrollable pattern). E: Type 3 patterns over S and Z folds showing opposite sense of asymmetry. F: Symmetric type 4 pattern with closure of U pointing in opposite directions over antiform (FA) and synform (FS). Pattern is more pronounced over synform.

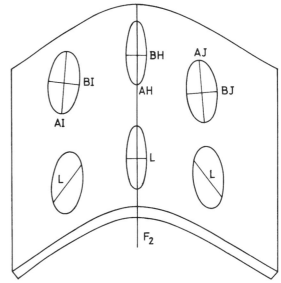

Figure 24. Schematic diagram showing orientations of sectional strain ellipses on hinge and limbs of fold. Note that minor axes of strain ellipses (BI and BJ) are larger on limbs than at hinge (BH). Major axes of ellipses (AI and AJ) make small angle with fold axis on limbs, while on hinge surface they (AH and F_2) are parallel. In lower part of figure, orientation of early lineation (L), which was initially at right angle to fold axis, shows rotation toward fold axis on limbs. Resulting pattern is symmetric U.

Halyburton, 1977; Ghosh and Chatterjee, 1985; Ghosh and Sengupta, 1987, 1990; Mukhopadhyay and Pati, 1997; Sengupta and Ghosh, 1997; Ghosh et al., 1999). The mathematical model of Ghosh and Chatterjee (1985) has shown that, depending on the initial angle between the early lineation (L_1) and the later fold axis (F_2), the different lineation patterns (from type 1 to type 6) may develop under coaxial deformation. The present series of experiments shows that similar lineation patterns (Fig. 23) may be produced in noncoaxial deformation as well.

In a type 1 pattern (with a constant angle between the lineation and the fold axis in all parts of the fold), the lineation can be straightened out when the form surface of the fold is unrolled. This pattern is generally explained by flexural folding without accompanying layer-parallel strain. The current experiments indicate, however, that the type 1 pattern can also develop over buckling and flattening folds when the initial angle between L_1 and F_2 is moderate to small. Because the angle between L_1 and the F_2 axis may change considerably due to layer-parallel strain, the initial L_1 cannot be obtained by unrolling of the later fold.

If the enveloping surface is parallel to a plane that is perpendicular to the vorticity vector of simple shear, a type 2 pattern developed when the angle between L_1 and F_2 has a moderate to large value (see Table 1). The fold axis initiates parallel to the long axis of strain ellipse on the enveloping surface. With progressive rotation of the fold axis toward the direction of the X-axis of the bulk strain ellipsoid, there is equal extension parallel to the fold axis both in the limbs and in the hinge zone. However, the minor axes of the strain ellipses have unequal lengths at the hinges and limbs (Fig. 24). Thus, for example, in the present series of experiments with the initial layering subparallel to the XZ-plane of simple shear, the aspect ratio of the strain ellipse on the subhorizontal hinge zone (parallel to the XZ-plane of the strain ellipsoid) is equal to X/Z. Over the limbs inclined to this plane, however, the minor axis (B) of the sectional strain ellipse must be larger than Z over the limbs (Fig. 24). Because the angle between the major axis of sectional strain ellipse (A) and the lineation on the layer changes according to the equation $\tan\theta' = (B/A)\tan\theta$ (cf. Ramsay, 1967, equation 3-34), the angle between the fold axis and lineation decreases to a greater extent on the hinge than on the limbs, thereby producing a type 2 lineation pattern. The pattern is accentuated by tangential longitudinal contraction over the inner arcs of synforms. The experiments indicate that, in the field, the orientation of early lineation should be studied on the limbs as well as on the hinges. Thus, for a type 2 pattern, if the orientation of the lineation is measured only on the limbs and not on the hinge, the lineation pattern may be wrongly interpreted as an unrollable type 1 pattern.

The U pattern of lineation developed on layers inclined to the XZ-plane when the orientation of the early lineation (L) was more or less at a right angle to the later fold axis. With such inclined enveloping surfaces, the major axes of sectional strain ellipses are not parallel in hinges and limbs, but occur at a small angle to each other (AI or AJ of Fig. 24). Because the fold axis (F_2) remains parallel to the major axis of the strain ellipse (AH) over the hinge zones, the near right angle between L and F_2 (i.e., also the angle between L and AH) either remains unchanged or changes to a much smaller extent than over the limbs where L is at an acute angle to the A axis (Fig. 24).

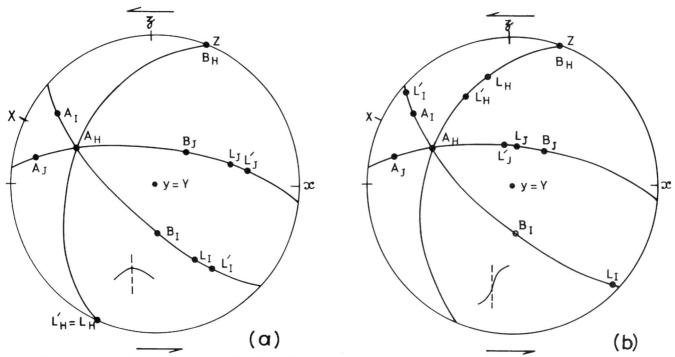

Figure 25. A, B: A_H, A_I, and A_J are major axes and B_H, B_I, and B_J are minor axes of sectional strain ellipses on hinge surface and two limbs (I and J). L_H, L_I and L_J are initial and L'_H, L'_I, and L'_J are deformed lineations after simple shear of $\gamma = 2$. In (A) early lineation was at right angle to fold axis. Over hinge lineation remained unchanged ($L_H = L'_H$). Over limbs I and J same lineation makes angle of 80° with fold axis (A_H). In (B) lineation was initially at angle of 45° with fold axis. After deformation lineation makes smaller angles on hinge than at limbs. Deformed lineation patterns are shown schematically in insets.

This point has been clarified by taking a simple situation (Fig. 25) where, after a simple shear of $\gamma = 2$, the enveloping surface at the hinge zone is inclined to the XZ-plane of the strain ellipsoid at an angle of 40°, and the axial surface (parallel to the XY-plane of the strain ellipsoid) makes an angle of 22.5° with the x-axis. The orientations of the two limbs are symmetrically inclined with respect to the axial surface. The orientations and the lengths of the semiaxes of the sectional strain ellipse from the hinge surface and on the limbs were determined from the equations of the Appendix 1. It is found that at $\gamma = 2$, the aspect ratio of the sectional strain ellipse over the hinge zone is $A_H/B_H = 1.68$, the major axis A_H being parallel to the fold axis. Over the two limbs I and J the aspect ratios of the strain ellipse is 1.43.

Let us now consider the rotation of a lineation that was initially perpendicular to the fold axis in all parts of the fold (Fig. 25A). On the hinge surface H, the initial angle of 90° between A_H and the lineation remains unchanged. Over the limbs, however, the initial angle of 65° between L_I and A_I or between L_J and A_J was reduced to 54°. As a result the angle between the fold axis (A_H) and the lineation (L'_I or L'_J) on the limbs changes from 90° to 80°. If the folded surface is unrolled about its axis, the lineation forms a gentle U pattern (type 4 lineation pattern). In this simple analysis, we have considered only a symmetrical fold and only a particular stage of deformation. In progressive deformation, however, the material planes parallel to the current axial plane and the current hinge surface will no longer remain at a right angle, and by hinge migration the new hinge will shift to another place. Hence the U pattern will have a type A asymmetry (type 3 lineation pattern), the turning point of U being shifted on one side of the hinge.

We apply the same model to a lineation, which makes an initial angle of 45° with the fold axis in both the limbs and the hinge zone (Fig. 25B). The lineation would rotate to a greater extent on the hinge zone than at the limbs in such a manner that the acute angle between the fold axis and the lineation is decreased to a greater extent on the hinge than on the limbs. At $\gamma = 2$, the lineation would make an angle of 31° with the fold axis at the hinge while the corresponding angles at the limbs I and J would be 40°. On the unrolled surface the lineation will show a type 2 pattern.

CONCLUSIONS

1. When buckle folding of a lineated layer is accompanied by layer-parallel strain, the angle between the lineation and the fold axis changes differently in different parts of the fold.

2. The U pattern of lineation, recorded from ductile shear zones, can develop in bulk simple shear under the conditions that the layer is inclined to the XZ-plane of simple shear, and the axis of the later fold is initially at a right angle to the early

lineation. The U pattern of lineation does not develop over folds in which the enveloping surface is parallel to the XZ-plane of simple shear.

3. With progressive simple shear, a symmetrical U pattern changes to an asymmetrical pattern.

4. The nature of the U pattern also depends on the symmetry of the folds. Over the parasitic M folds, the U pattern is symmetric; over S and Z folds, the pattern is asymmetric, with opposite sense of asymmetry over S and Z folds.

5. On inclined layers, the U pattern does not develop if the angle between the lineation and the fold axis is moderate or small. Under this condition a type 2 or a type 1 lineation pattern generally develops over buckling and flattening folds.

6. When the angle between the early lineation and the later fold axis is small the deformed lineation is nearly straight on the unrolled form surface of the fold. Again, the angle between the early lineation and the later fold axis may be considerably reduced by initial layer-parallel strain before buckle folding becomes significant. If the early lineation is brought to a low angle with the fold axis by initial layer-parallel strain, subsequent buckling may simply produce an unrollable lineation pattern, with a constant angle between the lineation and fold axis in all parts of the fold.

7. The pattern of deformed lineations is modified by tangential longitudinal strain. Over the inner arcs of synformal hinge zones there is contraction across the fold axis. As a result the U pattern of deformed lineation on the upper surface of the layer is more pronounced over the synformal hinges. The closure of U is oppositely directed in synforms and antiforms. The type 2 pattern also becomes more pronounced over the synformal hinges.

8. The experiments indicate that if the bulk deformation is essentially by simple shear, the U pattern of lineation is likely to be asymmetric. This conclusion is also in conformity with the theoretical model of Ghosh et al. (1999).

ACKNOWLEDGMENTS

The experiments were carried out at the Hans Ramberg Tectonic Laboratory of Institute of Earth Sciences, Uppsala University. We thank Chris Talbot, Richard Lisle, and Martin Burkhard for review and helpful suggestions, and Christina Wernstrom and T. Bhattacharya for drafting the line drawings. Comments and suggestions from Subir Ghosh significantly improved the manuscript. Sengupta acknowledges the financial support from Swedish Institute and Council of Scientific and Industrial Research (India). Koyi is funded by the Swedish Natural Sciences Research Council (NFR).

APPENDIX 1

If A and B are the major and minor semiaxes of the sectional strain ellipse, X, Y, Z are the semiaxes of the strain ellipsoid, and l_1, l_2, l_3 are the direction cosines of the normal to the section plane with reference to the principal axes of the strain ellipsoid, then the lengths of A and B are given by the following equations (McRea, 1953; Ghosh et al., 1996):

$$A = \sqrt{\frac{1}{2P}[Q + \sqrt{Q^2 - 4P}]}, \quad (1)$$

and

$$B = \sqrt{\frac{1}{2P}[Q - \sqrt{Q^2 - 4P}]}, \quad (2)$$

where

$$P = X^2 l_1^2 + Y^2 l_2^2 + Z^2 l_3^2, \quad (3)$$

and

$$Q = X^2 Z^2 [l_1^2 + l_3^2] + X^2 Y^2 [l_1^2 + l_2^2] + Y^2 Z^2 [l_2^2 + l_3^2]. \quad (4)$$

The direction ratios of the A-axis are:

$$a = \frac{X^2 l_1}{A^2 - X^2}, \quad (5)$$

$$b = \frac{Y^2 l_2}{A^2 - Y^2}, \quad (6)$$

$$c = \frac{Z^2 l_3}{A^2 - Z^2}. \quad (7)$$

The direction cosines of the A-axis are:

$$l_1 = \frac{a}{\delta}, \quad (8)$$

$$l_2 = \frac{b}{\delta}, \quad (9)$$

$$l_3 = \frac{c}{\delta}, \quad (10)$$

where

$$\delta = \sqrt{a^2 + b^2 + c^2}. \quad (11)$$

If θ and θ' are the initial and final angles between A-axis and the lineation L, then from Ramsay (1967, equation 3-34), we have

$$\tan\theta' = \left(\frac{B}{A}\right)\tan\theta \quad (12)$$

REFERENCES CITED

Ghosh, S.K., 1974, Strain distribution in superposed buckling folds and the problem of reorientation of early lineation: Tectonophysics, v. 27, p. 323–441.

Ghosh, S.K., 1993, Structural geology: Fundamentals and modern developments: Oxford, Pergamon Press, 598 p.

Ghosh, S.K., and Chatterjee, A., 1985, Patterns of deformed early lineations over late folds formed by buckling and flattening: Journal of Structural Geology, v. 7, p. 651–666.

Ghosh, S.K., and Sengupta, S., 1987, Progressive evolution of structures in a ductile shear zone: Journal of Structural Geology, v. 9, p. 277–288.

Ghosh, S.K., and Sengupta, S., 1990, Singhbhum shear zone: Structural transition and a kinematic model: Indian Academy of Sciences Proceedings (Earth and Planetary Sciences), v. 99, p. 229–247.

Ghosh, S.K., Deb, S.K., and Sengupta, S., 1996, Hinge migration and hinge replacement: Tectonophysics, v. 263, p. 319–337.

Ghosh, S.K., Hazra, S., and Sengupta, S., 1999, Planar, nonplanar and refolded sheath in the Phulad shear zone, Rajasthan, India: Journal of Structural Geology, v. 21, p. 1715–1729.

Hudleston, P.J., 1973, The analysis and interpretation of minor folds developed in the Moine rocks of Monar, Scotland: Tectonophysics, v. 17, p. 89–132.

McRea, W.H., 1953, Analytical geometry of three dimensions: Edinburgh, U.K., Oliver and Boyd. 144 p.

Mukhopadhyay, D., and Pati, D.P., 1997, Deformed lineation in ductile shear zones: A case study from the Proterozoic fold belt of Singhbhum, eastern India, in Sengupta, S., ed., Evolution of geological structures in micro to macro-scales: London, U.K., Chapman and Hall, p. 219–236.

Naha, K., and Halyburton, R.V., 1977, Structural pattern and strain history of superposed fold system in the Precambrian of central Rajasthan, India. I and II: Precambrian Research, v. 4, p. 39–84.

Ramsay, J.G., 1958, Moine-Lewisian relations in Glenelg: Geological Society of London Quaternary Journal, v. 113, p. 487–523.

Ramsay, J.G., 1960, The deformation of early linear structures in areas of repeated folding: Journal of Geology, v. 68, p. 75–93.

Ramsay, J.G., 1967, Folding and fracturing of rocks: London, U.K., McGraw-Hill, 568 p.

Ramsay, J.G., and Huber, M.I., 1987, The techniques of modern structural geology, 2: Folds and fractures: London, Academic Press, 700 p.

Sengupta, S., and Ghosh, S.K., 1997, The kinematic history of the Singhbhum shear zone: Indian Academy of Sciences Proceedings (Earth and Planetary Sciences), v. 106, p. 185–196.

Weiss, L.E., 1955, Fabric analysis of triclinic tectonite and its bearing on the geometry of flow in rocks: American Journal of Science, v. 253, p. 223–236.

Weiss, L.E., 1959, Geometry of superimposed folding: Geological Society of America Bulletin, v. 70, p. 91–106.

MANUSCRIPT ACCEPTED BY THE SOCIETY APRIL 12, 2000

Single-layer folds developed from initial random perturbations: The effects of probability distribution, fractal dimension, phase, and amplitude

Neil S. Mancktelow
Geologisches Institut, ETH-Zentrum, CH-8092 Zurich, Switzerland

ABSTRACT

Natural folds develop from small irregularities in the layer's surface that are probably best described by a self-affine fractal distribution, similar to topography. Such distributions differ from even or Gaussian normal distributions, which have commonly been assumed in analytical and numerical studies of single-layer folding, in that their amplitude spectrum is not flat but is biased toward longer wavelengths. This could lead to a corresponding bias in the measured wavelengths at finite amplitude in natural folds and thereby influence estimates of the rheological contrast between layer and matrix. Finite-element modeling of single-layer folds developed in viscoelastic materials with a viscosity contrast of 100:1 between layer and matrix demonstrates that the effects are only minor, despite the order of magnitude variation in amplitude of the different spectral components of the initial perturbations. Little difference is observed in the final amplified fold wavelengths whether the initial distribution is even, Gaussian normal (white noise) or self-affine fractal, with slopes to the spectral power distribution of 1, 2 (Brownian walk), or 3. Whether the perturbations in the upper and lower surface are completely independent or perfectly in phase also has little effect, as does variation in the Poisson's ratio to values approaching incompressibility. The only effect that does have a significant influence is the amplitude of the initial perturbations. Scaling the random distribution to smaller and smaller initial amplitudes promotes layer-parallel shortening, resulting in a decreased arc length and thicker layer and consequently in a lower arc-wavelength to layer-thickness ratio. This effect can be corrected if the shortening within the layer is known, but accurate estimates are seldom available for natural folds.

INTRODUCTION

The geometry of initial irregularities of small but finite amplitude in the layer's surface can have an important influence on the final shape of folds developed by buckling of a single layer embedded in a weaker matrix (e.g., Biot, 1961; Cobbold, 1976; Williams et al., 1978; Abbassi and Mancktelow, 1992). In the end-member case of a single sinusoidal perturbation, the initial waveform determines the wavelength of the perfectly periodic fold that develops, even when this is far removed from the dominant or potentially most rapidly growing wavelength for the material properties involved (Mancktelow, 1999). At the other extreme, a single isolated initial perturbation in a perfectly planar layer influences the location, shape (Cobbold, 1975; Abbassi and Mancktelow, 1992), and even the symmetry (Abbassi and Mancktelow, 1990) of folds developed, particularly at lower values of bulk shortening. In natural examples, however, the initial perturbations are more likely to approach a less correlated, near random distribution. Even in this case, the shape of the initial perturbation can be recognized in the final quasiperiodic fold shape (Mancktelow, 1999). What is not yet established is if the initial random distribution can exert a systematic influence on the measured wavelength of folds. Because the wavelength of natural folds represents one of the few potential measures of rheological contrast from field observation, determination of any systematic bias due to more realistic natural initial perturbation distributions is of considerable practical interest. Currently, it is generally assumed in analytical and numerical studies of fold development that perturbations in adjacent surfaces are in phase and that random distributions are either even or Gaussian normal. However, there is no basis for the in-phase assumption, and natural forms are also more likely to follow a fractal random distribution (e.g., Turcotte, 1997) with a consequent bias in the amplitude spectrum of the initial perturbation toward longer wavelengths. The aim of this chapter is to investigate if these effects have any consistent influence on measured fold wavelengths.

PROPERTIES OF SELF-AFFINE FRACTALS

This overview of self-affine fractals is a short summary of the relevant chapter of Turcotte (1997), which is based on the pioneering work of Mandelbrot (1977) and many subsequent workers. A statistically self-similar fractal is by definition isotropic, i.e., the scaling does not depend on orientation. Variation in topography in the horizontal plane is a good two-dimensional example. A statistically self-affine fractal is not isotropic. The scaling of one dimension is some constant proportion of the other (e.g., as in a vertical topographic profile). A formal definition of a self-affine fractal in a two-dimensional xy-space is that $f(rx, r^{Ha}y)$ is statistically similar to $f(x, y)$, where Ha is known as the Hausdorff measure and r is a scaling parameter. Using the Fourier transform, the distribution can be considered either in the physical domain $y(x)$ or the frequency domain, in terms of the spectrum $Y(f, X)$, where f is the frequency. In practice, for distributions where y is known at a discrete number of equally spaced points $y(n)$, the discrete Fourier transform is employed to obtain $Y(N)$, where N is the wavenumber (i.e., the integer number of wavelengths in the length of the data set considered). The spectral density is then defined by $S(N) = (1/n_m)|Y(N)|^2$ as the total number of data points n_m approaches infinity. A value of 16384 was used for n_m in the current study, sufficient to provide a good statistical approximation.

For a self-affine fractal, the power spectral density has a power-law dependence on wavenumber, i.e., $S(N) \simeq N^{-\beta}$. The relationship between this power exponent β, the Hausdorff measure Ha, and the fractal dimension D is given by $\beta = 2Ha + 1 = 5 - 2D$ (Turcotte, 1997, equation 7.48). For a self-affine fractal, $0 < Ha < 1$, $1 < D < 2$, and $1 < \beta < 3$. Even and Gaussian normal distributions (white noise) have a flat power spectrum ($\beta = 0$) and are not self-affine fractals. However, the cumulative sum of a distribution with a power exponent β has a power exponent $\beta + 2$. The cumulative sum of a white noise is therefore a self-affine fractal with a power exponent of 2. It is known as a Brownian walk, because it corresponds to the path of a molecule undergoing Brownian motion, in which each individual step is described by a Gaussian normal probability distribution. It was also called a white roughness spectrum by Kamb (1970) and has the important property that its limb-dip spectrum is flat. This follows from the relation between amplitude, tangent of limb dip, and wavenumber, $Y(N) = \tan\theta_{max}/N$ (e.g., Fletcher and Sherwin, 1978, equation 9) and the definition of the spectral density $S(N) = (1/n_m)|Y(N)|^2$. So-called fractional Brownian walks with β values other than 2 can be generated by scaling the amplitude components of an existing Gaussian normal distribution in the frequency domain followed by inverse transformation back into the physical domain. Measurements of natural topography return β values of ~2 (i.e., approximately Brownian walks; e.g., Gilbert, 1989; Fig. 7.17 of Turcotte, 1997), and rock surfaces $1.8 < \beta < 3$ ($1 < D < 1.6$; e.g., Brown and Scholz, 1985; Turcotte, 1997). In this study the complete range of self-affine fractals is investigated ($1 < \beta < 3$).

INITIAL PERTURBATION DISTRIBUTIONS

To maintain good statistics, the starting distribution in every case consists of 16384 data points. Equally spaced values were then selected from this distribution to represent the initial irregularity in the layer surface, discretized as 513 nodes, the initial and final nodes having no offset from a perfectly planar layer. For analysis, the right boundary node was dropped (because it represents the implied repeat distance of the whole sequence, with a value equal to the left boundary node) leaving 512 data points for efficient discrete fast Fourier transform (FFT, e.g., Press et al., 1986). Generation and modification of random distributions was effected with Mathematica® and the relevant notebooks can be downloaded from *http://www.erdw. ethz.ch/~neil/fractaldist.html*

The starting distribution was a Gaussian normal distribution with a variance of 0.05. The seed for the random distribution was 0 for the upper surface and 10^6 for the lower surface, to ensure there was no overlap in the random distributions generated. Even distributions (i.e., with equal probability over a given range) were also generated and scaled to have the same root-mean-square (rms) average amplitude as the Gaussian distribution over the same range of data values. Both these distributions have a flat spectral distribution ($\beta = 0$; Figs. 1 and 2). The discrete Fourier components of the Gaussian distribution were scaled to obtain a partial Brownian walk with $\beta = 1$ (Fig. 3). The cumulative sum of the initial Gaussian distribution produced a Brownian walk with $\beta = 2$ (Fig. 4) and the cumulative sum of the distribution with $\beta = 1$ a distribution with $\beta = 3$ (Fig. 5).

In the first instance, the amplitudes of the perturbation distributions were scaled so that they all have the same rms average amplitude in the spatial domain as the Gaussian distribution. As can be seen from Figure 6A, the relative amplitude of

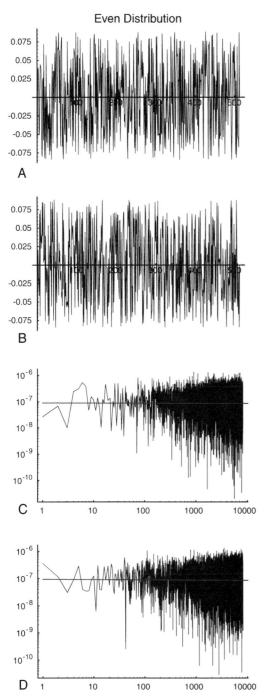

Figure 1. Initial even random distribution with same root mean square average amplitude as Gaussian normal distribution with variance 0.05. Represented are 512 equally spaced data points, selected from full set of 16384 values, that were assigned as initial perturbation in y-coordinate of nodes on (A) upper and (B) lower surfaces of layer. C, D: Corresponding log-log plots of spectral density against wavenumber, returned by discrete Fourier transform of this full data set (number of distinct spectral values is always one-half of number of data values analyzed). Least-squares best linear fits to values in (C) and (D) are also given, demonstrating flat spectral distribution.

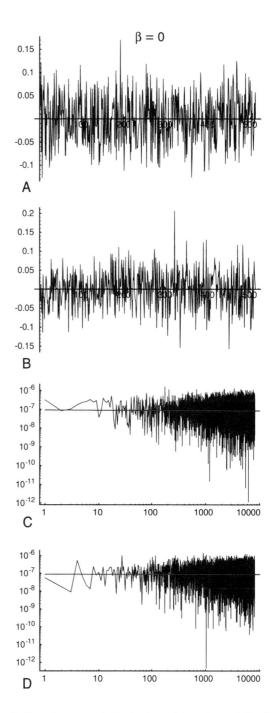

Figure 2. Gaussian normal distribution with variance 0.05. Equally spaced 512 values selected from 16384 values of full data set for (A) upper and (B) lower surface are shown, together with corresponding log-log plots (C) and (D) of spectral density distribution against wavenumber for full data set. Least-squares best linear fits to values in (C) and (D) highlight flat spectral distribution, similar to even distribution of Figure 1.

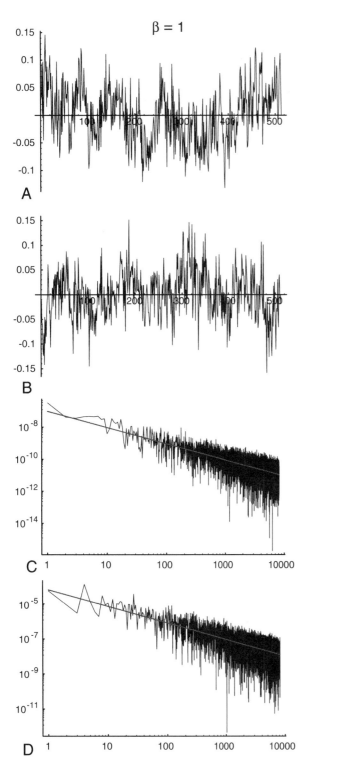

Figure 3. Self-affine fractal distribution with $\beta = 1$ for perturbation in (A) upper surface and (B) lower surface of layer, derived by scaling Fourier components of Figure 2 (C and D) to obtain required slope of 1 in corresponding log-log plots (C and D) of spectral components against wavenumber, as shown by best-fit line to spectral distribution. Magnitudes of values in distribution are scaled so that root mean square average of full data set is same as for normal distribution of Figure 2.

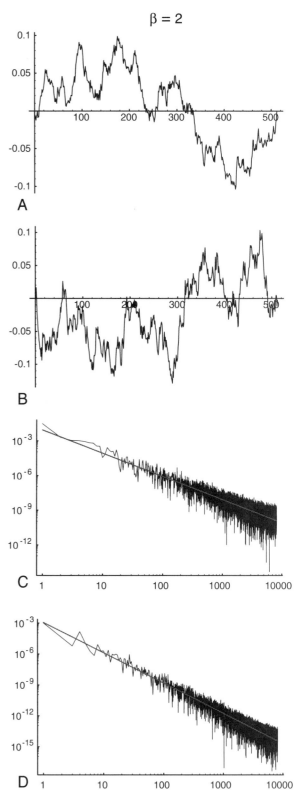

Figure 4. Brownian walk (self-affine fractal distribution with $\beta = 2$) derived from cumulative sum of Gaussian normal distribution of Figure 2 for (A) upper surface and (B) lower surface of layer. Corresponding log-log plots of spectral distribution against wavenumber in C and D have slopes for best-fit lines of -2.01 and -1.94, respectively. Magnitudes of values in distribution are scaled so that root mean square average of full data set is same as for normal distribution of Figure 2.

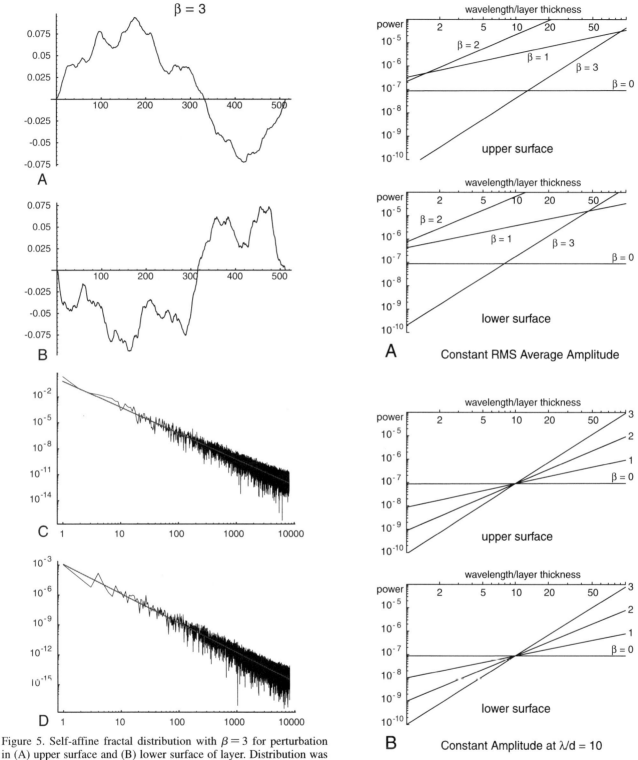

Figure 5. Self-affine fractal distribution with $\beta = 3$ for perturbation in (A) upper surface and (B) lower surface of layer. Distribution was derived from cumulative sum of distribution with $\beta = 1$ (see Fig. 3). Corresponding log-log plots of spectral distribution against wavenumber in (C) and (D) have slopes for best-fit lines of -3.01 and -2.94, respectively. Magnitudes of values in distribution are scaled so that root mean square average of full data set is same as for normal distribution of Figure 2.

Figure 6. A: Summary of best-fit lines through spectral distributions of Figures 2–5, plotted against wavelength/layer thickness, as appropriate to numerical models. As in Figures 2–5, each of these distributions has same root mean square (rms) average. B: The same distributions scaled so that they all have same amplitude at wavelength/layer thickness value of 10. Necessary scaling factors are listed in Table 1.

particular frequency components between the different distributions varies as a function of frequency. For $\beta \neq 0$, it is no longer possible to speak of a specific amplitude for a distribution or to say that the amplitude of one distribution is some constant multiple of another. Therefore, in a second set of experiments (Fig. 6B) the amplitudes were scaled so that in the frequency domain the distributions all have the same amplitude at a wavelength 10 times the layer thickness (a value chosen to be close to the dominant wavelength). The necessary scaling factors are listed in Table 1. In another series of experiments, the effect of initial perturbation amplitude was considered by multiplying the amplitude of the distribution with $\beta = 2$ in Figure 6B by factors of 0.01, 0.1, 10, 50, and 100. For out of phase initial irregularities, this results in important variations in initial layer thickness, which is most pronounced in the distributions with larger β values (≥ 2), where the persistent long wavelength components are more pronounced (Figs. 4 and 5).

MODEL GEOMETRY AND BOUNDARY CONDITIONS

All models were deformed in pure shear at a constant logarithmic strain rate of 10^{-14} s^{-1}. The width of the layer was 2 units and the length 396 units, twice that employed in Mancktelow (1999), to reduce boundary constraints on the amplification of longer wavelength components. The model width was initially 128 units. The converging sides were kept straight and the only other constraint was that the y-coordinates of the center line of the layer were fixed at the left and right sides (necessary to eliminate rigid-body rotational components). Models were usually deformed to a logarithmic strain of -0.6 (~45% shortening). Modeling was carried out with the commercial finite-element package MARC-Mentat, following the same general procedures as described in Mancktelow (1999). A total of 11264 elements was employed; the mesh was refined toward the layer, which was represented by 512×6 quadratic elements. The finite-element program automatically varies the time step to maintain input tolerances, namely that creep strain in any one increment does not exceed 0.5 of the elastic strain and that the stress change during one increment does not exceed 0.1 of the current stress. With these tolerances, models took from 900 to 3000 time increments to reach a logarithmic strain of -0.6.

TABLE 1. SCALING FACTORS FROM FIGURE 6 (A AND B)

β value	1	2	3
Upper Surface	0.1624	0.0633	1.4548
Lower Surface	0.1527	0.0365	0.7034

Scaling factors necessary to convert constant root mean square average amplitude distributions of Figure 6A into distributions of Figure 6B, with constant amplitude at wavelength to layer thickness ratio of 10.

MATERIAL PARAMETERS

The rheology considered follows a Maxwell viscoelastic model, taken as the in-series addition of a compressible elastic element and an incompressible linear viscous element. There is no standard definition of the Deborah number, De, which measures the relative importance of the two rheological components (i.e., elastic vs. viscous) for the total rheological response under particular boundary conditions. The definition used here follows Poliakov et al. (1993) and Mancktelow (1999), i.e., $De = \eta\dot{\varepsilon}/G$, where η is the viscosity, $\dot{\varepsilon}$ is the applied bulk strain rate, and G is the elastic shear modulus ($G = E/2[1 + \nu]$, where E is Young's modulus and ν is Poisson's ratio). For elements in series, the stress is the same in both elements. The normal stress in the viscous element is $4\eta\dot{\varepsilon}$, and therefore De represents a measure of the elastic strain stored in the system, with increasing De corresponding to an increasing elastic strain component. Alternatively, because the Maxwell relaxation time is $\eta/2G$ and a viscous strain of 0.5 is attained in the time $1/2\dot{\varepsilon}$, De can be considered as the ratio of the Maxwell relaxation time to the time required to attain a viscous strain of 0.5.

In this study, the same values of viscosity were used in all experiments, namely 10^{22} Pa s for the layer and 10^{20} Pa s for the matrix, respectively, to give a ratio of 100:1. Two different sets of Young's moduli were used: (1) 10^9 Pa for both layer and matrix; and (2) 5×10^{10} Pa for the layer and 5×10^9 Pa for the matrix. For the strain rate of 10^{-14} s^{-1} used for all the models and a Poisson's ratio of 0.25, this corresponds to (1) $De = 0.25$ in the layer and $De = 0.0025$ in the matrix and (2) $De = 0.005$ in the layer and $De = 0.0005$ in the matrix. Test models demonstrate that identical results are obtained for the same values of De and the same ratios in viscosities and elastic moduli whether it is the magnitude of the rheological parameters or the strain rate that is varied. In one run, a Poisson's ratio of 0.45 (i.e., near the incompressible value of 0.5) was used to test for any possible influence, because incompressibility is often assumed in numerical and analytical studies.

METHODS OF ANALYSIS

For the quasiperiodic fold trains developed naturally from initial random perturbations and modeled numerically in this study, it is not a straightforward matter to establish a measure of wavelength that is both unbiased and reproducible. Sherwin and Chapple (1968) employed an averaging technique that is simple in principle but difficult to apply unequivocally in practice. They divided the total arc length measured along the center line of a fold train by the number of folds (one fold consisting of one syncline plus one anticline), and normalized this value against the average layer thickness. As Sherwin and Chapple (1968) noted, however, determining the position of inflection points and hence the number of distinct folds is often a matter of subjective judgement. When there is only a small number of folds in the layer segment available for analysis, a difference of one or two in the number of folds distinguished results in a significant difference in the calculated average wavelength. Fletcher and Sherwin (1978) demonstrated that the arc wavelength measured in this way corresponds to the highest amplitude component in the limb-dip spectrum, assuming that it developed from an initial perturbation distribution with a flat limb-dip spectrum (i.e., a Brownian walk with $\beta = 2$). In general, the most amplified spectral component (i.e., the dominant wavelength) should correspond to the maximum in the amplitude spectrum if the initial amplitude spectrum was flat ($\beta = 0$) and the maximum in the limb-dip spectrum if the initial limb-dip spectrum was flat ($\beta = 2$).

An alternative method for determining the wavelength distribution in quasiperiodic fold trains is introduced here that is both more objective and gives greater information on the full spectral distribution in a particular fold sequence. First, the xy-coordinates of the midline of the layer are digitized for natural folds or analogue models, or obtained directly from nodal coordinates in the case of numerical models. Then the arc length is calculated as the cumulative sum of $\sqrt{x^2 + y^2}$. An interpolating cubic spline is determined that passes through the data points of y versus arc length to extract a set of equally spaced data points whose total number is a factor of 2 (in this study 4096 points). A discrete real FFT is then taken of these data points to acquire the spectral distribution, either relative to the x-coordinate (referred to here simply as the wavelength) or to the arc length (referred to as the arc wavelength), in both cases normalized against the average layer thickness. The results are smoothed by taking a three-point moving average, repeating the process four times. Plots of the spectral distribution are presented for clarity as curves, but in fact are closely spaced discrete values, as returned by the discrete Fourier transform.

RESULTS

Influence of probability distribution

The models in Figure 7 provide a comparison between the fold train geometry developed from the even distribution of Figure 1 and that developed from the Gaussian normal distribution of Figure 2, at a bulk logarithmic strain of −0.6. From inspection, the average wavelength developed from the initial even random distribution appears to be slightly longer, and this is confirmed in the plots of the spectral components in Figure 8, particularly for the plots against arc wavelength (Fig. 8, B and D). However, overall the differences are only small and unlikely to be significant for realistic examples. The tendency in the literature to be nonspecific about the actual random distribution employed is therefore of no real consequence.

Influence of initial spectral distribution (β value)

The model results for the range of β values from 0 to 3 are given in Figure 9 for $De = 0.25$ and Figure 10 for $De = 0.005$.

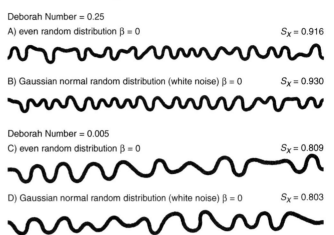

Figure 7. Effect of even versus Gaussian normal random initial perturbation distribution on geometry of single-layer folds. Viscosity ratio between layer and matrix is 100:1 and results are shown for bulk logarithmic strain of −0.6. Two examples, with Deborah number, $De = 0.005$ and $De = 0.25$ in layer, respectively, demonstrate effects of increasing elastic influence in rheology. S_x is amount of layer shortening, measured as stretch (i.e., ratio of current arc length to initial layer length).

The corresponding spectral distributions for amplitude and limb dip are given in Figures 11 and 12, respectively. These results are for the constant rms amplitude examples of Figures 2–5. For comparison, results for the input distributions scaled to constant amplitude at a wavelength 10 times the layer thickness (Fig. 6B) are given in Figures 13–16. In general, the influence of the β value of the initial distribution on the wavelength spectrum of the final fold train is small. The biggest shift occurs between $\beta = 0$ and $\beta = 1$, particularly for the constant rms examples (Fig. 11). As discussed by Fletcher and Sherwin (1978), for initial distributions with a flat initial amplitude spectrum ($\beta = 0$), the most strongly amplified component should correspond to the maximum in the final amplitude spectrum, whereas for initial distributions with a flat limb-dip spectrum ($\beta = 2$), it should be the corresponding maximum in the final limb-dip spectrum. This can be confirmed in the results presented here, the most amplified component at logarithmic strain −0.6 corresponding to an arc-wavelength: layer-thickness ratio of ~9 for $De = 0.25$ and ~13 for $De = 0.005$. Within the range of true self-affine fractals (i.e., $1 < \beta < 3$), the influence of the β value is small and, considering the other inaccuracies involved in analyzing fold geometry, can be ignored. The only possible exception is for $\beta = 3$ for the low De example in Figures 15, C and D, and 16, C and D, where there is a distinct shift to longer wavelength.

The examples with $De = 0.005$ approach linear viscous material behavior and can be tested against the existing relevant first-order theory for single-layer folding (e.g., Fletcher, 1974, 1977; Smith, 1975). For the relevant viscosity ratio of 100:1, the

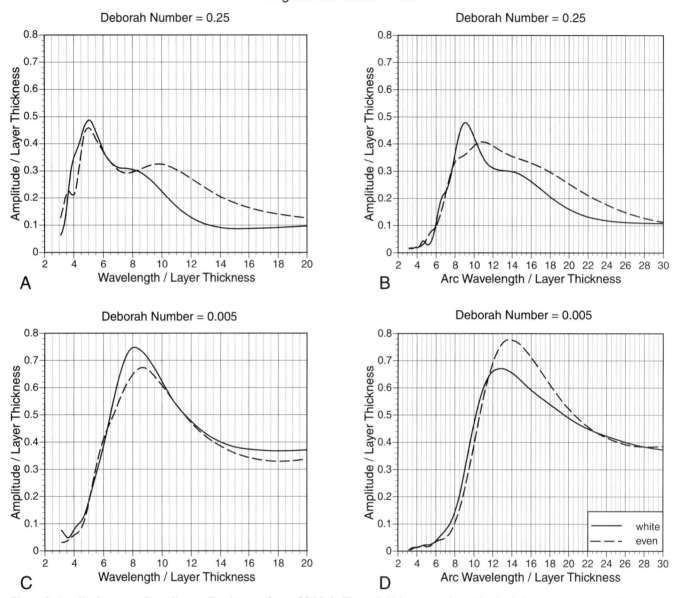

Figure 8. Amplitude spectra from discrete Fourier transform of folds in Figure 7. Values were determined relative to current x-coordinate in (A) and (C) and to arc length in (B) and (D); rms is root mean square.

theory predicts a dominant arc wavelength of 16.4 times the layer thickness during initial amplification at infinitesimal amplitude. This arc wavelength with the fastest dynamic growth rate is modified during continued progressive folding by the concomitant kinematic shortening of the layer. Johnson and Fletcher (1994, their equation 5.2.20a) numerically integrated the relevant equation including this kinematic shortening effect to derive the amplification distribution as a function of wavelength: thickness ratio (presented, for a viscosity ratio of 100:1,

as their Fig. 5.7). They also calculated the preferred wavelength, i.e., the most strongly amplified wavelength at any particular value of layer shortening. They established, however, that the preferred wavelength determined by this laborious procedure is effectively identical to that predicted by the much simpler relationship proposed by Johnson and Pfaff (1989), i.e., that the preferred wavelength:thickness ratio is equal to the theoretical dominant wavelength:thickness ratio multiplied by the stretch $S_x = S/S_o$ within the layer. The stretch within the layer in the nu-

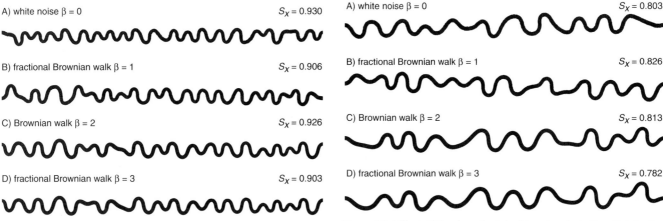

Figure 9. Effect of β value of original input perturbation distribution on geometry of single-layer folds. Examples presented here are for constant root mean square (rms) average for all distributions (see Fig. 6A). Viscosity ratio between layer and matrix is 100:1, Deborah number in layer is 0.25, and results are shown for bulk logarithmic strain of −0.6. S_x is amount of layer shortening, measured as stretch.

Figure 10. Effect of β value of original input perturbation distribution on geometry of single-layer folds. Examples presented here are for constant root mean square (rms) average for all distributions (see Fig. 6A). Viscosity ratio between layer and matrix is 100:1, Deborah number in layer is 0.005, and results are shown for bulk logarithmic strain of −0.6. S_x is amount of layer shortening, measured as stretch.

merical models of Figures 10 and 14 is ~0.8, giving a preferred arc wavelength to layer thickness ratio of ~13, in accord with the results presented here. Considering that the first-order theory is only strictly appropriate to very low amplitudes, whereas the folds in the numerical models at $\varepsilon = -0.6$ are strongly amplified, with high limb dips, the accuracy with which the theory predicts the observations is remarkable.

Influence of phase

It is generally assumed in theoretical and modeling studies of folding that the initial perturbations in the upper and lower surfaces of the layer are in phase. There is no reason that this should be the case in nature. Models in this study generally have completely independent initial perturbation distributions on the upper and lower surfaces. However, to test the influence of phase, two models were performed (for $De = 0.25$ and $De = 0.005$ in the layer) in which the same distribution was introduced in the two surfaces. A comparison of the results is presented in Figures 17 and 18. The differences in the results are very small, with perhaps a slight shift to longer wavelength and greater amplitude in the in-phase case. The two effects are interrelated. At low amplitude, for which the distinction between arc length and wavelength is unimportant, the rate of shortening due to buckling is approximated by (Ramberg, 1970, equation 34):

$$\frac{dx}{dt} \approx \frac{d\lambda}{dt} \approx 2\pi^2 \frac{y}{\lambda} \frac{dy}{dt} \quad (1)$$

It follows that for the same shortening rate accommodated by buckling dx/dt, a longer wavelength λ is associated with either a higher growth rate dy/dt or a larger current amplitude y (or both). Perfect phase coherence between the upper and lower surfaces aids the fold growth rate and increases the effective amplitude of the initial perturbation, both of which promote a longer wavelength. Equation 1 also explains the correlation between maximum amplitude and arc wavelength observed in Figures 8, 11, 15, and 18.

Influence of initial amplitude

The influence of the amplitude of the perturbation distribution was investigated in a series of experiments in which the y-coordinate of the initial distribution was multiplied by factors of 0.01, 0.1, 10, 50, and 100. The chosen starting distribution was that with $\beta = 2$ in Figure 6B (i.e., the Brownian walk or white roughness spectrum, with a statistically constant limb-dip spectrum of average value 0.00115°). Two series were considered, one with $De = 0.25$ (Fig. 19) and one with $De = 0.005$ (Fig. 20). For higher initial amplitudes (relative amplitude ≥50), the predominance of the high-wavelength components and the out-of-phase distribution leads to important variation in initial layer thickness. The shortening along the layer is correspondingly heterogeneous, being concentrated in the thinner parts of the layer. This effect, together with the shorter wavelength associated with the thinner segments, results in locally very high strains, particularly on the inner arc of appressed folds. For this reason, the example with relative amplitude = 100 and

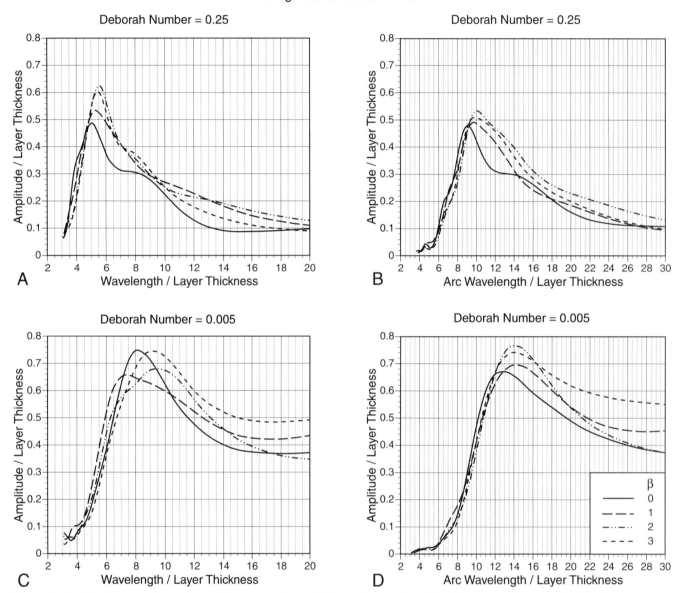

Figure 11. Amplitude spectra from discrete Fourier transform of folds in Figures 9 and 10. Values were determined relative both to current *x*-coordinate in (A) and (C) and to arc length in (B) and (D); rms is root mean square.

$De = 0.005$ failed to converge above a logarithmic strain of -0.47, due to local excessive distortion of the finite-element grid. The amplitude and limb-dip spectral distributions corresponding to Figures 19 and 20 are given in Figures 21 and 22, respectively. From these figures it is seen that, in contrast to cases considered so far, there is a marked difference between the spectral distributions relative to wavelength (Fig. 21, A and C) and those relative to arc wavelength (Fig. 21, B and D) with increasing initial perturbation amplitude. As would be expected, the amplitude at bulk logarithmic strain $= -0.6$ is consistently greater for greater initial amplitude. However, the shift to longer wavelength with increasing initial amplitude is only weak relative to the wavelength, but is quite marked relative to the arc wavelength. This directly reflects the relationship between the amount of layer shortening and the initial perturbation amplitude. In the extreme example, when there is no initial perturbation (zero amplitude), folds do not develop and the bulk shortening is entirely accommodated by layer shortening. Increased layer shortening de-

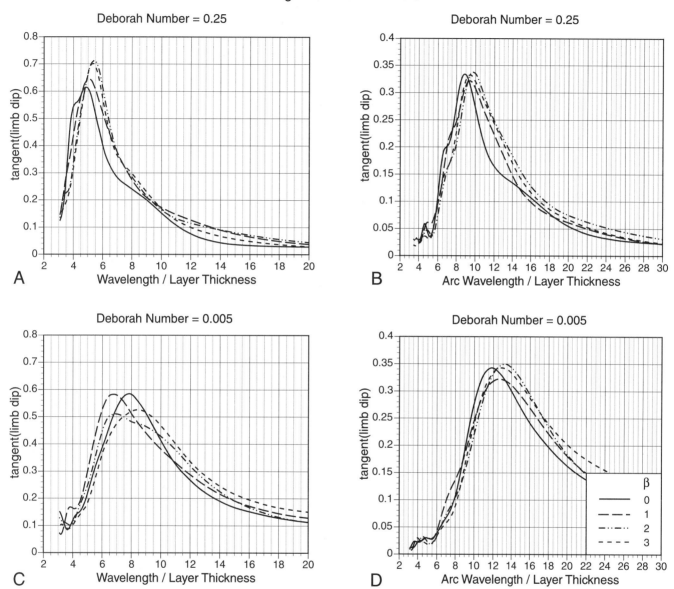

Figure 12. Tangent (limb dip) spectra from discrete Fourier transform of folds in Figures 9 and 10. Values were determined relative both to current x-coordinate in (A) and (C) and to arc length in (B) and (D); rms is root mean square.

creases the arc length and increases the layer thickness. The first effect shortens the wavelength relative to arc length, the second will increase it. The first effect dominates in Figures 21, B and D, as noted by Ramberg (1970, p. 92). Following Ramberg (1970, p. 92), it is possible to estimate this influence on the fold wavelength by considering, for low fold amplitude, the relative contributions of shortening due to buckle folding and layer-parallel shortening (termed respectively buckle shortening and layer shortening by Ramberg, 1970). For buckle shortening

$$\frac{y}{y_0} = \exp[-(1 + \gamma)\varepsilon], \quad (2)$$

where γ is the dynamic growth rate of folding (e.g., Fletcher, 1974; Smith, 1975) and ε the logarithmic strain. For layer shortening

$$\frac{S}{S_0} = S_x = \exp[\varepsilon], \quad (3)$$

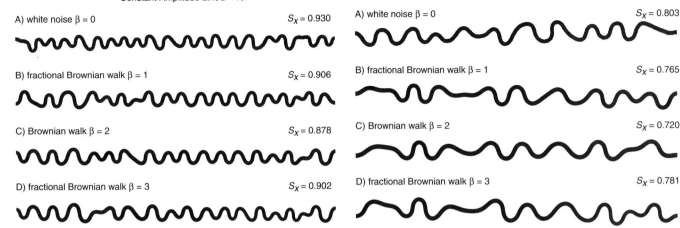

Figure 13. Effect of β value of original input perturbation distribution on geometry of single-layer folds. Examples presented here are for constant amplitude for all distributions at wavelength to layer thickness ratio of 10 (see Fig. 6B). Viscosity ratio between layer and matrix is 100:1, Deborah number in layer is 0.25, and results are shown for bulk logarithmic strain of −0.6. S_x is amount of layer shortening, measured as stretch.

Figure 14. Effect of β value of original input perturbation distribution on geometry of single-layer folds. Examples presented here are for constant amplitude for all distributions at wavelength to layer thickness ratio of 10 (see Fig. 6B). Viscosity ratio between layer and matrix is 100:1, Deborah number in layer is 0.005, and results are shown for bulk logarithmic strain of −0.6. S_x is amount of layer shortening, measured as stretch.

where S_x is the stretch in direction x parallel to the layer. Now consider an increase in y/S by some factor F, i.e.,

$$\frac{y}{S} = F \frac{y_0}{S_0}. \quad (4)$$

Now,

$$F = \frac{y}{y_0} \bigg/ \frac{S}{S_0} = \frac{\exp[-(1+\gamma)\varepsilon]}{\exp(\varepsilon)} = \exp[-(2+\gamma)\varepsilon] \quad (5)$$

and therefore

$$\varepsilon = \frac{\ln F}{-(2+\gamma)}. \quad (6)$$

Substituting in equation 3 then gives

$$S_x = \exp\left[\frac{\ln F}{-(2+\gamma)}\right] = F^{-[1/(2+\gamma)]}. \quad (7)$$

If we now consider the relative effect on S_x of two different F factors, where $F_a = (1/R) F_b$, then

$$\frac{S_{xa}}{S_{xb}} = \frac{F_a^{-[1/(2+\gamma)]}}{(RF_a)^{-[1/(2+\gamma)]}} = R^{[1/(2+\gamma)]}. \quad (8)$$

In other words, all other parameters being equal, a reduction in the initial perturbation amplitude by a factor R should produce a corresponding increase in the amount of layer-parallel shortening by a factor $R^{[1/(2+\gamma)]}$. For a dynamic growth rate of the fastest growing component $\gamma = 24.5$ (the theoretical value for viscosity ratio 100:1 from Fletcher, 1974), and values for R of 10, 100, 1000, and 5000 (the factors relative to the smallest initial amplitude presented in Figs. 19 and 20), the corresponding ratios in S_x are listed in Table 2. The measured ratios in S_x for the numerical models of Figure 20 are also listed in Table 2, and the correspondence between predicted and measured values is excellent (maximum difference <2.5%). Using the measured values of S_x, the preferred wavelength/current layer thickness ratios predicted from the relation $\lambda_p/d = S_x \lambda_d/d_0$ (Johnson and Pfaff, 1989), where λ_d is the initial dominant wavelength and d_0 is the initial layer thickness, are given in Table 2. These predictions are in close agreement with the most amplified arc wavelengths determined from the limb-dip spectra in Figure 22D. The correspondence is less good for the amplitude spectra of Figure 21D, confirming the observation of Fletcher and Sherwin (1978) that it is more appropriate to consider the limb-dip spectrum for initial perturbations with $\beta = 2$, because in this case it is the limb-dip spectrum that is initially flat. The numerical models establish that, provided the amount of layer shortening can be established, the existing first-order theory is a very good approximation for the range of initial random perturbation distributions expected in natural examples.

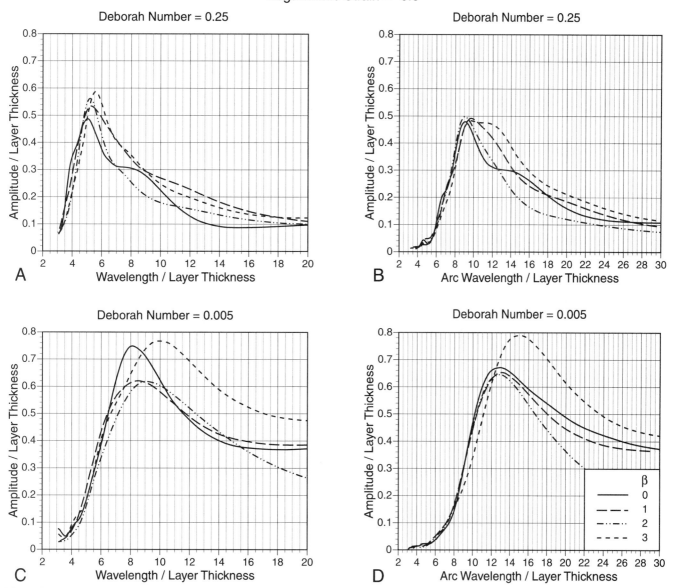

Figure 15. Amplitude spectra from discrete Fourier transform of folds in Figures 13 and 14. Values were determined relative both to current x-coordinate in (A) and (C) and to arc length in (B) and (D).

Influence of Poisson's ratio

To consider the possible effect of Poisson's ratio ν on fold development two experiments were run with the same parameters except that in one case $\nu = 0.45$ (i.e., the materials are nearly incompressible) rather than the usual $\nu = 0.25$ (normal range for rocks $\nu = 0.2$–0.3; e.g., Turcotte and Schubert, 1982). The example with the more important component of elastic behavior to the rheology (i.e., $De = 0.25$) was chosen because the effects of any variation in elastic parameters are likely to be more pronounced. However, as can be seen from Figure 23, the effect of changing the Poisson's ratio is negligible in these experiments.

DISCUSSION AND CONCLUSIONS

Overall, the influence of the initial probability distribution (e.g., even or Gaussian normal) and the fractal dimension of these distributions does not have a dramatic influence on the wavelength of folds developed to the finite amplitudes at which they can be studied. The Poisson's ratio of the materials also has

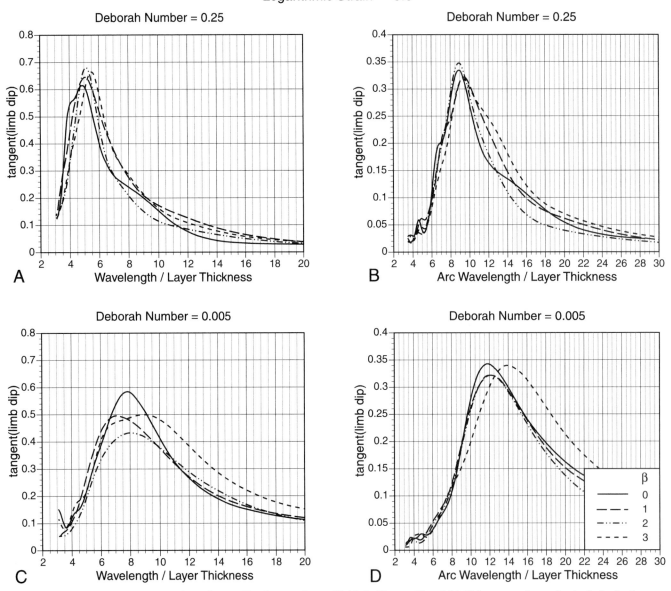

Figure 16. Tangent (limb dip) spectra from discrete Fourier transform of folds in Figures 13 and 14. Values were determined relative both to current *x*-coordinate in (A) and (C) and to arc length in (B) and (D).

little effect, so that assumptions of incompressibility should not have much influence on predicted wavelengths. The strongest effects are seen in the influence of initial perturbation amplitude. Increased initial perturbation amplitude is reflected in less layer shortening, with the result that measured arc wavelengths are longer for larger initial amplitude. The observed shift can be predicted well by consideration of this layer-shortening effect alone. The additional influence of the initial self-affine fractal distribution, with its bias to longer wavelength, is not of great importance. The wavelength determined relative to the horizontal *x*-direction (or the median line in natural folds) is much less sensitive to this initial amplitude effect. However, this wavelength progressively decreases with increasing bulk strain, in contrast to the arc wavelength, which remains more or less constant for an important part of the folding history. Obtaining some reliable estimate based on wavelength that can be directly related to rheology contrast is therefore difficult without additional information that is often unavailable: i.e., the bulk

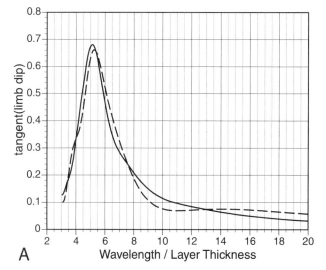

Deborah Number = 0.25

A) independent S_x = 0.878

B) in phase S_x = 0.878

Deborah Number = 0.005

C) independent S_x = 0.720

D) in phase S_x = 0.729

Figure 17 (left). Effect of phase coherency between perturbations on upper and lower layer surface on geometry of folds developed. Viscosity ratio between layer and matrix is 100:1 and results are shown for bulk logarithmic strain of −0.6. S_x is amount of layer shortening, measured as stretch.

Linear Viscosity Ratio = 100
Logarithmic Strain = -0.6
Brownian Walk β = 2
Constant Amplitude at λ/d = 10

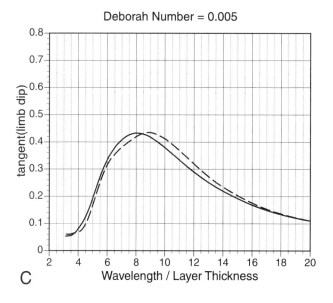

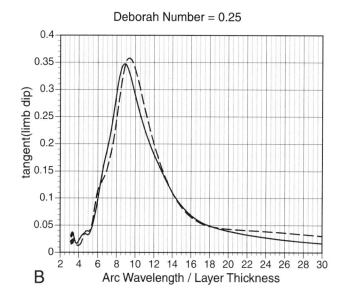

Figure 18 (below). Tangent (limb dip) spectra from discrete Fourier transform of folds in Figure 17. Values were determined relative both to current x-coordinate in (A) and (C) and to arc length in (B) and (D).

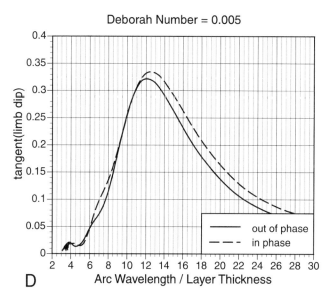

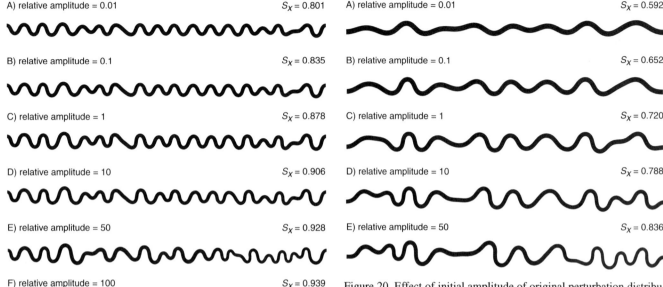

Figure 19. Effect of initial amplitude of original perturbation distribution on geometry of single-layer folds. Initial distribution corresponding to Brownian walk in Figure 6B was scaled by factors of 0.01, 0.1, 10, 50, and 100, respectively. Viscosity ratio between layer and matrix is 100:1, Deborah number in layer is 0.25, and results are shown for bulk logarithmic strain of −0.6. S_x is amount of layer shortening, measured as stretch.

Figure 20. Effect of initial amplitude of original perturbation distribution on geometry of single-layer folds. Initial distribution corresponding to Brownian walk in Figure 6B was scaled by factors of 0.01, 0.1, 10, and 50, respectively. Viscosity ratio between layer and matrix is 100:1, Deborah number in layer is 0.005, and results are shown for bulk logarithmic strain of −0.6. S_x is amount of layer shortening, measured as stretch.

TABLE 2. EFFECT OF LAYER SHORTENING

R	S_{xa}/S_{xb}*	S_{xa}/S_{xb}†	λ_p/d§
1	1	1	9.7
10	1.091	1.101	10.7
100	1.190	1.216	11.8
1000	1.298	1.331	12.9
5000	1.379	1.412	13.7

Effect of amplitude of initial perturbation on stretch factor S_x (ratio of current arc length to initial layer length) and preferred wavelength/current layer thickness ratio λ_p/d, using data from Figure 20. R is the factor by which initial perturbation amplitude is increased, relative to Figure 20A, in 20B, C, D, and E, respectively; S_{xa}/S_{xb} is the corresponding ratio in stretch factors.
* Theoretical values from equation 8.
† Measured values from numerical models.
§ Theoretical preferred wavelength/layer thickness.

shortening (ε), the layer shortening (S_x), or the initial perturbation amplitude.

The influence of initial amplitude is less for more elastic behavior and for increased fold growth rates in general (whether due to higher De, higher effective viscosity ratio, or nonlinear, power-law viscous behavior). This general observation was discussed in detail by Biot (1961). In this study, it has been shown that considering more natural random initial perturbations, which probably follow a self-affine fractal distribution with $\beta \simeq 2$, does not result in a consistent bias of the finite amplitude folds toward longer wavelengths, at least not for the viscosity contrast of 100:1 considered here. This result is in spite of the fact that, for such fractal distributions, there are order of magnitude differences in initial perturbation amplitude as a function of wavelength. As discussed by Fletcher and Sherwin (1978), for distributions with $\beta \simeq 2$ it is more appropriate to consider the limb-dip spectrum, because for $\beta = 2$ this spectrum is initially flat.

The major influence on the final fold geometry remains the interplay between layer shortening and buckle shortening (e.g., Ramberg, 1964, 1970; Hudleston and Stephansson, 1973), which is determined by initial perturbation amplitude and fold growth rate. Even for the largest initial amplitude distribution investigated in this study, for which variation in layer thickness becomes an appreciable effect (Figs. 19F and 20E), the bandwidth of the dynamic amplification is sufficiently narrow to effectively swamp the initial spectral distribution of the perturbation. Although the initial perturbation strongly influences the final shape of the fold train, this control is not reflected in the spectral dis-

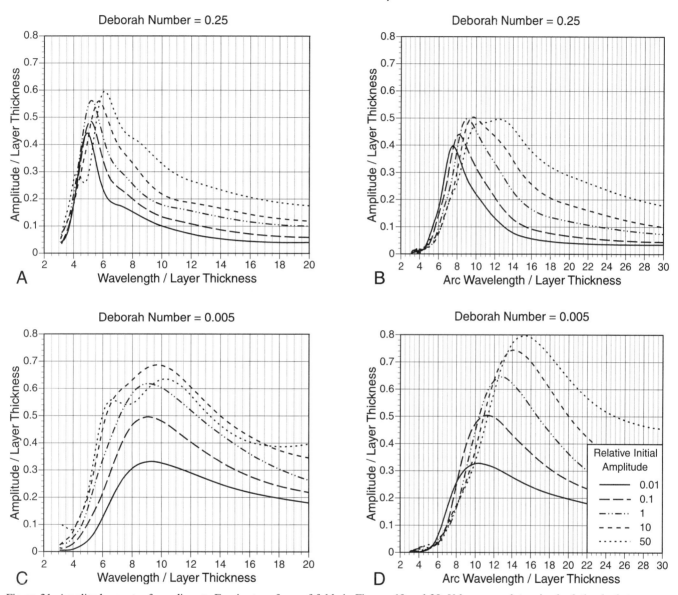

Figure 21. Amplitude spectra from discrete Fourier transform of folds in Figures 19 and 20. Values were determined relative both to current *x*-coordinate in (A) and (C) and to arc length in (B) and (D).

tribution. This is a very encouraging result for considering natural fold trains to establish viscosity contrasts between different rock layers, because one variable that cannot be easily estimated is shown to not introduce any consistent bias into the results. The current numerical models emphasize one problem that remains in any practical application to natural folds, i.e., the necessity to establish the amount of layer shortening during folding, because this parameter has a critical influence but is often very difficult to establish accurately in natural examples.

ACKNOWLEDGMENTS

I thank Yuri Podladchikov and Stefan Schmalholz for many discussions on folding in general and the potential influence of different initial perturbation distributions. Yuri's suggestions also helped greatly in establishing the code necessary to scale the Fourier transform components correctly for any required slope in the spectral distribution. Constructive reviews from Dazhi Jiang and Frederick Vollmer are also acknowledged.

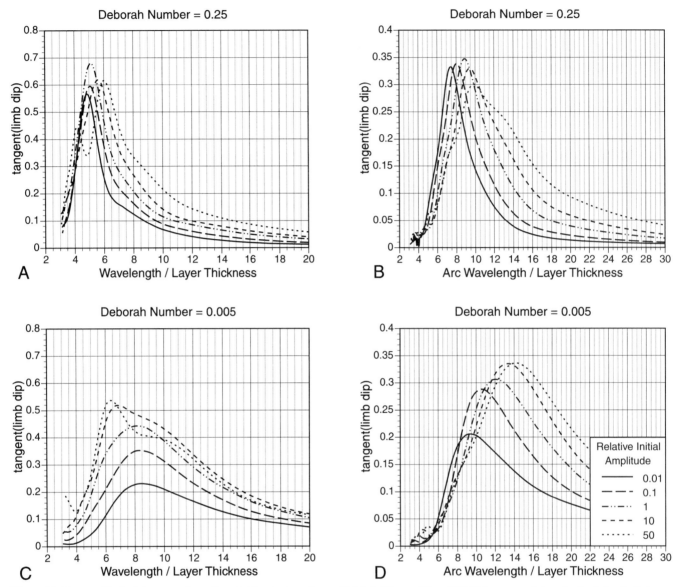

Figure 22. Tangent (limb dip) spectra from discrete Fourier transform of folds in Figures 19 and 20. Values were determined relative both to current x-coordinate in (A) and (C) and to arc length in (B) and (D).

Figure 23. Influence of Poisson's ratio ν on geometry of single-layer folds developed from initial Gaussian normal distribution (white noise, Fig. 2). All other examples in this study use $\nu = 0.25$ (value typical for natural rocks), but this figure demonstrates that assumption of effective incompressibility ($\nu = 0.5$) does not significantly influence results.

REFERENCES CITED

Abbassi, M.R., and Mancktelow, N.S., 1990, The effect of initial perturbation shape and symmetry on fold development: Journal of Structural Geology, v. 12, p. 273–282.

Abbassi, M.R., and Mancktelow, N.S., 1992, Single layer buckle folding in nonlinear materials—I. Experimental study of fold development from an isolated initial perturbation: Journal of Structural Geology, v. 14, p. 85–104.

Biot, M.A., 1961, Theory of folding of stratified viscoelastic media and its implication in tectonics and orogenesis: Geological Society of America Bulletin, v. 72, p. 1595–1620.

Brown, S.R., and Scholz, C.H., 1985, Broad bandwidth study of the topography of natural rock surfaces: Journal of Geophysical Research, v. 90, p. 12575–12582.

Cobbold, P.R., 1975, Fold propagation in single embedded layers: Tectonophysics, v. 27, p. 333–351.

Cobbold, P.R., 1976, Fold shapes as functions of progressive strain: Royal Society of London Philosophical Transactions, ser. A, v. 283, p. 129–138.

Fletcher, R.C., 1974, Wavelength selection in the folding of a single layer with power-law rheology: American Journal of Science, v. 274, p. 1029–1043.

Fletcher, R.C., 1977, Folding of a single viscous layer: Exact infinitesimal-amplitude solution: Tectonophysics, v. 39, p. 593–606.

Fletcher, R.C., and Sherwin, J.A., 1978, Arc lengths of single layer folds: A discussion of the comparison between theory and observation: American Journal of Science, v. 278, p. 1085–1098.

Gilbert, L.E., 1989, Are topographic data sets fractal?: Pure and Applied Geophysics, v. 131, p. 241–254.

Hudleston, P.J., and Stephansson, O., 1973, Layer shortening and fold-shape development in the buckling of single layers: Tectonophysics, v. 17, p. 299–321.

Johnson, A.M., and Fletcher, R.C., 1994, Folding of viscous layers: New York, Columbia University Press, 461 p.

Johnson, A.M., and Pfaff, V.J., 1989, Parallel, similar and constrained folds: Engineering Geology, v. 27, p. 115–180.

Kamb, B., 1970, Sliding motion of glaciers: Theory and observation: Reviews of Geophysics and Space Physics, v. 8, p. 673–728.

Mancktelow, N.S., 1999, Finite-element modeling of single-layer folding in elasto-viscous materials: The effect of initial perturbation geometry: Journal of Structural Geology, v. 21, p. 161–177.

Mandelbrot, B.B., 1977, Fractals: Form, chance, and dimension: San Francisco, Freeman, 365 p.

Poliakov, A.N.B., Cundall, P.A., Podladchikov, Y.Y., and Lyakhovsky, V.A., 1993, An explicit inertial method for the simulation of viscoelastic flow: An evaluation of elastic effects on diapiric flow in two- and three-layers models, in Stone, D.B., and Runcorn, S.K., eds., Flow and creep in the solar system: Observations, modeling and theory: Dordrecht, Holland, Kluwer Academic Publishers, p. 175–195.

Press, W.H., Flannery, B.P., Teukolsky, S.A., and Vetterling, W.T., 1986, Numerical recipes. The art of scientific computing: Cambridge, Cambridge University Press, 818 p.

Ramberg, H., 1964, Selective buckling of composite layers with contrasted rheological properties; a theory for simultaneous formation of several orders of folds: Tectonophysics, v. 1, p. 307–341.

Ramberg, H., 1970, Folding of laterally compressed multilayers in the field of gravity, II—Numerical examples: Physics of the Earth and Planetary Interiors, v. 4, p. 83–120.

Sherwin, J.-A., and Chapple, W.M., 1968, Wavelengths of single layer folds: A comparison between theory and observation: American Journal of Science, v. 266, p. 167–179.

Smith, R.B., 1975, Unified theory of the onset of folding, boudinage and mullion structure: Geological Society of America Bulletin, v. 86, p. 1601–1609.

Turcotte, D.L., 1997, Fractals and chaos in geology and geophysics (second edition): Cambridge, Cambridge University Press, 398 p.

Turcotte, D.L., and Schubert, G., 1982, Geodynamics—Applications of continuum physics to geological problems: New York, John Wiley & Sons, 450 p.

Williams, J.R., Lewis, R.W., and Zienkiewicz, O.C., 1978, A finite-element analysis of the role of initial perturbations in the folding of a single viscous layer: Tectonophysics, v. 45, p. 187–200.

MANUSCRIPT ACCEPTED BY THE SOCIETY APRIL 12, 2000

Experimental study of single layer folding in nonlinear materials

Tatiana Tentler
Hans Ramberg Tectonic Laboratory, Department of Earth Sciences, University of Uppsala, SE-75236 Uppsala, Sweden

ABSTRACT

Analogue scale-model experiments were performed to study the growth of folding instabilities in shortened single layers of nonlinear materials (mixtures of paraffin, Vaseline, and plaster of paris) embedded in a less competent power-law matrix (Dow Corning bouncing putty). In particular, the influence of the degree of strain hardening and/or softening on the shape, growth rate, wavelength and thickness selection, and degree of layer-parallel shortening of single-layer folds during their growth to finite amplitudes was studied. To analyze these factors, single-layer models were shortened along their length in pure shear box tests that produced a statistically valid number of waves. The rheological behavior of the materials was determined separately in a uniaxial pressure vessel in a series of cylinder tests. The materials of the competent layers show elasto-plastic strain hardening or softening rheologies; the stress-strain rate behavior of the matrix approximated a power law.

In both types of tests, folds in the elasto-plastic strain-hardening single layer have a characteristic sinusoidal wave form throughout the deformation and exhibit hinge thickening and limb thinning with progressive shortening. In the strain-softening single layers, folds form initially close to sinusoidal in shape and become angular after faults nucleated in their hinges in the early stage of shortening. These angular folds exhibit little or no layer-parallel shortening, the preferred arc length/thickness ratio is larger by a factor of 1.5, and the growth rate is higher by a factor of 1.8 than folds in an elasto-plastic strain hardening single layer shortened under otherwise similar conditions. The similarity of fold shapes in both the uniaxial pressure vessel and pure shear box tests suggests that rheological parameters can also be used to interpret the results of the pure shear experiments.

The degree of sharpness of the folds depends on the amount of plastic strain accumulated in the single layers prior to yielding at fold hinges. Folds have sharper hinges and longer, straighter limbs when the ductile strain is low prior to yielding at fold hinges. By studying fold shapes in analogue models it may be possible to infer the rheological properties of folded layers in rocks during fold growth to final amplitudes.

INTRODUCTION

Folding is one of the most common and spectacular expressions of rock deformation. Many natural folds are initiated by buckling, a mechanical instability in layered systems shortened along their layers by a compressive load. Previous work has shown that the geometry of folding instabilities depends on the viscosity and stiffness contrast between the layer and its matrix. Hudleston and Lan (1993) gave a broad analysis of factors controlling the shape of folds, such as the initial layer anisotropy and geometry, the principal stress direction, the dominant mechanism of folding, and the rheological properties. Most of the theoretical models based on linear viscous or elastic materials can only predict the growth rates and spacing of folds along a train in the initial stages of deformation (Ramberg, 1960; Biot, 1961; Sherwin and Chapple, 1968). There are some theories available to make predictions of fold growth and shape to significant amplitudes. Thus, Fletcher (1974) and Smith (1977) showed

Tentler, T., 2001, Experimental study of single layer folding in nonlinear materials, *in* Koyi, H.A., and Mancktelow, N.S., eds., Tectonic Modeling: A Volume in Honor of Hans Ramberg: Boulder, Colorado, Geological Society of America Memoir 193, p. 89–99.

that quantitative agreement between theory and the observed folds requires the use of a non-Newtonian model. Fletcher developed the theory to third order. Smith's theory is more general, does not depend on any specific rheology, and takes account of nonlinear behavior. Smith (1979) predicted theoretically that, in a strain-softening material, resistance to additional normal straining is significantly reduced from the background level, whereas resistance to tangential straining is unchanged. This increases the growth rates of folds and alters their dominant wavelengths. Nevertheless, numerical or analogue models are necessary to extend the study of fold growth to finite amplitudes in material with complex rheologies. Although rocks show both short-term elastic and long-term viscous or plastic creep, theoretical and numerical models usually consider only one or other of these particular rheological behaviors. Only recently have limited combinations of these complex rheologies been combined in numerical experiments (Zhang et al., 1996; Mancktelow, 1999). Models assuming linear rheologies are based on the assumption that viscous behavior dominates over elastic behavior during slow natural deformation (strain rates on the order of 10^{-14} s^{-1}). Fletcher (1974) derived linearized constitutive relations for an incompressible power-low fluid to show that the rheology of a limestone layer embedded in shale was strongly nonlinear in folding. Rocks exposed in the Earth's crust are clearly not viscous materials, and field studies as well as available experimental data on natural rock deformation suggest that rocks deform with complex nonlinear rheologies and exhibit transitions from brittle to ductile behavior with increasing confining pressure (Goodman, 1989).

Folds in nature and experiment exhibit a wide variety of geometric shapes ranging from sinusoidal forms with rounded outer arcs, to angular forms with straight limbs (Hudleston 1973; Ramberg, 1964; Hudleston and Lan, 1994). Angular folds, e.g., chevron-shaped or kink bands, are particularly common in multilayers (Ramsay and Huber, 1987). The shape of folds is known to depend on a number of parameters such as initial shape and thickness of layers, their spacing, competent contrast, the degree of nonlinearity of flow law (strain-rate hardening or softening) and type of flow (strain hardening or softening). This chapter considers how the type of flow law in the layer material affects the fold shape. The range of fold forms exhibited in nature may be an expression of the range of strain-dependent behavior, from strain hardening, through steady state, to strain softening of rocks deformed under different geological conditions. Experimental rock deformation studies indicate a very complex range of strain-rate dependent behavior from linear (for diffusion-controlled deformation mechanisms) to nonlinear with different types of flow law. Thus experiments on natural rocks deformation at 5 kbar and strains of 15% show that dunite, Yule marble, and pyroxenite deform steadily or strain harden (Griggs et al., 1960; Heard and Rayleigh, 1972), whereas basalt, dolomite, volcanic tuff, and granite exhibit strain softening (Ranalli, 1987; Schultz and Li, 1995; Zhao et al., 1999). Rocks near the Earth's surface, e.g., in sedimentary environments, are also expected to behave as elasto-plastic strain-hardening or softening materials (Fuchu et al., 1999; Samieh and Wong, 1998).

In this chapter I investigate the shapes of single layer folds using elasto-plastic strain-hardening and strain-softening single layers embedded in a bouncing putty matrix with a power law behavior. The experiments study how strain-hardening and/or softening rheological properties of single layers embedded in a matrix influence general fold shapes, hinge sharpness, and growth rate, and establish criteria based on shape that may give information about the rheologies of rocks deformed in natural conditions. The experiments simulate small-scale folding of single layers where the effect of gravity is negligible. Analogue elasto-plastic materials were prepared specially for this modeling with the aim of providing new insights on the reasons for complex deformational patterns. However, the extent to which any particular natural fold fulfils the rheological assumptions made in the present study is not evaluated.

METHOD AND MATERIALS

Method

This experimental study was carried out in three parts. First, the stress and strain response of two types of competent materials for single layer was investigated independently by a series of tests where cylinders of each material (4 cm in diameter and 6 cm high) were compressed uniaxially in a pressure vessel (Fig. 1, A and B) at a constant strain rate (10^{-4} s^{-1}), room temperature (22 °C), and confining pressure (3×10^5 Pa). Second, cylindrical models consisting of single planar layers of the same strain-hardening or softening materials were embedded in a non-Newtonian matrix and compressed parallel to layering in the same pressure vessel to study fold shapes (Fig. 1C). Third, rectangular parallelepipeds were prepared from the same materials for layer and matrix. In each experiment, a single competent layer was sandwiched between two planar layers of matrix material so that cohesion between layer and matrix was maximized (Fig. 2B). Models were compressed parallel to layering in a pure shear box under constant strain rate to study the evolution of fold shapes (Fig. 2A).

During tests in the uniaxial pressure vessel, the stress and strain rate response of the folding layer was monitored by a continuous display of the stress-time and strain-time curves on a computer screen during the experiment and stored in a data file. The uniaxial stress was computed by dividing the measured force by the instantaneous value of the cross-sectional area of the cylinder. Strain rate was determined by dividing the ram speed by the instantaneous height of the cylinder. Measured forces were converted to stresses by assuming a homogeneous increase in cross-sectional area of the cylinders.

The cylinder tests and the pure shear tests were conducted at identical shortening rates (parallel to the layer) and confining pressures. However, the strain rate in the direction perpendicular to the layer (direction of growth of the folds) is not the same, because one case is uniaxial shortening and the other is plane strain. These considerations can make the fold amplification different in the two cases (Fletcher, 1974). However, in this work

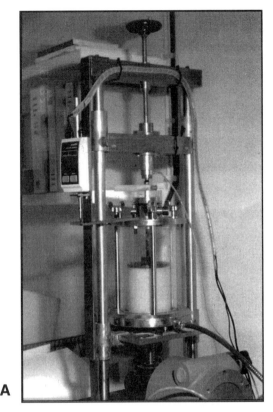
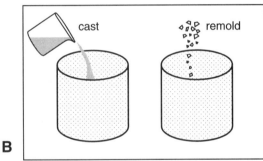
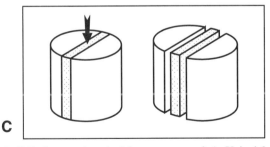

Figure 1. Cylinder tests in uniaxial pressure vessel. A: Uniaxial deformation rig. B: Two different modes of preparation of analogue materials for competent layer. Material with ductile (strain-hardening) rheology was prepared by remolding solidified mixture of molten and stirred paraffin, Vaseline, and plaster of paris to homogeneous cake. Material with semibrittle (strain softening) rheology was cast by allowing same melt to cool and solidify slowly in appropriate mold and then cutting body of cooled material to required dimensions. C: Arrangement of model cylinders with planar single layers embedded in cylindrical matrix for subsequent uniaxial compression parallel to layering.

the shape of fold produced in the two types of experiments is the subject of comparison. The results of rheological tests are capable of providing a basis for interpretation of fold characteristics under a wide range of experimental conditions. The similarity of fold shapes obtained by applying uniaxial shortening and plane strain suggests that the given rheological conditions produce consistent results.

Folds were developed in plane competent single layers of uniform thickness without any initial perturbations being intentionally imposed. The thickness of the encasing medium is three to four times the wavelengths of the folds. Photographs were taken at regular time intervals throughout each experiment. Deformed models were removed from the pressure vessel or pure shear box and sectioned. Although layers generally remained well bonded to their matrix, in shear box experiments the host separated from the strain-softening layer near the hinges of mature folds after the phase of wavelength selection. Special tests were performed to check the effect of the cylinder height/diameter ratio and the value of confining pressure on fold shapes without producing significantly different results.

To study the evolution of the strain distribution in folds, strain grids in the form of initially circular and square markers were printed on the upper surface of each model. For every experiment, the ends of cylinders and the walls of the pure shear box were copiously lubricated with Vaseline to minimize boundary friction. Pressure vessel cylinders and pure shear models deformed fairly homogeneously.

Materials

The competent single layer consisted of a mixture of 39% by weight paraffin, 33% Vaseline, and 28% plaster of paris, homogenized by thorough stirring when molten. Two different analogue materials were produced by molding or casting this initial mixture (Fig. 1B). A material with ductile (strain hardening) rheology was prepared by remolding the solidified mixture to a homogeneous cake. A material with semibrittle (strain softening) rheology was cast by pouring the molten mixture into an appropriate form and allowing it to cool and solidify slowly before cutting the body of cooled material to the required dimensions. These two different techniques for material preparation resulted in materials with distinct rheological properties. Changing the mechanical properties of paraffin and other components during preparation leads to either a disordered material with ductile rheology or a more ordered structure with semibrittle rheology. The rheological behavior of the different analogue materials was measured separately in the uniaxial pressure vessel. Flow curves of layer and matrix materials show different variation of flow stress with applied strain (Fig. 3). Tests of cast cylinders of the semibrittle strain-softening material for layers show that this material yielded or strain softened (curve *a* of Fig. 3). The stress drops on the stress-strain curve correspond to particular brittle failures in the test cylinders. Tests of cylinders of the same components molded to a homogenous cake show strain-hardening behavior (curve *b* of Fig. 3).

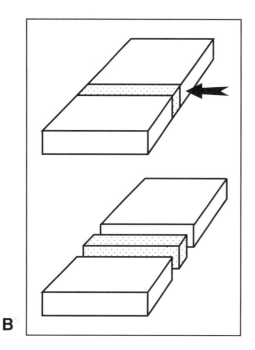

Figure 2. Planar models in pure shear box. A: Pure shear box. B: Arrangement of model with planar single layer prepared for shortening by pure shear along layering.

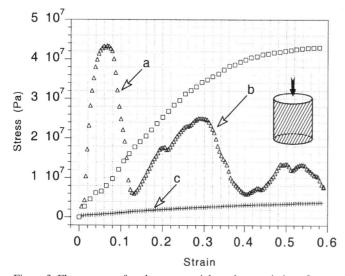

Figure 3. Flow curves of analogue materials to show variation of stress with strain: a is semibrittle strain-softening (cast cylinder) material for layer; b is ductile strain hardening (molded cylinder) material for layer; c is silicone putty used as matrix material for embedded layer.

Dow Corning compound DC3179 was used to represent the less competent matrix enclosing the folding layers (curve c of Fig. 3). Previous tests have shown that this compound flows with a power-law constitutive relationship of the form (Dixon and Summers, 1986; Weijermars, 1986):

$$\dot{\varepsilon} = k\,\sigma^n, \qquad (1)$$

where k is a material constant and n is the stress exponent.

The steady-state creep equation does not provide a good fit for the deformation of this material (Hailemariam and Mulugeta, 1998). The same authors determined that at room temperature the viscosity of this material is $\eta \approx 10^6$ Pa s with a power law exponent ($n > 3\text{--}7$) in the strain rate range 10^{-4} to 10^{-2} s^{-1}.

MODELING

Pressure vessel tests

To interpret the results of the layered model experiments, the pressure vessel (otherwise used for material testing) was also employed to study differences in fold shape between the strain-hardening and strain-softening single layers compressed uniaxially at the same confining pressure as in the cylindrical material tests.

In models deformed in the pressure vessel, folding developed under uniaxial loading along the layer and radial expansion of the matrix. Figure 4 shows photographs of fold shapes produced in these tests. The ductile strain hardening single layer develops folds of sinusoidal shape with little thickening in the hinges (Fig. 4A). The strain-softening single layer develops folds with an angular chevron shape (Fig. 4B). However, the dimensions of the cylinders in the pressure vessel did not allow the growth of many waves. A pure shear box was therefore used to ensure statistically valid spectra of wavelength selection and growth rate for both types of layer.

Pure shear box tests

A pure shear box that provides a wide range in convergence rates (5×10^{-4} to 5×10^{-1} cm s^{-1}) was used to shorten single competent layers along their length and study the initiation and development of folds as well as their shape characteristics. In this paper two representative experiments of folding of elasto-plastic hardening or softening single layers shortened by pure shear at a strain rate of 10^{-4} s^{-1} are described. The growth of the regular fold train from a uniform layer to finite amplitudes (Figs. 5 and 6) was monitored continuously during the experiment. The folds were not seeded, and both types of folding instabilities developed simultaneously all along their length rather than by serial sideways propagation from a preimposed nucleation site, as in the experiments of Cobbold (1975) and Abbassi and Mancktelow (1992).

Figure 5 shows the progressive evolution of folds in the elasto-plastic strain-hardening single layer embedded in the matrix material. Folds develop with sinusoidal forms and the hinges thicken slightly while the limbs thin with increasing shortening. Figure 6 shows progressive stages of fold development in an elasto-plastic strain-softening single layer. Folds were smoothly sinusoidal in the early stage of shortening, but acquired sharp hinges and straight limbs as deformation proceeded. This change in fold shape is attributed to the complex rheology of the semibrittle material. The single layer behaved elastically until each hinge underwent brittle failure at a critical load. Bending strains concentrated in the fold hinges led to their breaking at about 10%–15% bulk shortening.

During the late stages of compression, the layer separated from the weaker matrix and voids developed near the hinges, but this did not affect the preceding wavelength selection and growth rates. This assumption was checked by additional tests in which the layer interfaces remained perfectly bonded. The model was constructed with a thin film of supplementary material (polydimethyl-siloxane, PDMS) between layer and matrix, preventing void formation. In these tests the progressive evolution of folds was the same as for the illustrated semibrittle single-layer model. The different styles of folds (Figs. 5 and 6) are clearly related to the behavior of the materials constituting the layers. The following section examines in more detail the rheological control of fold shapes and, in particular, the role of strain hardening and/or softening in controlling growth rate and wavelength selection during evolution of folds to finite amplitudes.

RESULTS AND DISCUSSION

Fold shape

The principal difference in fold shape, strain distribution, and growth rate of the structures is attributed to the differences in the flow laws of the material making up the layer. The most distinctive geometric characteristic of the model folds is the sharpness of the fold hinge. The inner and outer interfaces of fold hinges in the ductile strain-hardening layer are smooth concentric curves, whereas in the semibrittle strain-softening layer they develop angular hinges that further sharpen as the folds amplify.

Folds in the strain-hardening single layer show sinusoidal shapes that thicken in the hinges and thin in the limbs at a moderate stage of deformation. Many authors have investigated the influence of material rheology on the geometry of amplifying folds. Chapple (1969) studied the finite-amplitude folding of a thin viscous-plastic layer and inferred that an increase in stress during fold growth will induce plastic yielding at the fold crest.

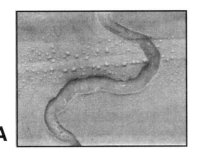 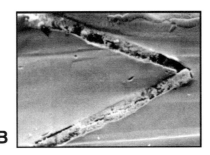

Figure 4. Photographs of fold shapes in pressure vessel tests. A: Cylindrical model with sinusoidal folds in strain-hardening layer. B: Cylindrical model with angular fold in strain-softening layer.

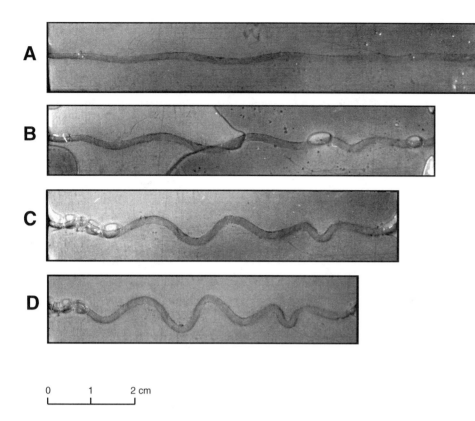

Figure 5. Progressive stages of folding in pure shear model with strain-hardening single layer. Photographs taken throughout experiments. A: 10% bulk shortening. B: 20% bulk shortening. C: 30% bulk shortening. D: 40% bulk shortening.

The shape of a viscous-plastic fold will depend primarily on when in the folding history plastic yielding occurs. Chapple found that, if this fraction is large, the resulting fold will have longer straight limbs, a higher amplitude, and a stronger curvature at the fold crest than will a viscous fold with the same interlimb angle. Folds with the same kind of geometry as presented in this chapter have also been described in layered viscous materials (Johnson and Fletcher, 1994) and non-Newtonian and strain-softening materials (Neurath and Smith, 1982). This implies that it is difficult to determine, from analysis of the final fold form alone, whether a rock layer deformed as a viscous or a plastic material.

In the experiments described here, folding in the strain-softening layer initiates with a nearly sinusoidal wave form, but the hinges sharpen and the limbs straighten as deformation proceeds. A possible explanation for this behavior is strain softening in the layer, which yields plastically near the hinges when the layer deflection reaches a critical value. Brittle failure occurs in the hinges when limb dips reach about 30°. Previous works have attributed angularity of folds to a number of causes. Biot (1961) showed theoretically that buckle folds in stratified viscoelastic media start with sinusoidal profiles, but become increasingly angular as progressive shortening leads to yielding of the material in the hinges. This is in agreement with the strain-softening folds in the experiments reported here. Other rheological factors that could account for sharp hinges and straight limbs in the layer are pseudo-plastic behavior (Johnson, 1970), strain-softening behavior (Neurath and Smith, 1982), or an increase in the layer's power-law exponent or material anisotropy (Lan and Hudleston, 1996). Hudleston and Lan (1994) gave quantitative analyses of differences in fold shape for layers of different rheological properties, although they showed that these differences are minimal if the wavelength/thickness ratio is <~10. They also noted that particular shapes—such as angular chevron folds—can be produced in a number of different ways. On the basis of numerical modeling, Lan and Hudleston (1996) found that increasing the degree of strain softening has almost the same effect on fold shape as increasing the power-law exponent (as predicted by Neurath and Smith, 1982). Folds systematically become increasingly angular as the power-law exponent increases. As in the models described here, the degree of strain hardening or softening strongly affects the amplification rate and wavelength selection of folds (as discussed in the following).

Growth rate

The rate at which the dominant and other wavelengths amplified as a function of the hardening or softening behavior of the layer was investigated. In the absence of a competence contrast, folds grow by passive amplification of initial irregularities in the layer, a process unlikely to produce regular folds. There are several reasons for studying growth rates in preference to wavelengths. First, the amplitude of a growing disturbance is a more direct measure of the growth mechanism than fold wavelength (Neurath and Smith, 1982). Second, the normalized

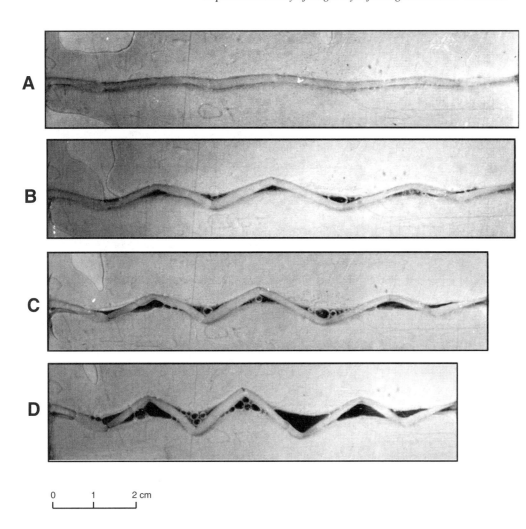

Figure 6. Progressive stages of folding in pure shear model with strain-softening single layer. Photographs taken throughout experiments. A: 15% bulk shortening. B: 20% bulk shortening. C: 30% bulk shortening. D: 35% bulk shortening.

growth rate of a disturbance is theoretically independent of the initial amplitude as long as the amplitude is small. Third, the growth rate varies much more with changing material rheology than the dominant wavelength. Growth rates are, therefore, more characteristic of a particular rheological combination.

Figure 7 shows the growth rates of sinusoidal folds in a strain-hardening layer and of angular folds in a strain-softening layer. The fold amplitude *(A)* is normalized against the amplitude of the initial preferred fold wavelength *(Ao)*, and plotted as a function of bulk shortening. The initial preferred fold wave length is defined as the first recognizable stable fold wavelength developed during the layer shortening. The growth curves for folds in layers of both materials, obtained from analysis of a number of folds in a single successful experiment, are sigmoidal. Folds do not amplify at a constant rate, but growth rate decreases after a particular amplitude. Early in the folding history (to 10% bulk shortening), the layers shorten by uniform thickening rather than folding. As the deformation develops, sinusoidal folds show a stable increase in amplitude, whereas the growth of angular folds is more irregular. Angular folds increase in amplitude to about 23% bulk shortening and then decelerate in growth until they reach a maximum amplitude at about 30% bulk shortening.

Comparison of amplification rates for both types of folds (Fig. 8) shows that the amplitude of angular folds in the strain-softening layer increases rapidly to 20% bulk shortening, while the amplitude of folds in the strain-hardening layer increases more slowly, to about 30% bulk shortening. The amplification rate of folds in the softening layer increases until about 20% bulk shortening and then falls rapidly. The amplification rate of folds in the hardening layer increases until about 30% bulk shortening and thereafter gradually decreases. Theory predicts exponential amplification of folds, consistent with the first part of the plots in Figure 7.

The distinctive features of sinusoidal and angular folds can be attributed to differences in the rheologies of the single layers. The greater initial stiffness of the strain-softening layer accelerates the growth of the selected preferred wavelength of angular folds at the earlier stage compared to the slower growth of sinusoidal folds in the strain-hardening material. The subsequent decrease in growth rate for both types of folds can be explained as a consequence of the contrast in growth imposed by

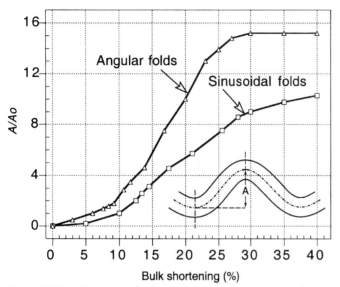

Figure 7. Variations in amplitude during progressive growth of sinusoidal folds in strain-hardening layer and angular folds in strain-softening layer. Normalized fold amplitude (*A/Ao*) is plotted as function of bulk shortening, where *A* is measured amplitude and *Ao* is amplitude of initial preferred fold wavelength. Method of measuring amplitude is shown.

fold geometry. Angular folds with straight limbs generally have larger amplitudes compared to sinusoidal folds for the same amount of shortening. Sinusoidal folds in the strain hardening layer increased in amplitude steadily throughout the experiment. Amplification of angular folds in the strain-softening layer slows more rapidly when brittle failure at the hinges separates the two fold limbs, which then rotate passively as two disconnected layers.

Wavelength selection

Using available theories it is difficult to infer the competence contrast from wavelength selection of structures in materials with such complex rheologies as the layers of strain-hardening and/or softening materials embedded in non-Newtonian media employed here. Therefore, the data obtained from rheological tests in the pressure vessel are used, together with measurements of the geometries of folds produced in the shear box, to discuss the wavelength selection process qualitatively.

In single layers of simple viscous materials it is possible to predict the dominant wavelength or the wavelength that amplifies most rapidly. The competence contrast of a Newtonian viscous single layer with viscosity η embedded in a viscous medium of lower viscosity η_o is usually found using the Biot-Ramberg equation for the dominant wavelength λ divided by T, the layer thickness:

$$\lambda/T = 2\pi(\eta/6\eta_o)^{1/3}. \qquad (2)$$

Hudleston (1973) produced a series of experiments on single-layer folding using viscous materials. He noticed that there is little change in the arc length of the folds once limb dips of 10°–20° have been reached regardless of the viscosity contrast. Models presented in Figure 9 also show a rapid decrease in arc length until the limb dip exceeds 20° for both types of materials. In Figure 9 the fold arc length *(S)* is normalized by the arc length of the initial preferred fold wavelength *(So)* and plotted as a function of limb dip *(θ)*. The curves of variations in normalized arc length for folds in layers of both materials are obtained from analysis of a number of folds from one successful

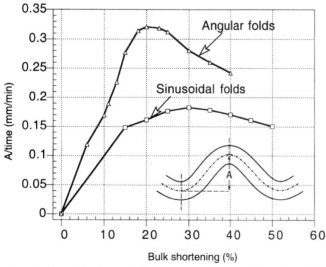

Figure 8. Amplification rate during progressive growth of sinusoidal folds in strain-hardening layer and angular folds in strain-softening layer. Change of amplitude with time (*A*/time) is plotted as function of bulk shortening. Method of measuring amplitude (*A*) is shown.

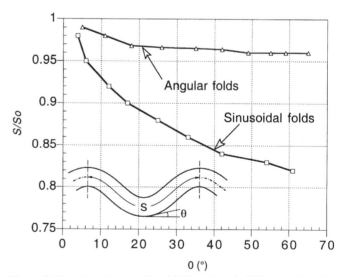

Figure 9. Variations in normalized fold arc length (*S/So*) are plotted as a function of limb dip (*θ*) for sinusoidal and angular folds, where *S* is measured arc length and *So* is arc length of initial preferred fold. Method of measuring arc length (*S*) and limb dip (*θ*) is shown.

experiment. The initial preferred fold wavelength is measured at a limb dip of 3.5° for sinusoidal folds and 5° for angular folds. In the strain-softening layer there was no significant change in arc length beyond limb dips of about 20°. Arc length of the sinusoidal folds in the strain-hardening layer decreased more rapidly throughout the experiment because more layer-parallel shortening preceded the folding than in the chevron folds.

Variations in arc length/thickness ratio during progressive shortening for sinusoidal and angular folds are shown in Figure 10. The arc length (S) is measured in succession for each amount of shortening, and the average thickness (T) is obtained by measuring different parts of the layer from hinge to limb and calculating the mean number for each amount of shortening. Arc lengths decrease more slowly at the later stage of fold development in both strain-hardening and strain-softening layers. However, the curve for sinusoidal folds is relatively smooth compared to the curve for angular folds, which displays two stages of development. There is an abrupt change in the folding process at 25% bulk shortening when the decrease in the mean ratio of the arc length to thickness slows markedly.

Most fold wavelength/thickness ratios measured in rocks in the field are between 4 and 8. Smith (1979) predicted theoretically that in highly nonlinear materials, fold wavelength approaches a small and constant value regardless of viscosity contrast. He found that the competent layer does not act mechanically as a coherent unit, but that irregularities on one interface produce motions that deform the other interface and vice versa. Sherwin and Chapple (1968) studied wavelength selection in folded quartz veins in sandy and phyllitic matrices. They found that the mean wavelength/thickness ratio ($L/h \approx$ 4–6) was much smaller, and the range of arc length/thickness values ($S/T \approx$ 3–9) inferred from their theoretical considerations was much larger than predicted by the theoretical buckling equations. These authors modified the theory to take into account the layer-parallel shortening during progressive folding. In the experiments reported here, the arc length/thickness ratio varies from the start of the folding process to the end in the range $S/T \approx$ 4.5–9.5. This is in agreement with the wavelength selection observed in natural rocks if layer-parallel shortening is taken into consideration. Although the results presented here are not proof, they are used to infer that competent layers in natural rocks (e.g., quartz veins) deform as elasto-plastic materials.

Strain distribution

The strain distribution in the models has been analyzed, particularly the strain variations in layer and matrix developed in response to the rheological stratification. Strain markers were painted on top of the model for direct strain measurement, to establish further differences between the two modes of folding. In this way an inhomogeneously strained folded layer embedded in the matrix has been divided into areal units small enough to be considered to have deformed homogeneously. The distribution of strain during folding depends on the manner in which the mechanical properties and shape of the layers change with progressive deformation. This is shown in Figure 11, where maximum layer shortening in the layer occurs in the inner arcs of both types of folds, while the outer arc region accommodates the deformation by layer-parallel extension. This mode of deformation is in agreement with previously reported experimental results (Cobbold, 1975; Neurath and Smith, 1982; Abbassi and Mancktelow, 1992). The grid markers show local homogeneity of strains within layers, while near the boundaries distortion occurred by slip along the layer-matrix interface and local layer-perpendicular compression.

The strain gradient steepens considerably as the folds in the strain-softening layer become increasingly angular (Fig. 11B). Layer shortening is concentrated in the inner arcs of the fold hinges of angular folds in this layer. The fold limbs undergo minor, relatively homogeneous layer-parallel shortening. This strain distribution is essentially similar to the tangential longitudinal strain of Ramsay and Huber (1987). Nonlinear flow results in increasingly greater strain rate as bending strain increases in the inner and outer arcs of the folds. The higher strain near the hinges produces sharper fold hinges and straighter limbs. Strain concentration eventually leads to brittle failure in the hinge zones.

For the layer with strain-hardening behavior (Fig.11A), bending strain is more penetrative, resulting in the migration of layer material from fold limbs toward the hinge zones at moderate stages of deformation, leading to hinge thickening. The strain-hardening layer in Figure 11A is thicker than in other experiments in order to show the strain variation in more details. A zone of no apparent strain in the middle of the layer separates the outer and inner arc regions.

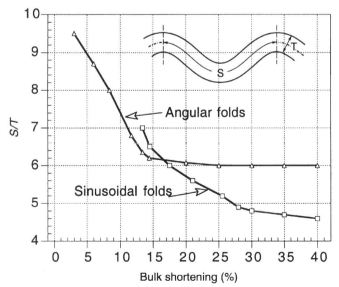

Figure 10. Variations in arc length/thickness ratio (S/T) during progressive shortening for sinusoidal and angular folds. Method of measuring arc length (S) and thickness (T) is shown.

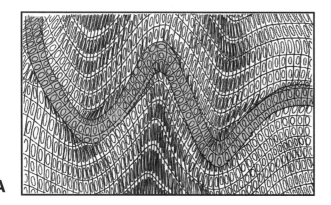

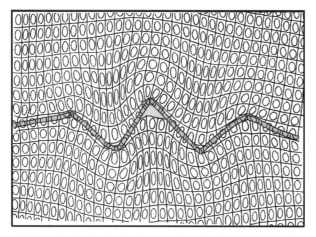

Figure 11. Strain distribution in layer and matrix in single-layer fold models. A: Sinusoidal folds in strain-hardening layer. B: Angular folds in strain-softening layer.

The matrix material accumulates heterogeneous strain during both types of folding. Material tends to migrate from the outer arc of one fold to the inner arc of the next fold. The highest strains in the matrix occur close to the inner arcs of folds, similarly to earlier reported experimental results (Hudleston, 1973; Cobbold, 1975; Abbassi and Mancktelow, 1992). Strain ellipses in the matrix fan around the outer arc of the folds and converge within the inner arcs. Strain intensity decreases with increasing distance from the layer boundaries.

CONCLUSIONS

Natural folds in rocks exhibit a wide spectrum of shapes, from wave forms with smoothly rounded hinges and short limbs to angular folds with narrow hinges and long straight limbs. The experiments described here demonstrate the rheological control of these variations. Elasto-plastic strain-hardening single-layer folds develop sinusoidal wave forms and exhibit limb thinning and hinge thickening with progressive strain. By comparison, folds in the elasto-plastic strain-softening single layer, which also initiate with sinusoidal wave form, become increasingly angular with sharper hinges and straighter limbs as deformation progresses. The amplification rate of folds prior to 25% bulk shortening is influenced by initial competence contrast and is higher in the strain-softening layer. At later stages of shortening, folding is controlled by complex rheological properties and the amplification rate is larger in the strain-hardening layer. The change in the ratio of arc length to layer thickness during folding is higher for folds in the strain-softening layer ($S/T \approx 6\text{--}9.5$) than in the strain-hardening layer ($S/T \approx 4.5\text{--}7$) for the same bulk shortening. Folds in the strain-hardening single layer continuously decrease their arc length during compression. In the strain-softening single layer, changes in arc length were insignificant after fold limbs reached dips of 25°. In sinusoidal folds, penetrative bending strain results in the migration of layer material from fold limbs toward the hinge zones. In contrast, strain in the limbs of the angular folds remains relatively small and homogeneous as strain is focused in the hinges. Evolution of folds in the strain-hardening layer takes place gradually with regular modification of shape characteristics under compressive strain. Folds in the strain-softening layer show an abrupt change in shape when strain distribution leads to the brittle failure of the fold hinges as critical loads are reached. Structures in models evolve through distinctive stages characterized by individual parameters. Folds undergo progressive changes in growth rate, wavelength selection, and strain distribution, probably in response to variations in the nonlinear properties and complex strain-dependent rheologies of the materials used.

ACKNOWLEDGMENTS

The experimental study was carried out under the guidance of G. Mulugeta. Constructive and thoughtful reviews of an earlier version by G. Mulugeta and C. Talbot, and by P. Hudleston, L. Skjernaa, R. Walter, and N. Mancktelow of the final manuscript, are gratefully acknowledged.

REFERENCES CITED

Abbassi, M.R., and Mancktelow, N.S, 1992, Single layer buckle folding in nonlinear materials–I. Experimental study of fold development from an isolated initial perturbation: Journal of Structural Geology, v. 14, p. 85–104.

Biot, M.A., 1961, Theory of folding of stratified viscoelastic media and its implications in tectonics and orogenesis: Geological Society of America Bulletin, v. 72, p. 1595–1620.

Chapple, W.M., 1969, Fold shape and rheology: The folding of an isolated viscous-plastic layer: Tectonophysics, v. 7, p. 97–116.

Cobbold, P.R., 1975, Fold propagation in single embedded layers: Tectonophysics, v. 27, p. 333–351.

Dixon, J.M., and Summers, J.M., 1986, Another word on the rheology of silicone putty: Bingham: Journal of Structural Geology, v. 8, p. 593–595.

Fletcher, R.C., 1974, Wavelength selection in the folding of a single layer with power-law rheology: American Journal of Science, v. 274, p. 1029–1043.

Fuchu, D., Lee, C.F., and Sijing, W., 1999, Analysis of rainstorm-induced slide-debris flows on natural terrain of Lantau Island, Hong Kong: Engineering Geology, v. 51, p. 279–290.

Goodman, R.E., 1989, Introduction to rock mechanics: New York, J. Wiley and Sons, 562 p.

Griggs, D.T., Turner, F.J., and Heard, H.C., 1960, Deformation of rocks at 500°–800°C, *in* Griggs, D., and Handin, J., eds., Rock deformation: A symposium, Geological Society of America Memoir 79, p. 39–104.

Hailemariam, H., and Mulugeta, G., 1998, Temperature-dependent rheology of bouncing putties used as rock analogs: Tectonophysics, v. 294, p. 131–141.

Heard, H.C., and Rayleigh, C.B., 1972, Steady-state flow in marble at 500° to 800°C: Geological Society of America Bulletin, v. 83, p. 935–956.

Hudleston, P.J., 1973, An analysis of "single-layer" folds developed experimentally in viscous media: Tectonophysics, v. 16, p. 189–214.

Hudleston, P.J., and Lan, L., 1993, Information from fold shapes: Journal of Structural Geology, v. 15, p. 253–264.

Hudleston, P.J., and Lan, L., 1994, Rheological controls on the shapes of single-layer folds: Journal of Structural Geology, v. 16, p. 1007–1021.

Johnson, A.M., 1970, Physical processes in geology: San Francisco, Freeman, Cooper and Company, 577 p.

Johnson, A.M., and Fletcher, R.C., 1994, Mechanical analysis and interpretation of structures in deformed rock: New York, Columbia University Press, 461p.

Lan, L., and Hudleston P., 1996, Rock rheology and sharpness of folds in single layers: Journal of Structural Geology, v. 18, p. 925–931.

Mancktelow, N.S., 1999, Finite-element modeling of single-layer folding in elasto-viscous materials: The effect of initial perturbation geometry: Journal of Structural Geology, v. 21, p. 161–177.

Neurath, C., and Smith, R.V., 1982, The effect of material properties on growth rates of folding and boudinage: Experiments with wax models: Journal of Structural Geology, v. 4, p. 215–229.

Ramberg, H., 1960, Relationships between length of arc and thickness of ptygmatically folded veins: American Journal of Science, v. 258, p. 36–46.

Ramberg, H., 1964, Selective buckling of composite layers with contrasted rheological properties, a theory for simultaneous formation of several orders of folds: Tectonophysics, v. 1, p. 307–341.

Ramsay, J.G., and Huber, M.I., 1987, The techniques of modern structural geology, Volume 2: Folds and fractures: London, Academic Press, 700 p.

Ranalli, G., 1987, Rheology of the Earth: London, Allen and Unwin Inc., 366 p.

Sherwin, J., and Chapple, W.M., 1968, Wavelengths of single layer folds: A comparison between theory and observation: American Journal of Science, v. 266, p. 167–179.

Samieh, A.M., and Wong, R.C.K., 1998, Modelling the responses of Athabasca oil sand in triaxial compression tests at low pressure: Canadian Geotechnical Journal, v. 35, p. 395–406.

Schultz, R.A., and Li Qizhi, 1995, Uniaxial strength testing of non-welded Calico Hills Tuff, Yucca Mountain, Nevada: Engineering Geology, v. 40, p. 287–299.

Smith, R.B., 1977, Formation of folds, boudinage, and mullions in non-Newtonian materials: Geological Society of America Bulletin, v. 88, p. 312–320.

Smith, R.B., 1979, The folding of a strongly non-Newtonian layer: American Journal of Science, v. 279, p. 272–287.

Weijermars, R., 1986, Flow behavior and physical chemistry of bouncing putties and related polymers in view of tectonic laboratory applications: Tectonophysics, v. 124, p. 325–358.

Zhang, Y., Hobbs, B.E., Ord, A., and Muhlhaus, H.B., 1996, Computer simulation of single-layer buckling: Journal of Structural Geology, v. 18, p. 643–655.

Zhao, J. Li, H.B., Wu, M.B., and Li, T.J., 1999, Dynamic uniaxial compression tests on a granite: International Journal of Rock Mechanics and Mining Sciences and Geomechanics Abstracts, v. 36, no. 2, p. 273–277.

MANUSCRIPT ACCEPTED BY THE SOCIETY APRIL 12, 2000

Sheath fold development in bulk simple shear: Analogue modeling of natural examples from the Southern Iberian Variscan fold belt

Filipe M. Rosas,* Fernando O. Marques, Sara Coelho, and Paulo Fonseca
Departamento de Geologia, Faculdade de Ciências, Universidade de Lisboa, Edifício C2, Piso 5, 1700 Lisboa, Portugal

ABSTRACT

Experimental analogue modeling in bulk simple shear was carried out to study the evolution of natural sheath folds developed from noncylindrical deflections of foliation, associated, or not, with competent boudins. Results of experiments with competent inclusions embedded in a less viscous matrix show that sheath folds exhibit different development when formed from deflections around nonrotating or rotating rigid inclusions—the greater the rotation (and rotation velocity), the lesser the development. Nonrotating boudins are obvious obstacles to the progress of the incompetent (less viscous) matrix, whereas rotating (especially fast rotating) boudins inhibit sheath fold development. Experimental results also suggest that sheath fold development around rigid inclusions is inhibited during active boudinage (layer-parallel extension) concomitant with relatively high rotation velocity of boudins.

Experimental investigation of sheath fold development from noncylindrical deflections in a foliation (no rigid bodies present) illustrates their variable geometric patterns observed in crosssections—from a mushroom-like pattern close to the root zone, to an eye-like pattern close to the apex of the sheath fold. The recognition of the root zone (i.e., a mushroom section) is crucial for shear sense determination because one can attribute polarity to the fold.

INTRODUCTION

Natural rotated inclusions and sheath folds in high strain shear zones have been the object of study and experimental analogue modeling by several authors (e.g., Ghosh and Ramberg, 1976; Cobbold and Quinquis, 1980; Van Den Driessche and Brun, 1987), with the aim of understanding their evolution and usefulness as shear criteria. According to the quoted authors, those structures are thought to develop mainly during bulk simple shear deformation. To study how competent elliptical and rectangular inclusions rotate when embedded in an incompetent matrix subjected to bulk simple shear, Ghosh and Ramberg (1976) carried out theoretical and experimental investigations and concluded that, during homogeneous simple shear, particles with axial ratios >1 rotate synthetically with variable velocity, and particles with axial ratios = 1 rotate synthetically at constant velocity. However, Marques and Cobbold's (1995) experiments with bulk simple shear revealed that, under certain circumstances, ellipsoidal inclusions, instead of behaving as theoretically predicted, rotate antithetically, even when starting from a position with their longest axis parallel to the shear plane. Other published experimental work has dealt with rotated inclusions (e.g., rolling structures of Van Den Driessche and Brun, 1987), sheath fold development from noncylindrical deflections in foliations or associated with boudinage (Cobbold and Quinquis, 1980), and sheath fold development associated with nonrotating rigid inclusions (Marques and Cobbold, 1995). No published results have dealt with simultaneous rotating competent inclusions and associated sheath fold generation. Thus, Cobbold and Quinquis' (1980, model 3) and Marques and Cobbold's (1995) work with nonrotating boudins fails to account for the structural associations observed in our natural examples. Then, in the present

*E-mail: frosas@fc.ul.pt

study we worked with rotating and extending trails of boudins, and rotating individual boudins, as observed in the natural example, to evaluate the effects of rotation on the development of noncylindrical folds associated with them.

Building on the theoretical and experimental work on sheath fold development and rigid particle rotation mentioned in the preceding, and motivated by our analysis of natural examples of sheath folds associated with rotating inclusions (and/or trails of inclusions), we used analogue experimentation in bulk simple shear to simulate, and try to understand, the mechanics of: (1) differential development of sheath folds as a function of their position relative to folded sets of competent boudins within an incompetent matrix; (2) differential sheath fold development depending on variable shape and rotation of rigid bodies; and (3) development of sheath folds from noncylindrical deflections in the foliation surface, illustration of their variable geometry in crosssections along different parts of the fold, and use of geometry to determine polarity and thus sense of shear.

Skjernaa (1989) defined a sheath fold as a noncylindrical fold with $\omega < 90°$ and x:y $> 1/4$ (we followed this definition in our study). Sheath fold shape depends mostly on amount of strain and shape of original deflection (Skjernaa, 1989; Marques and Cobbold, 1995); e.g., from deflections elongated parallel to shear direction one obtains a tubular fold for relatively low values of shear strain (γ), whereas from spherical deflections a sheath fold develops for the same amount of strain (see Marques and Cobbold, 1995).

GEOLOGICAL SETTING AND NATURAL EXAMPLES

The motivation for the experimental work presented here was based on field observations carried along a geotraverse in the western Ossa-Morena Zone, Southern Iberian Variscan fold belt, southern Portugal, which includes a unit of marbles. The earliest recognizable tectonometamorphic event (D1) in this unit is characterized by a pervasive mylonitic foliation (coincident with a metamorphic layering) and by a north–south trending mineral stretching lineation (currently mostly erased by recrystallization, due to later gabbroic intrusions, but still recognizable by north–south elongated crystals of wollastonite). The most interesting structures occur in the marbles as a result of (1) contrasting rheological behavior between incompetent marbles and competent intrusive mafic dikes and (2) noncylindrical deflections of the foliation.

We carried out analogue modeling in bulk simple shear because all D1 structures observed along the geotraverse are interpreted to have been formed in a noncoaxial deformation regime; this, together with the suspected allochthonous character of all units (Fonseca and Ribeiro, 1993; Araújo, 1995), suggests that the deformation regime was bulk layer-parallel simple shear, with a subordinate component of layer-normal pure shear. Rotations like the one observed in the mafic boudin on the left in Figure 1 would not be possible if a vertical (layer-normal) significant component of flattening was present, because then it would stabilize at a low angle to the shear plane after synthetic rotation (cf. Ghosh and Ramberg, 1976).

The marbles present a metamorphic layering of mostly calcite (white layers) and impure (yellow, green, gray to dark gray) siliceous marble that can contain as accessory phases dolomite, forsterite, wollastonite, chlorite, and talc. The mineral association of the mafic boudins comprises plagioclase + green hornblende + actinolite + epidote + chlorite + sphene.

Natural structures observed along north–south sections (slightly oblique to the stretching lineation) in the marble quarries (Fig. 1) are characterized by the presence of several rigid mafic boudins inside the marbles, and by deflection of the metamorphic layering around them. It is possible to observe fracturing of the boudins, as a result of a bookshelf mechanism, and folds associated with both boudin and fractured pieces. Eye-like patterns of folds in oblique sections are characteristic of a sheath geometry. This kind of structural association was experimentally studied by Cobbold and Quinquis (1980, model 3) and Marques and Cobbold (1995). Comparison of their experimental results with the natural example of Figure 1 suggests that boudins must have remained almost nonrotated during sheath fold development, and that the fractured pieces have rotated only slightly. The same has not happened with the vertical boudin, which does not show associated fold generation (besides the deflection of the metamorphic layering around it). This is the topic of a set of experiments to investigate fold development associated with rotating boudins.

Along other north–south sections, sets of mafic boudins are rotated and coherently folded with the metamorphic layering of the marbles. In contrast to the structures illustrated at the bottom of Figure 1 (nonrotated boudins and associated sheath folds), the section of the boudins is approximately square, the trail of boudins is folded, and individual boudins are considerably rotated and show no associated folds. It is only possible to observe one isoclinal, highly stretched fold associated with the boudin located at the hinge zone. Another purpose of our first set of experiments was the investigation of sheath fold development associated with folded and rotated sets of boudins.

Mushroom-like geometric patterns defined by the folded metamorphic layering are observed along east–west sections in the marbles. Experimental investigation on the variable geometric patterns observed in cross sections of sheath folds, developed from noncylindrical deflections of the foliation, under bulk simple shear, was the aim of our third set of experiments.

EXPERIMENTAL PROCEDURE

The viscosity contrast between marble (incompetent matrix) and amphibolite (competent boudin) is obviously large from direct observation of outcrops: the foliation in the marble is heavily folded (isoclinal and eye-like folds in Figure 1), and the boudin is fractured. From recent literature, where methods for evaluation of viscosity contrast are reviewed (Treagus, 1999; Talbot, 1999), there seems to be no method suitable for

Figure 1. A: North–south section in marble quarry. Inside layered marbles three sets of fractured mafic boudins are recognizable. Left: Fractured boudin standing in vertical position relative to shear plane, deflecting metamorphic layering in its vicinity but without associated fold patterns. Central bottom: Fractured boudin with fold patterns associated with boudin and fractured pieces. Right: Boudins disrupted by bookshelf-like mechanism and showing less evident associated fold pattern. B: Schematic illustration of central bottom fractured boudin: a, fold pattern associated with boudin; b, fold pattern associated with fractured pieces of boudin. Interpreted shear sense is top to right.

that evaluation in the studied rocks: they do not have cleavage refraction (Treagus, 1999), folds are not appropriate because of wave length, amplitude, and unfavorable thickness relations, and pinch and swell structures are missing from the boudins (Smith, 1977; Talbot, 1999). These structures are fractured instead of pinched and swelled, which suggests a substantial viscosity contrast between mafic rock and marble. Thus, our choice of analogue materials followed previous work dealing with identical rock types and viscosity contrasts. Typically, high viscosity contrast in rocks has been simulated using silicone putty and plasticine (e.g. Ramberg, 1959; Ghosh and Ramberg, 1976; Van Den Driessche and Brun, 1987). Marques and Cobbold (1995), facing a natural example similar to ours, used transparent and pink silicone putty as analogues of an incompetent marble matrix, and plasticine for the competent mafic boudins. Moreover, Weijermars (1986) showed the advantages of the use of bouncing putties as analogues of incompetent rocks at strain rates typical of laboratory experiments. As noted by Ramberg (1959), plasticine behavior is strongly dependent on thickness: thin strips embedded in silicone putty can fold ptygmatically, while thick slabs behave as stiff solids. Appropriately, plasticine parallelepipeds remained unstrained in our experiments, and their edges and corners remained sharp throughout the experiments (as did the natural broken pieces of the mafic boudin at the bottom of Figure 1 during deformation of the marbles).

In our analogue experiments, we used transparent silicone putty (polydimethyl-siloxane [PDMS], manufactured by Dow Corning of Great Britain under the trade name SGM 36) for physical properties and convenience as geological model materials (see Weijermars, 1986) to simulate the incompetent marble matrix, pink silicone putty (Rhodorosil gomme speciale 70009, manufactured by Rhone-Poulenc, France) as a passive strain marker (viscosity contrast between SGM36 and RGS70009 is not significant; Weijermars, 1986), and plasticine (standard yellow plasticine, manufactured by Omyacolor SA-F51240, France) as our analogue for the competent mafic boudins.

The experiments were carried out in a Perspex simple shear rig which allows shear strains (γ) to 12. Our shear box (Fig. 2) is an improved version of those described by Ramberg (1959) and Van Den Driessche and Brun (1987): it comprises two fixed upper and lower walls (normal to the shear plane), two lateral walls (parallel to the shear plane), one fixed and the other driven by a motor, and two top walls articulated with the side walls to maintain the simple shear geometry of the model. The model dimensions (coincident with X, Y, and Z in Fig. 2B) were $500 \times 30 \times 70$ mm. To ensure simple shear flow in the model, PDMS must be perfectly adherent to the lateral walls (which is easy to achieve if the Perspex is perfectly clean) and a neutral liquid soap was used to guarantee that friction was minimum on all other walls. The perfect adherence of the PDMS to the lateral walls ensured that the shear was integrally transmitted into the model. The used coordinate system was x, y, and z for the inclusions and X, Y, and Z for the matrix: the shear plane was coincident with the XY plane (parallel to the side walls) and X was the shear direction.

EXPERIMENTAL RESULTS AND DISCUSSION

In the first and second set of experiments we distinguished between the rotation of the layer of preformed boudins and the rotation of individual boudins. The preformed trail of boudins behaved, as a whole, as a passive line marker: it rotated very slowly toward the shear plane as theoretically predicted for simple shear (only at $\gamma = \infty$ would the layer of preformed boudins and shear plane would become parallel). Rotations of individual boudins are shown in the graphs of Figure 3. Although boudins started abutted, they rotated almost as theoretically predicted by Ghosh and Ramberg (1976), because they separated in the early stages of the experiments.

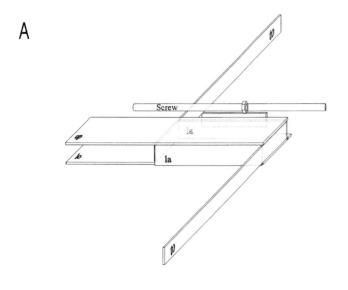

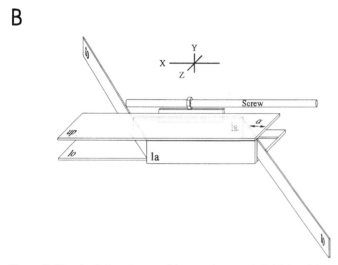

Figure 2. Sketch of shear box used in experiments. A: Initial position. B: After $\gamma = 1$. to—top walls; la—lateral walls; up—upper wall; lo—lower wall; X, Y, and Z—kinematic axes; α—angular shear strain.

First set of experiments

The model was made up of transparent silicone putty, inside which we placed several adjacent plasticine boudins (10 mm < x < 20 mm, y = 15 mm, z = 10 mm and 1 < R = x/z < 2) according to an original fold geometry with a long limb–short limb relationship defining a top-to-the-left geometrical verging structure (Fig. 4A). The y direction of the boudins coincides with the Y kinematic axis for simple shear and Y direction in the matrix, whereas their x and z directions vary along the XZ plane, defining the trace of the original fold geometry. On both sides of the plasticine boudins, thin pink silicone layers were introduced, in contact with the transparent silicone putty, to allow observation of the interference effects between the rigid inclusions and the matrix during deformation. Because the y dimension of inclusions is smaller than the Y dimension of the shear box, the deflection of the pink silicone is noncylindrical.

The results of the top-to-the-left simple shear for different stages of deformation are illustrated in Figure 4. Due to initial different orientation of the fold limbs (defined by the alignment of boudins; Fig. 4A) relative to the shear plane and because the shear sense is top-to-the-left, the long and short limb show different behaviors: the long limb extends and rotates toward the shear plane throughout deformation; the short limb starts by shortening, in the early stages as it rotates synthetically toward an orientation orthogonal to the shear plane, after which it extends.

For low shear strain values ($\gamma < 1$), the plasticine boudins in the long limb rotate synthetically as they separate. The combined effect of boudin rotation and necking of the thin pink silicone layer around the plasticine inclusions originates a stair-stepping geometry as deformation proceeds (Fig. 4B). In the short limb, a similar but somewhat delayed evolution is observed, because extension only starts to occur when the short limb is reversed. In our experiments extension does not result from a component of pure shear, but from the position of the layer of preformed boudins relative to the shear plane.

We now focus on what happens in the hinge zone of the original folded set of boudins. At relatively low γ values (<1), it is possible to observe a geometric accommodation of the pink silicone marker to the edges of the boudins at the hinge zone of the fold; the boudins define a noncylindrical geometric anisotropy. The shapes of the sheath folds generated from these anisotropies depend on the y dimension of each boudin. For $\gamma \cong 6$, the result is $\omega \cong 14°$ and x:y $\cong$ 4, which is, according to Skjernaa (1989), a tubular fold. This shape develops gradually from the noncylindrical deflection of the pink silicone in the boudin occupying the hinge zone of the folded set. Thus, in the initial stages, the frontal fold is slightly noncylindrical; for higher γ values it develops into a sheath fold and finally into a tubular fold. Every one of these sheath folds correlates with one of the edges of the boudins that, by turns and for higher γ values, occupy the hinge zone of the folded set (Fig. 4C).

In these experiments, it is important to note the following: (1) At the final stage ($\gamma \cong 6$), most boudins rotated synthetically and significantly (over 45°), especially in the reverse limb. (2) Initiation of sheath fold development at the hinge zone of folded row of boudins began at relatively low γ values (~2) when the individual boudins rotated <45° (Fig. 4B). (3) Well-developed sheath folds did not form in association with boudins in the limbs of the initial fold. This result is identical to our observations in the marbles (Fig. 5), where sheath folds only formed near the hinge zone of the folded set of boudins, but is in clear contrast to the well-developed sheath folds associated with the nonrotating boudins in Figure 1, and with previous results by Cobbold and Quinquis (1980) and

Marques and Cobbold (1995), in whose experiments the boudins remained nonrotating.

We propose that deviation of our results from what is predicted by those authors could be due to (1) significant synthetic rotation of the boudins, which started from a position at a low angle (~10°) to the shear plane (Fig. 4A); (2) behavior of both fold limbs, which lie in the extension field and thus undergo extension that inhibits accumulation of matrix around the boudins and so leads to the progressive development of sheath folds; or (3) our maximum shear strain of $\gamma = 7$ was obviously not enough to rotate both limbs to the shear plane, where extension comes to an end. From this orientation onward, the only controlling factor remaining would be synthetic rotation at variable velocity.

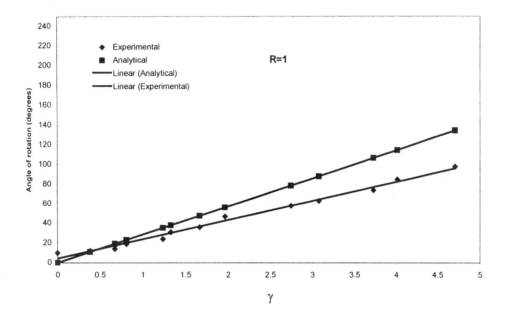

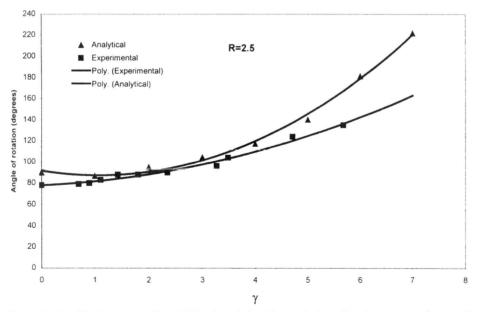

Figure 3. Graphical representation of behavior of plasticine inclusions. For shear strain values used in our experiments, there was no significant deviation of experimental results from analytical predictions of Ghosh and Ramberg (1976).

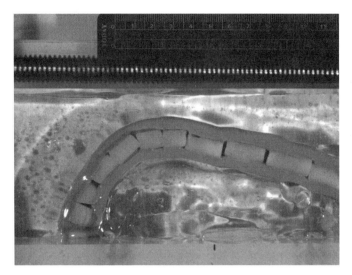

A

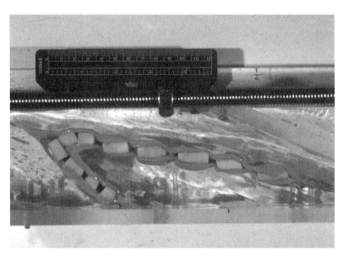

B

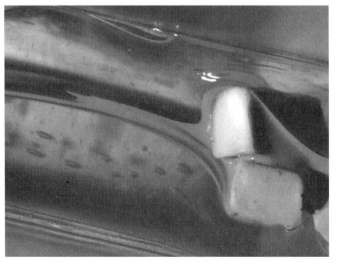

C

A

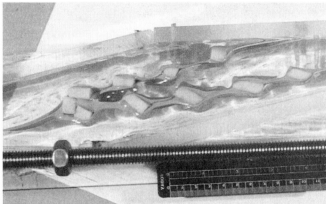

B

Figure 5. Illustration of similarity between natural example and the end result of our first set of experiments. A: Folded set of boudins, defining short limb/long limb top-to-right geometrical sense of shear; one highly stretched isoclinal fold is associated with hinge zone of folded trail of boudins, but fold patterns associated with individual boudins of limbs are either absent of very poorly developed. B: End result of our first set of experiments, characterized by conspicuous similarity with natural example.

Figure 4. XZ plane photographs of first set of experiments (shear sense is top to left). A: Initial undeformed state. B: ($\gamma \cong 3$) Long limb—boudinage and necking of pink silicone layer around synthetically rotated plasticine inclusions, and development of stair-stepping geometry; short limb—boudinage and initial necking around inclusions; hinge zone—development of highly stretched isoclinal noncylindrical fold. Note absence of sheath fold development around inclusions in both limbs of folded set of inclusions. C: ($\gamma \cong 4, 5$) Detail of hinge zone of folded trail of inclusions: second-generation sheath fold associated with upper edge of lower boudin (note curvature of fold hinge).

Second set of experiments

Previous work (Ghosh and Ramberg 1976) has established that the rotation velocity of rigid particles in a less viscous medium during simple shear depends on their axial ratios (and original orientation). Thus, a second set of experiments using boudins with an equidimensional xz section ($R = x/z = 1$) was carried out.

A straight row of plasticine boudins ($x = 10$ mm, $y = 15$ mm, $z = 10$ mm, and $R = x/z = 1$) was embedded in transparent silicone putty (Fig. 6A). This layer started with an initial angle (θ) of 10° with the shear plane so that the trail of boudins would rotate as inferred in the natural example. The geometric interference between the rigid bodies and the matrix is shown by two parallel thin pink silicone layers, separated by PDMS, introduced on each side of the row of boudins. The results of top-to-the-left simple shear deformation of the model are illustrated in Figure 6 (B and C).

1. The layer of boudins rotated synthetically toward the shear plane, so that θ decreased as γ increased.
2. The pink silicone layers necked between the boudins as they separated due to progressive extension (Fig. 6B).
3. Small-scale sheath folds developed, at high γ values (~5) and total boudin rotation of ~135°, in the innermost thin pink silicone layer, associated with frontal and rear edges of the boudins (Fig. 6C).

The major result of this set of experiments is the almost complete lack of sheath fold generation associated with deflections on the foliation around plasticine inclusions. This agrees with the natural example (Fig. 5A), but is in clear contrast with the experimental results of Cobbold and Quinquis (1980), and Marques and Cobbold (1995). We interpret our results similarly to the first set of experiments, because we cannot separate the individual effects of rotation and extension, which occurred simultaneously in our experiments. The major difference between the two sets of experiments is in the rotation velocity of individual boudins: according to Ghosh and Ramberg (1976), boudins with an axial ratio = 1 (square section) rotate at a constant velocity, and boudins with an axial ratio >1 rotate with a variable velocity, with an orientation of minimum velocity at which the longer axis parallels the shear plane. In this set of experiments, boudins with square sections rotated at a constant velocity (Fig. 3), without nonrotating positions that could favor sheath fold development.

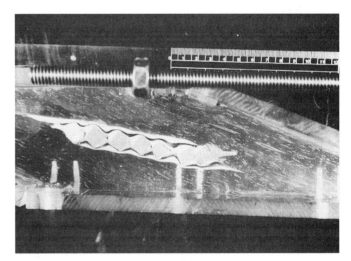

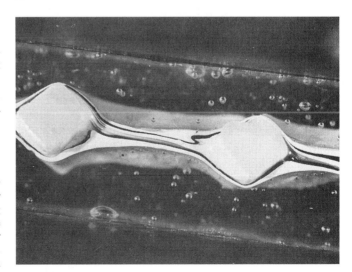

Figure 6. XZ plane photographs of second set of experiments (shear sense is top-to-left). A: Initial undeformed state. B: ($\gamma \cong 3$) Necking of innermost pink silicone layer around synthetically rotated plasticine inclusions, and development of stair-stepping geometry. Note that for this γ value no sheath fold pattern exists around rigid inclusions. C: ($\gamma \cong 4, 5$) Detail of deformed set of boudins: longitudinal section of minor noncylindrical fold associated with frontal edge of boudin (on right) that rotated ~135°.

Interpretation of our natural examples based on these experimental results suggests the following.

1. In the natural example of Figure 1, the absence of a fold pattern around the vertical boudin on the left is due to a large amount of rotation. From the rectangular section, one would expect well-developed sheath folds if the boudin had remained nonrotating and parallel to the shear plane (as in the boudins in the lower part of the outcrop). According to Ghosh and Ramberg (1976), the orthogonal position of the boudin relative to the shear plane could not be nonrotating (and significant rotation must have occurred), because this position corresponds to the one of maximum velocity.

2. Our results and those of Marques and Cobbold (1995) indicate that the boudins in the bottom part of Figure 1 rotated only slightly.

3. The considerable rotation of individual boudins in the limbs of the folded layer in the natural example of Figure 5A could be responsible for the absence of sheath fold development around boudins.

4. In the natural fold of Figure 5A, the long limb, as a whole, behaved as an almost nonrotating (or very slowly rotating) boudin, and thus only development of a frontal sheath fold was favored.

Third set of experiments

The third set of experiments was based on the analysis of a natural example, illustrated in and aimed at analogue modeling development of multilayer sheath folds by deformation of pre-existing noncylindrical surfaces, under bulk simple shear conditions (e.g., models 1 and 2 of Cobbold and Quinquis, 1980). Knowledge of geometry of sheath folds in sections perpendicular to the x axis is important because it can help the structural geologist to determine polarity and sense of shear (if sense of apex can be found).

The construction of the multilayer model, and its XYZ reference frame, is illustrated in Figure 7. Between two thick transparent silicone layers we introduced two 2-mm-thick pink silicone slices separated by a 2-mm-thick PDMS layer, all lying in the XY plane. Before covering this multilayer set with the uppermost PDMS layer, we imprinted, with spherical objects, a downward convex noncylindrical irregular deflection into the upper pink silicone layer and then filled the depression with PDMS.

Top-to-the-left simple shear deformation resulted in a well-developed sheath fold (with $\omega \cong 29$ and x:y $\cong 2.3$). Second-order sheath folds developed from irregularities in the initial noncylindrical deflection.

Several sections were cut parallel to the YZ plane at the end of the experiments; the sections show the different geometric patterns of a well-developed sheath fold, from the apex to its root zone (Fig. 8). These patterns vary between two end members: the eye-like shape found in sections near the sheath fold apex, and the wide-open mushroom-like shape found in sections near the root zone of the sheath fold. It is thus possible to distinguish four main types of geometric patterns: (1) an eye pattern defined only by the outermost pink silicone layer; (2) a double eye pattern defined by both the inner and outer pink silicone layers; (3) a mixture of an eye pattern in the inner pink layer and mushroom pattern in the outermost pink layer; and (4) a double mushroom pattern defined by both pink silicone layers (Fig. 9).

This set of experiments confirms that passive sheath folds develop from noncylindrical deflections in foliations, during bulk simple shear, according to model 1 of Cobbold and Quinquis (1980). Moreover, such results illustrate the typical geometric patterns for different cross sections of sheath folds in multilayers (Fig. 8), which is crucial for shear-sense determination. If we only know the sense to which the apex points (by measuring hinge attitudes; Fig. 10A), we cannot determine the sense of shear because we do not know whether it is a syncline or anticline (we know geometry but not polarity). However, if we can recognize the root zone (by finding mushroom sections), then we can determine shear sense because we are able to determine sheath fold polarity: if a mushroom section of a sheath fold is rooted downward (Fig. 10B), it is an anticline and the apex indicates a top-away sense of shear; if it is rooted upward (Fig. 10C), it is a syncline and the apex indicates a bottom-away shear sense.

CONCLUSIONS

The end results of our analogue experiments are strikingly similar to our natural examples (Figs. 5 and 9). This allows us to draw the following conclusions.

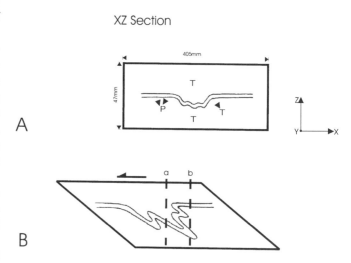

Figure 7. Schematic illustration of XZ section of constructed multilayer analogue model (A) and end result (B) of third set of experiments. T—polydimethyl-siloxane filling entire volume of shear box. P—2-mm-thick pink silicone layers. a and b—locations of cross sections, from root zone to apex of sheath fold, respectively. X, Y, and Z—used coordinate system.

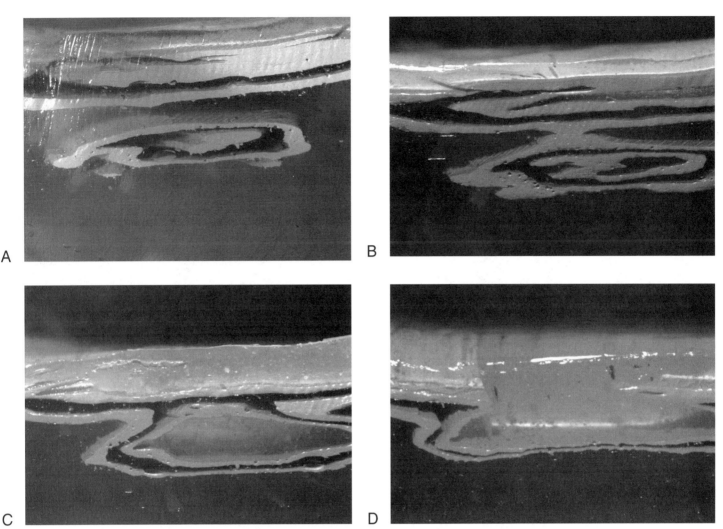

Figure 8. Illustration of different geometric patterns observed along experimental sheath folds, in cross sections parallel to YZ plane and from apex to root zone. A: Double-eye pattern defined by both inner and outer pink silicone layers. Note that, beneath main eye, two minor ones correspond to second-order sheath folds. B and C: Mixture of eye pattern in inner pink layer and mushroom pattern in outermost pink layer. D: Double mushroom pattern defined by both pink silicone layers.

1. Sheath folds do not show equal development when formed from deflections around nonrotating or rotating boudins; nonrotating rigid bodies are usually surrounded by a well-developed pattern of sheath folds, but rotating boudins only show associated minor (or none) frontal sheath folds.

2. Sheath folds are even less likely to form if extension and boudinage occur concomitantly with high rotation velocity.

3. Folded trails of boudins behave, as a whole, as a slowly rotating (or nonrotating) rigid body, giving rise to well-developed frontal sheath folds.

4. The geometric patterns of sheath folds in cross sections vary along their length, from a mushroom pattern close to the root zone, to an eye-like pattern closer to the apex. Knowledge of the root zone (mushroom section) is critical for shear-sense determination.

5. As proposed by Marques and Cobbold (1995), the distribution of folds around rigid inclusions can be used as a kinematic indicator. Our experimental results did not show the rear rim-like folds, but the frontal folds developed and were congruent with the imposed sense of shear.

ACKNOWLEDGMENTS

The experiments were performed in the Experimental Tectonics Laboratory of LATTEX, a research unit funded by PLURIANUAL, Project 125/N/92. Field work of F. Marques was supported by the same project. F. Rosas acknowledges a Ph.D. scholarship of Fundação para a Ciência e a Tecnologia. We thank the reviewers C.J. Talbot and A.R. Cruden for their critical comments.

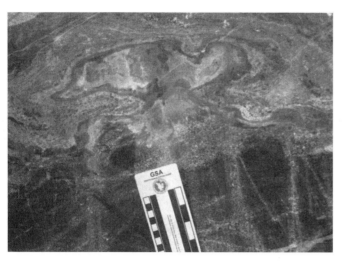

Figure 9. Illustration of similarity between natural example and end result of our third set of experiments. A: Mushroom-like geometric pattern of folding, defined by metamorphic layering and corresponding to an east–west cross section of root zone of sheath fold. B: End result of our third set of experiments.

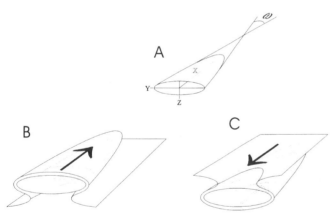

Figure 10. Illustration of determination of shear sense from sheath fold cross sections. A: Impossible to determine shear sense because polarity is unknown, although one can find sense of apex by measuring hinge attitudes. B: Solution with downward-rooted sheath fold (anticline), which indicates top-to-right sense of shear. C: Solution with upward-rooted sheath fold (syncline), which indicates top-to-left sense of shear (opposite to B). In any case, one must always know sense of apex. X, Y, and Z are principal axes of sheath fold and ω is angle proposed by Skjernaa (1989) to define sheath fold.

REFERENCES CITED

Araújo, A., 1995, Estrutura de uma geotransversal entre Brinches e Mourão (Zona de Ossa Morena): Implicações na evolução geodinâmica da margem sudoeste do Terreno Autóctone Ibérico. [Ph.D. thesis]: Faculty of Sciences, University of Lisbon, 200 p.

Cobbold, P.R., and Quinquis, H., 1980, Development of sheath folds in shear regimes: Journal of Structural Geology, v. 2, p. 119–126.

Fonseca, P.E., and Ribeiro, A., 1993, The tectonics of Beja-Acebuches Ophiolite: A major suture in the Iberian Variscan Fold Belt: Geogische Rundschau, v. 3, p. 440–447.

Ghosh, S.K., and Ramberg, H., 1976, Reorientation of inclusions by combination of pure shear and simple shear: Tectonophysics, v. 34, p. 1–70.

Marques, F.G., and Cobbold, P.R., 1995, Development of highly noncylindrical folds around rigid ellipsoidal inclusions in bulk simple shear regimes: Natural examples and experimental modeling: Journal of Structural Geology, v. 17, p. 589–602.

Ramberg, H., 1959, Evolution of ptygmatic folding: Norsk Geologisk Tidsskrift, v. 39, p. 99–161.

Skjernaa, L., 1989, Tubular folds and sheath folds: Definitions and conceptual models for their development, with examples from the Grapesvare area, northern Sweden: Journal of Structural Geology, v. 11, p. 689–703.

Smith, R.B., 1977, Formation of folds, boudinage and mullions in non-Newtonian materials: Geology Society of America Bulletin, v. 88, p. 312–320.

Talbot, C.J., 1999, Can field data constrain rock viscosities: Journal of Structural Geology, v. 21, p. 949–957.

Treagus, S.H., 1999, Are viscosity ratios of rocks measurable from cleavage refractions?: Journal of Structural Geology, v. 21, p. 895–901.

Van Den Driessche, J., and Brun, J.-P., 1987, Rolling structures at large shear strain: Journal of Structural Geology, v. 9, p. 691–704.

Weijermars, R., 1986, Flow behavior and physical chemistry of bouncing putties and related polymers in view of tectonic laboratory applications: Tectonophysics, v. 124, p. 325–358.

MANUSCRIPT ACCEPTED BY THE SOCIETY APRIL 12, 2000

Modeling the role of erosion in diapir development in contractional settings

Maura Sans*
Departament de Geologia Dinàmica i Geofísica, Universitat de Barcelona, Barcelona 08071, Spain
Hemin A. Koyi
Hans Ramberg Tectonic Laboratory, Department of Earth Sciences, Villavagen 16. SE-752 36 Uppsala, Sweden

ABSTRACT

The Cardona diapir of Eocene salt (Cardona salt formation) crops out in the deformed foreland of the Pyrenees. The diapir pierces 300 m of sediments on the northern limb of the northeast-southwest trending Cardona-Pinós fold close to the hinge area, and is 2 km long and 0.7 km wide at the surface. Dynamically scaled models were laterally shortened in a centrifuge in order to investigate the mechanism that triggered the Cardona diapir, and the role of postshortening erosion in its development. The models consisted of a microlaminate overburden simulating nonevaporitic sediments overlying a layer of silicone putty simulating rock salt. Model results show that moderate folding of overburden units causes the underlying ductile layer to flow from under syncline troughs to accumulate in the anticline cores. Postshortening erosion of one of the models thinned and weakened the overburden above the salt-cored anticlines and aided piercement of salt walls along the crests of the anticlines. Our models show that anticlines need to be open structures in order to allow accumulation of sufficient salt to rise diapirically when the overburden is thinned by erosion. Minimum ascent velocities, which were calculated using the deformed Pliocene-Pleistocene deposits around the Cardona diapir, support a postshortening age for salt piercement after 2.2 km of overburden was eroded from the crest of the anticline 2 Ma ago.

INTRODUCTION

Compressional tectonics may inhibit the formation of salt diapirs (Vendeville, 1991; Koyi, 1998) as opposed to extension, which is an important triggering factor (Vendeville and Jackson, 1992; Koyi et al., 1993; Jackson and Vendeville, 1994). Thin- and thick-skinned extension trigger diapirism by differentially thinning preexisting overburden units, which will load the underlying salt (Vendeville and Jackson, 1992; Koyi et al., 1993). Faulting in overburden also creates the necessary occupation space for the diapiric material (Vendeville and Jackson, 1992; Koyi, 1998). Lateral shortening inhibits initiation of diapirism mainly by thickenning the overburden and by laterally squeezing any preexisting diapir (Vendeville, 1991; Koyi, 1998). Prediapiric shortening thickens overburden above a salt detachment by folding and thrusting, and therefore suppresses piercement (Koyi, 1998). When shortening takes place simultaneously with formation of salt diapirs, the diapirs are laterally squeezed and suppressed by their overburden, which is thickened tectonically (Witschko and Chapple, 1977; Koyi, 1998). Postdiapiric shortening may squeeze or cut the feeding stem and starve preexisting diapirs by sequeezing salt back to the source layer or into the air (Vendeville and Nilsen, 1995; Roca et al., 1996; Koyi, 1998).

Despite all these mechanisms inhibiting salt diapirs in laterally shortened overburdens, diapirs are nonetheless common in areas that have undergone compression, such as the toe of gravitational gliding systems (e.g., Vendeville and Nilsen, 1995) and fold and thrust belts (e.g., Koyi, 1988; Serrano and Martínez del Olmo, 1990; de Ruig, 1992; Roca et al., 1996; Cotton and Koyi, 2000). Moreover, diapirs in fold and thrust

---

*E-mail: maura@geo.ub.es

Sans, M., and Koyi, H.A., 2001, Modeling the role of erosion in diapir development in contractional settings, *in* Koyi, H.A., and Mancktelow, N.S., eds., Tectonic Modeling: A Volume in Honor of Hans Ramberg: Boulder, Colorado, Geological Society of America Memoir 193, p. 111–122.

belts described in the literature can be preshortening, synshortening, or postshortening. Preshortening diapirs have been described in the Pyrenees (Serrano et al., 1994; Gil, 1998), the Betics (de Ruig, 1992; Roca et al., 1996), Zagros (Koyi, 1988), the Atlas (Vially et al., 1994; Letouzey et al., 1995), and the Hellenides (Underhill, 1988), among others. Synshortening diapirs are less common and have been described in the Zagros (Koyi, 1988; Talbot et al., 1998) and the Atlas (Vially et al., 1994). Postshortening diapirs have been identified in the Zagros (Koyi, 1988) and the Betics (de Ruig, 1992; Roca et al., 1996). In all these examples, preshortening diapirs are attributed to an earlier extensional phase that triggered diapirism in the area. These diapirs are later modified by lateral compression (Koyi, 1988; Serrano and Martínez del Olmo, 1990; Roca et al., 1996), or can be passively transported within thrust sheets without deformation (Serrano and Martínez del Olmo, 1990). Synshortening diapirs can be related to tear faults across salt-cored anticlines (Vially et al., 1994; Letouzey et al., 1995; Talbot and Alavi, 1996), whereas postshortening diapirs are either related to new extensional phases (Jackson and Vendeville, 1994; Roca et al., 1996) or to differential loading (Jackson and Vendeville, 1994).

In this chapter, we present the results of scaled centrifuge models designed to investigate the role of erosion in triggering diapirism in areas that have already been folded. One of the models is designed to conform to field data collected from the Cardona diapir. This diapir is located in the deformed foreland basin of the central southern Pyrenees and has been the subject of intense study; it has been mined for potash from Roman time until 1991, and has been recently proposed as a waste disposal site. However, the age and triggering mechanism of the diapir are still not well understood. Deformation of the surrounding sediments and internal structures of the diapir have been studied in the field and used to prepare an analogue model to determine the influence of erosion in the evolution history of the diapir. Results of eroded models are compared with models that did not undergo a phase of erosion.

GEOLOGICAL SETTING

Salt diapirs are common in the southern Pyrenees. Most of them are of Triassic salt and crop out, highly deformed, in the external parts of the south-central and southwestern Pyrenees, or, less deformed, in a more external position in the southwestern Pyrenees (Fig. 1). These diapirs developed during an extensional phase that took place during the Late Jurassic to Albian (Serrano and Martínez del Olmo, 1990; Gil, 1998). During Pyrenean compression (Oligocene-Miocene), these diapirs were transported about 20 km southward in a thrust sheet without any deformation except in the thrust front (Serrano and Martínez del Olmo, 1990). By contrast, in the southeastern Pyrenees foreland, Triassic salt diapirs in the external areas of the fold and thrust belt are fewer. In addition to these, a diapir of Eocene salt (Cardona Formation) crops out in the deformed foreland. The Cardona Formation of Priabonian age is the intermediate of the three horizons that define the basal detachment of the south Pyrenean triangle zone (Fig. 2). The presence of these evaporitic formations (Beuda, Cardona, and Barbastro) determines the thrust-wedge geometry and location of the deformation front (Sans et al., 1996).

The sedimentology and petrography (Pueyo, 1975; Ayora et al., 1995; Rosell and Pueyo, 1997) of the Cardona Formation

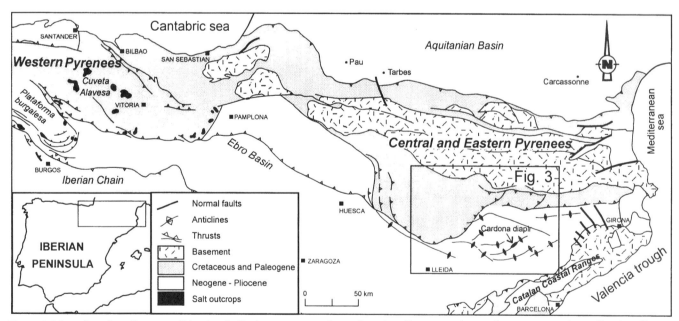

Figure 1. Map of southern Pyrenees. Triassic salt diapirs are common in southwestern Pyrenees, whereas they are scarcer in southeastern Pyrenees. In the southeastern Pyrenees Eocene salt diapir (Cardona diapir) crops out in deformed foreland. See location of Figure 3. Redrawn from Serrano and Martínez del Olmo (1990), Vergés et al. (1992), Vergés (1993), and Sans et al. (1996).

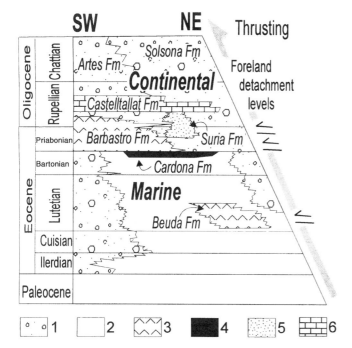

Figure 2. Schematic stratigraphic section of Ebro basin infill. Cardona salt formation is last marine infill of basin. Its overburden consists of continental, alluvial, fluvial, and lacustrine deposits (modified from Sans et al., 1996). To right of section three main detachment levels in basin are indicated. Arrow shows younging age of thrusts. 1: Sandstones and conglomerates. 2: Sandstones and marls. 3: Gypsum. 4: Salt. 5: Sandstones. 6: Limestones.

have been studied (Wagner et al., 1971; Riba et al., 1983; Sans, 1999; Miralles et al., 2000) together with its structure using underground mining data, mining exploration wells, and the excellent outcrops in the western termination of the Cardona diapir. The Cardona Formation, the thickness of which increases northward to reach 300 m in the center of the basin, is the last marine infill of the Ebro basin (Fig. 2). Three units can be distinguished in the Cardona salt formation that allow mapping of the internal structure: (1) a laminated 4–5-m-thick basal anhydrite unit; (2) a 130–200-m-thick lower massive salt unit, which consists of decimeter thick white and gray diffuse bands of halite (Busquets et al., 1985); and (3) an upper 50–100-m-thick potash-bearing banded salt unit (Pueyo, 1975; Ayora et al., 1995; Rosell and Pueyo, 1997). This unit is characterized by centimeter-thick beds of halite and sylvinite.

The structures developed in the Pyrenean triangle zone are detached fault-related folds cored by evaporites (Sans and Vergés, 1995) that have different trends above each of the three detachment horizons that form the basal decollement (Sans et al., 1996). In the central part of the triangle zone, the folds and thrusts that are detached on the Cardona salt formation have a northeast-southwest trend and merge to the west with the Sanaüja anticline, which represents the ramp to a higher evaporitic level (Fig. 3). The folds are open structures with large wavelength/amplitude ratios ranging from 52 in the El Guix anticline to 8 in the Pinós-Cardona anticline (Fig. 4). The synclines are wider than the anticlines and are flat bottomed (Fig. 4). The frontal anticlines, El Guix in the east and the Súria in the west, are double structures characterized by a south-directed fold or thrust in the north and a north-directed fold or thrust in the south (Fig. 4) (Sans and Vergés, 1995; Sans et al., 1996). To the north, the Cardona, l'Estany, and Vilanova de l'Aguda anticlines are south-verging detachment anticlines. Only the Pinós-Cardona anticline is pierced by diapiric salt. Bulk shortening across the El Guix, Súria and Cardona anticlines ranges from 21% to 33%, of which only 5%–10% is accommodated by folding and thrusting; the remaining 16%–23% is taken up by layer-parallel shortening, as determined by the analysis of strain markers (Sans et al., 1999).

Above the Cardona Formation, the initial thickness of the overburden involved in the folds and thrusts of the triangle zone was a maximum of 2.5 km. This value is estimated from the thickness of sediments preserved in the syncline located to the north of the Pinós-Cardona anticline (Fig. 4), vitrinite maturation measurements (Vergés et al., 1998), and the total thickness measured in a magnetostratigraphic section in the footwall of the main emergent thrust of the southern Pyrenees (Meigs et al., 1996). All the sediments are continental (Fig. 2) and consist of a lacustrine system and related alluvial successions to the north and southeast (Anadón et al., 1989). The lacustrine system is composed of the Barbastro and Castelltallat Formations. The Barbastro Formation consists of gypsum and grades upward and laterally to siltstones to the southwest (Sáez, 1987). The upper part of this lacustrine system is the Castelltallat Formation, which consists of lacustrine limestones. This lacustrine system passes transitionally to the alluvial terminal fan sediments of the Súria Formation and to the fluvial deposits of the Solsona Formation to the north. To the southeast, the lacustrine system grades to the fluvial deposits of the Artés Formation.

The age of the anticlines detached above the Cardona Formation in the triangle zone cannot be precisely determined be-

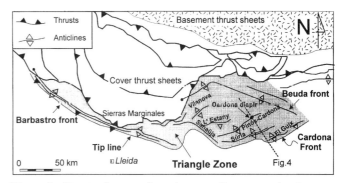

Figure 3. Structural map of southeastern Pyrenean triangle zone (shaded in gray). Three different fold trends correspond at depth with three detachments (Beuda, Cardona, and Barbastro Formations) that form basal detachment of triangle zone. Above Cardona detachment, folds and thrust have northeast-southwest trend. Cardona diapir pierces Pinós-Cardona anticline. Thrust wedge developed at thrust front of each detachment. See also location of Cardona diapir and section in Figure 4 (after Sans et al., 1996).

cause of the absence of related syntectonic sediments. However, regional sedimentological and paleontological data, paleomagnetically derived ages for syntectonic units located to the north of the studied area, and the structural chronology allow the age of the anticlines to be bracketed. The initiation age of the Súria and Cardona anticlines is younger than deposition of the Barbastro and Súria Formations (late Eocene–early Oligocene, 34.4 Ma). Because of lack of thickness or facies changes through the anticlines (Vergés, 1993), and on the basis of magnetic data and correlation with the Sierras Marginales thrust, Meigs et al. (1996) proposed an age of 32 Ma for the onset of folding, and suggested that folding must have ended at 27.8 Ma. This is in agreement with the prefolding early Oligocene age (Rupelian, 29 Ma) based on mammal fauna (Agustí et al., 1987) and carofites flora (Sáez, 1987) in the youngest deformed sediments preserved in the synclines in the Ebro basin. The folds were obviously growing after deposition of these sediments.

However, field evidence suggests that these anticlines also grew over a brief interval during the Quaternary. Old Quaternary river terraces on the limbs of the El Guix anticline are cut by small thrusts with offsets <1 m and are folded in the Súria anticline (Solé-Sabarís and Masachs; 1940; Masachs, 1952; Masana and Sans, 1995). In addition, the mining shaft in the Súria anticline has been narrowing in a northwest-southeast direction at an average rate of 12 mm/yr (Masachs, 1952).

An important Neogene erosional phase in the Ebro basin has been documented (Vergés, 1993; Coney, 1993; Vergés et al., 1998). The Ebro basin was filled by Oligocene synorogenic sediments to a level that is at present higher than 1500 m above sea level. During the Neogene, the river system reached the Ebro basin after crossing the Catalan Coastal Ranges. The older age for the start of the erosion in the Ebro basin is estimated to be synchronous with the formation of the Catalan Coastal Ranges during Aquitanian-Burdigalian (20.5 Ma). The preservation of unfolded 16.4 Ma sediments dated as upper Burdigalian in the center of the Ebro basin provides the younger estimate for the start of the erosion (E. Roca, 1992, personal commun.). Erosion of the crest of the Pinós-Cardona anticline has removed ~2000 m of strata. Taking the uncertainty on the onset of erosion as between 20.5 Ma and 16.4 Ma gives a minimum and a maximum rate of erosion of 0.1 mm/yr and 0.12 mm/yr if this was constant. During this period, the Ebro basin was no longer under compression. The extension that created the Valencia trough during the Neogene did not affect this region.

CARDONA DIAPIR

The Cardona diapir pierces the Pinós-Cardona anticline (Figs. 4 and 5). This anticline can be mapped in a N55E direction for ~30 km from where it merges with the northwest-southeast trending Sanaüja anticline to a few kilometers to the east of Cardona (Fig. 3). The Pinós-Cardona anticline is a south-vergent detachment anticline with a wavelength, measured between syncline troughs, ranging between 14 km in the east and 6 km in the west, and a maximum amplitude of 1 km. The northwestern limb has a relatively constant gentle dip along the strike of the anticline (20°–30°NW), whereas the southeastern limb dips 50°–70°SE in the eastern and central parts and becomes overturned and cut by a south-directed thrust in the west. The anticlinal hinge, which is well defined in its western and central parts, becomes broader toward the east. The Cardona diapir is located close to the eastern termination of the anticline and marks the transition between the narrow and wide hinge zones. Where the diapir pierces the anticline, the total shortening in the anticline is 23%. Folding accounts for 6%, whereas 17% is accommodated by layer-parallel shortening prior to folding (Sans et al., 1999).

The overburden in the diapir area consists of 100 m of gray marls at the bottom of the Barbastro Formation, 450 m of sandstones and marls from the Súria Formation, and 1500 m of sandstones and conglomerates from the Solsona Formation. The base of the unit is interbedded with thin limestones from the Castelltallat Formation (Fig. 5). The thickness of the overburden surviving erosion in the anticline crest is only 400 m, whereas it is ~2000 m in the northern syncline trough and 1200 m in the southern one.

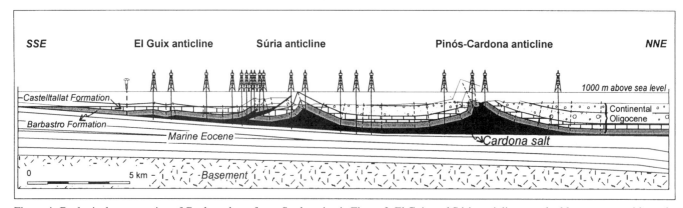

Figure 4. Geological cross section of Cardona thrust front. See location in Figure 3. El Guix and Súria anticlines are double structures with north-directed fold or thrust in south and south-directed fold or thrust in north. Cardona anticline is detachment anticline. Mining exploration and production wells projected into section are from <1 km.

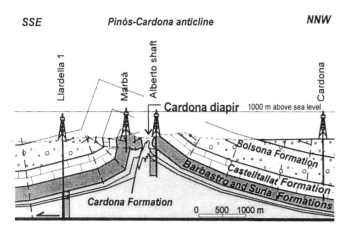

Figure 5. Cross section of Pinós-Cardona anticline and Cardona diapir. Pinós-Cardona anticline is south-vergent detachment anticline. Cardona diapir pierces back limb of anticline close to hinge. Three lithological distinct units in salt allow mapping of internal structure of diapir.

The contact between the overburden and the diapir corresponds to a 2–6-m-thick shear zone formed by a melange of country rock and sheared salt. Other than this zone, there is no other significant deformation in the overburden units close to the diapir (Iglesias, 1970; Wagner et al., 1971; Riba, 1975).

Subsurface geometry of the Cardona diapir can be constrained from well data. The diapir rises from the salt-cored detached anticline and develops a 250–500-m-wide and ~250-m-high stem (Fig. 5). The present-day erosion reveals outward-verging folds in the emergent salt, indicating remnants of either a diapiric bulb or salt extrusion (Fig. 6). The structure of the salt layers, the complete stratigraphic sequence in a normal position, and especially the horizontal fold axial planes in the highest parts of the diapir suggest that the bulb is almost completely preserved close to the diapir walls (Fig. 7). The diapiric bulb is small. It is ~60–80 m high and 400–700 m wide. Its internal structure consists of a major anticline parallel to the Pinós-Cardona anticline cut by a steep shear zone of the same trend (Fig. 6). The surfaces of axial planes and fold axes are vertical or steeply dipping near the western and eastern terminations of the diapir, whereas in the central part of the diapir the axial planes are vertical and the fold axes are subhorizontal (Wagner et al., 1971; Sans et al., 1999). The boudins have steep axes, indicating that the stretching direction was perpendicular to the fold axes in the center of the diapir and parallel to them at both terminations (Sans, 1999). This suggests that the folds are large sheath folds elongated in a northeast-southwest direction (Fig. 7). The significance of the shear zone dividing the diapir into two parts can be compared to that of the internal shear zones of Kupfer (1968, 1976), and could indicate different rates of salt movement between both sides of the shear zone. In this regard, two main spines in the Cardona diapir could be defined. The southwestern spine has risen relatively higher than the northeastern one.

The initiation of the Cardona diapir cannot be well constrained from its geometrical relationships with the surrounding sediments. The oldest record of diapiric movement is provided by the observed drag of the post-Oligocene sediments (Fig. 6). These sediments are poorly consolidated sandstones that flex upward from horizontal, 25 m away from the diapir, to subvertical close to it. The difference in height between the still-horizontal part and the highest points of any single layer gives the best estimate of the minimum rise of the salt since deposition of the layer. The age of these sediments has been attributed to middle Villafranquian (Riba, 1975), on the basis of its correlation with a river terrace containing an association of fauna

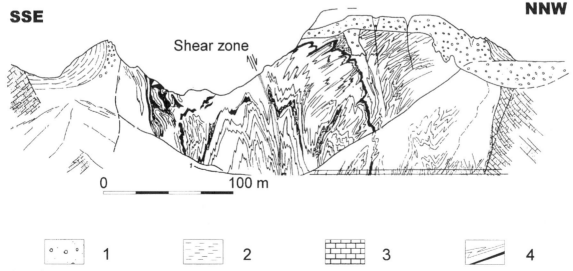

Figure 6. Drawing of Cardona diapir western outcrop (Muntanya de Sal). Shear zone is marked in center of diapir separating two spines. Modified from Wagner et al. (1971). 1: Cap rock and melange of salt and overburden fragments. 2: Pliocene-Quaternary deformed sediments. 3: Tertiary overburden. 4: Salt. Published with permission of Springer-Verlag.

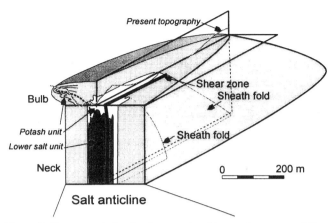

Figure 7. Transversal sketch of external and internal structure of Cardona diapir. Diapir has small bulb and short stem that rise from crest of salt anticline. Present-day erosion permits observation of almost complete bulb close to diapir walls.

with *Elephas meridionalis* (Solé-Sabarís and Masachs; 1940, Masachs, 1952). The Villafranquian period was defined by Bonifany (1975) and has been divided into Calabrian and Plasencian time (Riba and Reguant, 1986). The middle Villafranquian ranges between 0.9 and 1.7 Ma.

Current upward movement of the diapir is demonstrated by the presence of salt at the surface, which creates a local relief in the present climatic conditions. The Cardona diapir crops out in a subordinate valley, which drains to the Cardener River, and only in the western part of the valley is the Cardona salt exposed 100 m above the valley floor giving a local topographic relief of 100 m (Muntanya de sal). This relief is ~50–100 m lower than the highest surrounding topography. The presence of this relative relief in a climatic regime dominated by dissolution suggests that the diapir is still active. Upward movement of the diapir is also supported by topographic measurements carried out in the period 1979–1986. These measurements indicate that, away from the influence of the underground mining galleries in the deeper levels of the diapir and the salt-cored anticline, the leveling points have an average upward vertical motion of 1 mm/yr.

PREVIOUS MODELS

Salt tectonics has been studied by the use of a great number of analogue and numerical models (e.g., Jackson and Talbot, 1987; Talbot and Jackson, 1987; Vendeville and Jackson, 1992; Daudré and Cloetingh, 1994; Alsop, 1996; Waltham, 1997; Gugliemo et al., 1997; Gil, 1998). Differential sedimentary loading and regional extension have been the mechanisms most explored and are thought to be the most efficient triggers to diapirism. Previous models dealing with regional compression focused on the development of folds and thrusts above a ductile detachment and mostly fail to trigger diapirs. In these experiments, either a brittle granular overburden (e.g., sand, glass beads) or plastic microlaminates consisting of alternating laminae of plasticine and silicone putty are used. Granular overburdens laterally shorten to form box folds and conjugate thrusts. These oppose diapirism by thickening the overburden units and ejecting the viscous layer downward out of the cores of the anticlines (Vendeville, 1991). In such models, diapirs can only form if tear faults cut the anticlines (Vially et al., 1994). In models with plastic overburdens, upright folds can keep their cores of ductile material throughout the section (Fig. 8). Further deformation tightens the folds and the salt is injected back into the layer (Talbot et al., 1988). The presence of a thick ductile layer retards the nucleation of thrust faults (Dixon and Tirrul, 1991).

The effect of compression on preexisting diapirs has also been modeled. Contraction after depletion of the source layer in the toe of a gravity gliding system can rejuvenate preshortening diapirs by forcing salt upward instead of deforming the surrounding overburden (Vendeville and Nilsen, 1995). However, regional compression shows upward expulsion of salt in the diapir stems as well as a reaccumulation of the salt toward the area of thinned overburden (Roca et al., 1999).

Although some models have investigated overburden thinning related to extension (Vendeville and Jackson, 1992; Schultz-Ela et al., 1993; Jackson and Vendeville, 1994), and some numerical models have included erosion during diapir development (Poliakov et al., 1996), there are no analogue models exploring the effect of erosion on diapirism. Differential erosion unloading could trigger salt diapirism in the external areas of a fold and thrust belt. The models described in this chapter are a first attempt to explore the role erosion plays in triggering diapirs from contractional pillows.

MODEL PREPARATION AND KINEMATICS

Two centrifuge models investigating the effects of erosion in triggering diapirism in a previously shortened area are described here. The first model simulated shortening of an anisotropic microlaminate overburden over a ductile substrate (Fig. 8). This model was not directly scaled to any specific natural example. The second model was geometrically similar to the Cardona thrust front in the south Pyrenean triangle zone and postfolding erosion was induced to allow comparison between the results from the model and the real example (Figs. 9 and 10). The second model was also compared with the first model, which had a similar deformation history but was not eroded after it was shortened in the centrifuge.

In the models, rock salt was simulated by a Newtonian silicone putty (density 1.1 g cm^{-3}, viscosity = 2×10^4 Pa s) and the overburden by anisotropic microlaminates of plastilina (McClay, 1976) (effective viscosity = 3×10^6 Pa s at model strain rate) and Dow Corning silicone (Dixon and Summers, 1985) (viscosity = 6×10^5 Pa s). Silicon putty is a Newtonian viscous fluid, which deforms in a ductile manner under the experimental conditions. Dow Corning silicone is a non-Newtonian material. Plasticine is also a non-Newtonian mate-

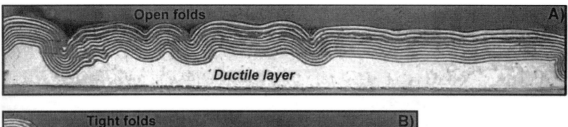

Figure 8. Profiles of first model (A) after 12% shortening and (B) after 35% shortening. Note that anticlines are open and ductile layer has accumulated in their cores in A, whereas open folds have tightened and squeezed ductile material out of their cores in B. On left side of model, synclines are totally closed. Anticlines have very narrow cores where most of ductile layer has been expelled.

rial that deforms in a ductile manner at low strain rates, but fails on discrete shear surfaces at higher strain rates (Dixon and Tirrull, 1991). Alternating laminae of Dow Corning silicone and plasticine microlamina are considered reasonable analogs for limestone-shale or sandstone-shale interlayers undergoing brittle deformation at shallow crustal levels (Dixon and Summers, 1985). Previously, models built with microlaminates were used to simulate anisotropy in sedimentary units in fold and thrust belts (Dixon and Liu, 1992) and as overburden to salt layers (Koyi, 1996). In both models here, the ductile layer/overburden thickness ratio in the model was 1/5, close to the 1/8 ratio in the field example. Both models were shortened during centrifuging by allowing a collapsing ductile mass to drive a rear wall that shortened the model (Dixon and Summers, 1985).

The first model was 8 cm long and consisted of a 1-mm-thick layer of silicone putty simulating salt and 5 mm of microlaminates simulating the overburden. The model was shortened first to 12%, then to 18%, and finally to 35% for a total of 2 min at 2000 g. After each increment of shortening, the model was sliced and photographed. The slice was then welded back by contact. No erosion was simulated in this model. Instead, after 35% of shortening, the model was spun in the centrifuge without being shortened so as to study the behavior of the ductile material that had accumulated in the cores of anticlines (Fig. 8).

The second model was initially 7.2 cm wide, 12 cm long, and 1.8 cm thick (Figs. 9 and 10A). The ductile layer consisted of a 3-mm-thick layer of silicone putty and the overburden of a 1.5-cm-thick layer of 0.5-mm-thick anisotropic microlaminates. The model was shortened by 12% in the centrifuge for a total of 7.5 min between 450 and 500 g, until the geometry of the folds was similar to the Pinós-Cardona anticline, and the low-density ductile layer had accumulated in the anticline core to 3 times its initial thickness. The model was then frozen and sliced (Fig. 10B). A slice was cut, photographed, and welded back to the model. A 1.2-cm-thick horizontal slice was removed from the top of the model to simulate the present-day erosion of the crest of the Cardona anticline (Fig. 10C). The model was then run in the centrifuge for an additional 5.5 min at 500 g without shortening.

MODEL RESULTS

Unlike the second model, no diapirs were initiated in the first model, which did not undergo any postshortening erosion, despite a second phase of centrifuging. In this model, the density inversion between the ductile substrate and the overburden was not sufficient to overcome the shear strength of the microlaminate overburden. The second model, in which a postshortening erosional phase was simulated, developed 6 symmetric buckle folds after 12% shortening in the centrifuge (Fig. 10B).

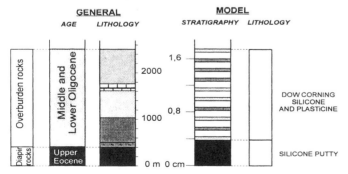

Figure 9. Initial set-up of second model, and comparison between startigraphic section in field example and materials used in model.

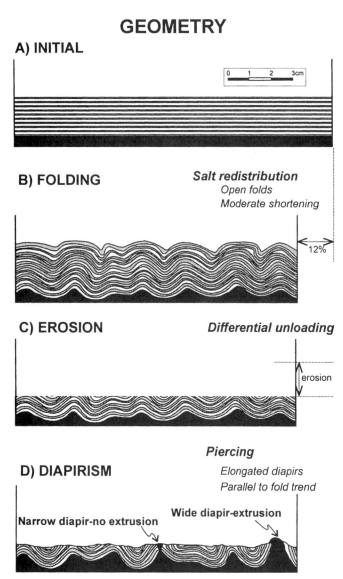

Figure 10. Line drawing of evolution history of second model. A: Initial configuration. B: Fold geometry after 12% of shortening. C: Erosion of top of model. D: Final geometry of folds and of resulting diapirs. At this stage several diapirs have pierced anticlines. Diapirs that have narrow neck and small bulb have not yet extruded at surface, whereas diapirs that extruded spread preferentially in direction of fold vergence.

The wavelengths of these folds, which were characterized by round hinges, ranged between 1.5 and 2.3 cm and the amplitudes were 1.7 and 2.5 mm. The wavelength/amplitude ratio in these anticlines ranged between 4/1 and 7.6/1 and the dip of the limbs was 30°–40°. At the salt-overburden interface, synclines and anticlines had similar sizes. However, anticlines at the top free surface of the model were much wider than the synclines. The folds were rectilinear in trend and had similar profiles. No significant layer-parallel shortening was observed in the model. These structures were similar to those in the first model (Fig. 8).

After the eroded model was centrifuged for a second time, diapirs rose through the anticlinal cores and extruded along their crests. This was due to differential loading that drove the ductile layer from beneath areas with higher overburden load (under the synclines) to the areas with less loading (the cores of anticlines). As the diapirs rose, the anticlines became slightly asymmetric and their limbs steepened to 40°–50°. In plan view, all the diapirs were elongate parallel to the fold axes and pierced the anticlines in the hinge area. In profile, some of the diapirs had oblique stems and pierced the back limb of the asymmetric anticlines close to the crest. The diapirs had short stems and small bulbs (Fig. 10D). When they reached the surface, they spread asymmetrically in the direction of the anticline vergence and the stems widened (Fig. 10D).

DISCUSSION

Role of erosion in the model

This modeling approach is the first attempt to evaluate the role of erosion as a triggering mechanism for diapirism in a previously compressed area, based on an excellent field example in the southeastern Pyrenees. Erosion of anticlines rising on the top free surface increases differential loading on the salt layer and weakens their overburden, and therefore promotes diapirism without any need to change the tectonic regime. In a previously compressed area, two conditions are necessary to create a diapir. First, the anticlines in the base of the overburden have to be sufficiently open structures to allow accumulation of ductile material (Fig. 10B) and second, erosion has to be sufficiently deep to thin the overburden for the diapirs to be able to pierce (Fig. 10C). Open folds, which have interlimb angles >70° (Twis and Moores, 1992), allow salt to accumulate in their cores rather than to be ejected downward, as in the tight folds (Figs. 8B and 10). With increasing shortening, open folds tighten and expel salt from their cores back into the source layer (Wiltschko and Chapple, 1977). If salt remains in anticlinal cores when erosion thins the overburden over the anticline crests, the resulting differential loading between the synclinal troughs and the anticline crests promotes salt flow toward the areas of lower pressure. In the crestal area the salt/overburden thickness ratio is greater than both the initial ratio and that under the synclinal troughs. Piercing takes place when the critical overburden thickness is reached. The diapirs formed in this way pierce anticlines, close to the hinge of symmetric anticlines or in the back limb of asymmetric anticlines, and both are expected to be elongate parallel to the fold axis. In section, the diapirs have short and narrow necks and small bulbs that widen as they extrude at the surface.

Cardona diapir

A precompressional age for the Cardona diapir can be ruled out because the Cardona salt was deposited in the late Eocene after compression had already started in the Pyrenees (Upper Cretaceous; Muñoz, 1992). In addition, deep erosion in the Ebro basin is documented (Coney, 1993; Vergés et al., 1998). The Cardona diapir pierces close to the hinge of the asymmetric anticline, is elongate parallel to the fold trend, and has a short and narrow neck and a small bulb. The geometry and evolution history of the Cardona diapir is therefore similar to that of the model diapirs.

Estimating the piercing age of the Cardona diapir

The age of the Cardona diapir cannot be well defined by its geometrical relationships with the surrounding sediments because of the lack of syndiapiric sediments. Whereas syndepositional diapiric piercement and growth can be determined by the geometry of syndiapiric sediments, the timing of diapirs triggered by erosion can only be determined by the critical thickness of the overburden. However, as displayed by the model, erosion triggers diapirism by thinning the overburden and causing differential loading between the synclines and the anticlines above the salt layer. Therefore the piercing onset can be estimated by calculating the piercing critical thickness.

A simple analytical approach is used here to estimate the critical thickness for piercing the salt-cored anticlines of the Ebro basin. The folds were modeled as an elongated ridge with a trapezoidal profile (Fig. 11). The difference in height and thickness of the overburden due to erosion created a pressure difference that drove deformation. Following Bishop (1978) and Schultz-Ela et al. (1993), simplified assumptions were made: the viscous flow resistance within the salt and retardation along the diapir walls is insignificant; and any elastic deformation is neglected. Resistance to flow on the diapir walls, however, is not negligible in the field example because there are marginal shear zones and the limbs are oversteepened. This would increase the thickness of the lid that can be lifted, because the driving force will be greater in this analytical approach than it is in nature.

In the model, the initial configuration was achieved during folding without synchronous erosion. After folding, several scenarios in which different amounts of instantaneous erosion occurred were studied. In each case the driving force for piercing and the resisting force were calculated. Piercing takes place when the driving force is larger than the resisting force.

To calculate the driving force we modified the analytical approach of Schultz-Ela et al. (1993), which relates the variables that govern piercement. We have added another parameter (erosion) to study its effect on the piercement of a salt diapir. The pressure at the base of the overburden column at point a (Fig. 11) is:

$$P_0 = \rho_0 g h_0, \quad (1)$$

where ρ_0 is the density of the overburden, g is the acceleration due to gravity, and h_0 is the initial thickness of the overburden. For a unit dimension normal to the section, the volume of the roof block (V_r) is:

$$V_r = w(h_r) + (h_r)^2/\tan \phi, \quad (2)$$

where w is the width of the anticline crest at the top of the salt layer, h_r is the roof block thickness, and ϕ is the dip of the faults

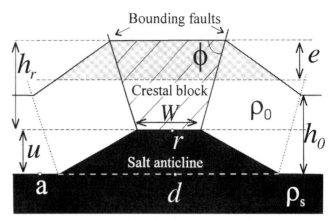

Figure 11. Configuration and variables for force balance analysis. Driving force results from differential unloading by erosion between anticline crest and syncline trough. Resisting force results from friction on fault planes and extra mass added to the roof block by nonvertical, inward dipping fault planes. Analysis assumes geometry extends infinitely perpendicular to section drawn. ρ_0 is density of overburden, h_0 its initial thickness, e amount of erosion, h_r roof block thickness, w width of the anticline crest at top of salt layer, and ϕ dip of faults limiting roof block.

limiting the roof block. The amount of erosion (e) relates h_r to the initial thickness (h_0), i.e.,

$$h_r = h_0 - e. \quad (3)$$

Therefore, the pressure (P_r) at the top of the salt ridge is the mass under the influence of gravity in the roof block divided by the basal area:

$$P_r = g[\rho_0(h_0 - e) + (h_0 - e)^2 \rho_0/\tan \phi \, w], \quad (4)$$

and the pressure (P_d) at the base of the ridge is:

$$P_d = g[\rho_0(h_0 - e) + (h_0 - e)^2 \rho_0/\tan \phi \, w] + v\rho_s, \quad (5)$$

where ρ_s is the density of the salt and u is the fold uplift. The driving force (F_d) is the pressure difference acting parallel to the fault planes, i.e.,

$$F_d = g \, w \sin \phi [\rho_0 e - u\rho_s - (h_0 - e)^2 \rho_0/\tan \phi \, w]. \quad (6)$$

If the amount of erosion exceeds the fold uplift, equation 1 is modified to

$$P_0 = \rho_0 g(h_0 - e + u). \quad (7)$$

In this case, the driving force (F_d) is

$$F_d = g \, w \sin \phi [u(\rho_0 - \rho_s) - (h_0 - e)^2 \rho_0/\tan \phi \, w]. \quad (8)$$

The opposing force (F_s) is the friction on the fault planes. Following Schultz-Ela et al. (1993), we calculate the shear resistance by assuming that the stress in the overburden is entirely lithostatic despite the shortening. Assuming also a unit dimension normal to the section, the total shear force (F_s) for both faults is the mean stress multiplied by the coefficient of internal friction (μ), assuming cohesion to be negligible. Cohesion can be considered negligible because the anticline crest was fractured during folding. Introducing the amount of erosion (e), the opposing force is calculated as

$$F_s = \mu\, g\, \rho_0 \cos \phi (h_0 - e)^2. \qquad (9)$$

Figure 12 shows graphical solutions of the two equations for the different folds detached above the Cardona salt formation. The values used to calculate the driving and opposing forces in these anticlines are 2000 m, 1200 m, 950 m, and 100 m of fold uplift for the Cardona, Vilanova, Súria, and El Guix anticlines, respectively. The bounding fault angle is 80°, and the diapir width is 400 m, which is the mean width for the Cardona diapir. Salt density (ρ_s) is 2300 g m^{-3} and the overburden density (ρ_o) is 2600 g m^{-3}, based on mean values from mining data. The initial thickness of the overburden (h_0) is 2500 m and the coefficient of friction (μ) is 0.85 for the numerical calculations plotted in the graphs shown in Figure 12. Although the calculations done in these sections are simplifications of nature, and "their correspondence with the analytical results and field example are somewhat fortuitous" (Schultz-Ela et al., 1993, p. 287), they show in a broad sense the relationship between erosion and piercing capability of the salt-cored anticlines. According to the calculation, the Cardona diapir pierced when the overburden thickness was ~300 m. If the erosion rate has been constant (0.15 m/yr), the age of the diapir would be 2 Ma. The calculated thickness of the overburden is in agreement with the field observations. If the Cardona diapir formed in a way similar to that of the diapirs in the analogue model, the height of the diapir (~300 m) would correspond to the thickness of the overburden at the crest of the anticline. The critical overburden thickness to create a diapir with characteristics similar to the Cardona diapir for the Vilanova and Súria anticlines ranges between 100 m and 200 m and, in the El Guix anticline, <100 m. The results show that only the Pinós-Cardona anticline has undergone enough erosion to allow piercing. If the erosion rate is constant, the Vilanova, Súria, and El Guix anticlines need between 2 and 3 m.y. more to be eroded to their critical thickness, which the diapirs can overcome to pierce.

CONCLUSIONS

1. External areas of fold and thrust belts detached above salt are suitable areas for diapir development if erosion thins the overburden to open salt-cored anticlines to the critical thickness. Erosion causes differential unloading between the anticline crests and the syncline troughs.

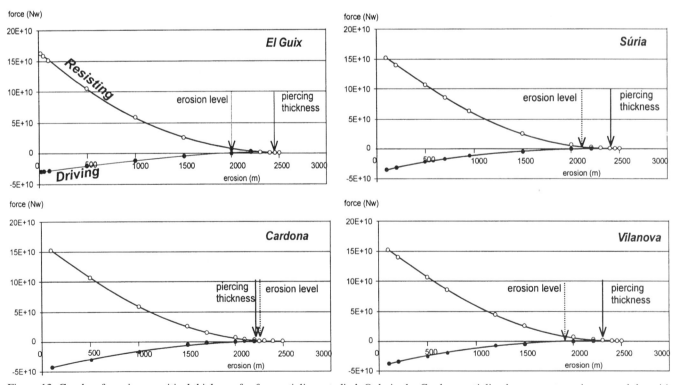

Figure 12. Graphs of maximum critical thickness for four anticlines studied. Only in the Cardona anticline has present erosion passed the critical thickness. Solid circles indicate driving force curve and white circles indicate resisting force.

2. Diapirs in the study area and our models have short stems and small bulbs that spread when they reach the surface. In map view they are elongated parallel to the fold trend and pierce the hinges of anticlines.

3. The Pinós-Cardona anticline in the southeastern Pyrenees has a salt accumulation in its core that is five times greater than the initial thickness, and differential unloading is eight times smaller in the synclines than in the anticline crest.

4. The age of piercing of the Cardona diapir can be bracketed by estimating the critical thickness of the overburden. The age is younger than 2 Ma, giving a postcompressional origin for the diapir.

5. Estimation of the critical thickness suggests that the Vilanova and Súria anticlines will pierce when the overburden reaches between 200 m and 100 m and <100 m for El Guix anticline.

ACKNOWLEDGMENTS

This work has been partially financed by Comissió Interdepartamental de Reurcai Innovació Tecnológica (CIRIT) grant 1997BEA/200109 from the Generalitat de Catalunya, and the Swedish Natural Research Council (NFR) funded Koyi. We thank C.J. Talbot for his useful comments on the final version of the manuscript and two reviewers, I. Alsop and I. Davison, and editor N. Mancktelow, for their suggestions and comments; A. Marcuello, F. Sabat, and A. Amilibia for comments on the numerical approach; and E. Roca for his input in the dating of the Ebro basin sediments. One of the models discussed here was prepared at the Bureau of Economic Geology at Austin during a postdoctoral stay. We also thank Springer-Verlag and F. Mauthe for permission to publish the section of the Cardona diapir.

REFERENCES CITED

Agustí, J., Anadón, P., Arbiol, S., Cabrera, L., Colombo, F., and Sáez, A., 1987, Biostratigraphical characteristics of the Oligocene sequences of northeastern Spain (Ebro and Campins Basins): Münchner Geowissenschaftliche Abhandlungen, v. A10, p. 35–42.

Alsop, G.I., 1996, Physical modelling of fold and fracture geometries associated with salt diapirism, in Alsop, G.I., et al., eds., Salt tectonics: Geological Society [London] Special Publication 100, p. 227–242.

Anadón, P., Cabrera, L., Colldeforns, B., and Sáez, A., 1989, Los sistemas lacustres del Eoceno superior y Oligoceno del sector oriental de la cuenca del Ebro: Acta Geologica Hispánica, v. 24, p. 205–231.

Ayora, C., Taberner, C., Pierre, C., and Pueyo, J.J., 1995, Modelling the sulphur and oxygen isotopic composition of sulphates through a halite-potash sequence: Implications for the hydrological evolution of Upper Eocene South-Pyrenean Basin: Geochimica et Cosmochimica Acta, v. 59, p. 1799–1808.

Bishop, R., 1978, Mechanism for emplacement of piercement diapirs: American Association of Petroleum Geologists Bulletin, v. 62, p. 1561–1581.

Bonifany, E., 1975, L'Ere quaternaire: Bulletin de la Société Géologique de France, v. 17, p. 380–393.

Busquets, P., Ortí, F., Pueyo, J.J., Riba, O., Rosell, L., Sáez, A., Salas, R., and Taberner, C., 1985, Evaporite deposition and diagenesis in the saline (potash) Catalan Basin, upper Eocene, in Milá, D., and Rosell J., eds., European Regional Geology Meeting (IAS), 6th: Lleida, Institue d' Estudis Ilerdencs, Excursion Guidebook, p. 13–59.

Coney, P., 1993, Syntectonic burial and post-tectonic exhumation of an active foreland thrust belt, southern Pyrenees, Spain, in Ortega-Gutierrez, F., et al., eds., First Circum-Pacific and Circum-Atlantic Terrane Conference: Guanajuato, Universidad Nacional Autónoma de México, p. 34–36.

Cotton, J., and Koyi, H.A., 2000, Modelling of thrust fronts above ductile and frictional décollements: Examples from the Salt Range and Potward Plateau, Pakistan: Geological Society of America Bulletin, v. 112, p. 351–363.

Daudré, B., and Cloetingh, S., 1994, Numerical modeling of salt diapirism: Influence of the tectonic regime: Tectonophysics, v. 240, p. 59–79.

de Ruig, M., 1992, Tectono-sedimentary evolution of the Prebetic fold belt of Alicante (SE Spain) [Ph.D. thesis]: Amsterdam, Brije Universiteit, 207 p.

Dixon, J.M., and Liu, S., 1992, Centrifuge modeling of the propagation of thrust faults, in McClay K.R., ed., Thrust tectonics: London, Chapman and Hall, p. 53–70.

Dixon, J., and Summers, J.M., 1985, Recent developments in centrifuge modeling of tectonic processes: Equipment, model construction techniques and rheology of model materials: Journal of Structural Geology, v. 7, p. 83–102.

Dixon, J., and Tirrul, R., 1991, Centrifuge modeling of fold-thrust structures in a tripartite stratigraphic succession: Journal of Structural Geology, v. 13, p. 3–20.

Gil, J., 1998, Modelización geodinámica y numérica de estructuras evaporíticas (cuencas Surpirenaica y Cantábrica) [Ph.D. thesis]: Barcelona, Universitat de Barcelona, 200 p.

Gugliemo, G., Jackson, M.P.A., and Vendeville, B., 1997, Three-dimensional visualization of salt walls and associated fault systems: America Association of Petroleum Geologists Bulletin, v. 81, p. 46–61.

Iglesias, M., 1970, Estudio estratigráfico y tectónico del area de Súria-Cardona [Ms. thesis]: Barcelona, Universitat de Barcelona, 77 p.

Jackson, M.P.A., and Talbot, C.J., 1987, External shapes, strain rates and dynamics of salt structures: Geological Society of America Bulletin, v. 97, p. 305–323.

Jackson, M.P.A., and Vendeville, B., 1994, Regional extension as a geologic trigger for diapirism: Geological Society of America Bulletin, v. 106, p. 57–73.

Koyi, H., 1988, Experimental modeling of role of gravity and lateral shortening in Zagros mountain belt: American Association of Petroleum Geologists Bulletin, v. 72, p. 1381–1394.

Koyi, H., 1996, Salt flow by aggrading and prograding overburdens, in Alsop, G.I., et al., eds., Salt tectonics: Geological Society [London] Special Publication 100, p. 243–258.

Koyi, H., 1998, The shaping of salt diapirs: Journal of Structural Geology, v. 20, p. 321–338.

Koyi, H., Jenyon, M.K., and Petersen, K., 1993, The effect of basement faulting on diapirism: Journal of Petroleum Geology, v. 16, p. 285–312.

Kupfer, D.H., 1968, Relationship of internal to external structure of salt domes, in Braunstein, J., and O'Brien, G.D., eds., Diapirism and diapirs: American Association of Petroleum Geologists Memoir 8, p. 79–89.

Kupfer, D.H., 1976, Shear zones inside Gulf Coast stocks help to delineate spines of movement: American Association of Petroleum Geologists Bulletin, v. 60, p. 1434–1447.

Letouzey, J., Coletta, B., Vially, R., and Chermette, J.C., 1995, Evolution of salt-related structures in compressional settings, in Jackson, M.P.A., et al., eds., Salt tectonics; a global perspective: American Association of Petroleum Geologists Memoir 65, p. 41–60.

Masachs, V., 1952, La edad y origen de los movimientos de las sales paleógenas de la cuenca del Ebro: Memoria y Comunicaciones del Instituto Geológico de Barcelona, v. IX, p. 51–65.

Masana, E., and Sans, M., 1995, Deformación neotectónica relacionada con anticlinales de núcleo salino (NE de la cuenca del Ebro, Barcelona): Geogaceta, v. 20, p. 846–849.

McClay, K., 1976, The rheology of plasticine: Tectonophysics, v. 33, p. T7-T15.

Meigs, A., Vergés, J., and Burbank, D.W., 1996, Ten-milion-year history of a thrust sheet: Geological Society of America Bulletin, v. 108, p. 1608–1625.

Miralles, L., Sans, M., Pueyo, J.J., and Santanach, P., 2000, 3D fabric characterization of an Internal shear zone (Cardona Diapir, Southern Pyrenees, Spain), in Vendeville, B., ed., Salt, shale and igneous intrusions in and around Europe: Geological Society [London] Special Publication 174, p. 149–167.

Muñoz, J.A., 1992, Evolution of a continental collision belt: ECORS-Pyrenees crustal balanced cross-section, in McClay, K., ed., Thrust tectonics: London, Chapman and Hall, p. 235–246.

Poliakov, A.N.B., Podladchikov, Y., Dawson, E.C.H., and Talbot, C.J., 1996, Salt diapirism with simultaneous brittle faulting and viscous flow, in Alsop, G.I., et al., eds., Salt tectonics: Geological Society [London] Special Publication 100, p. 291–302.

Pueyo, J.J., 1975, Estudio petrológico y geoquímico de los yacimientos potásicos de Cardona, Súria, Sallent (Barcelona, España) [Ph.D. thesis]: Barcelona, Universitat de Barcelona. 300 p.

Riba, O., 1975, Cardona, mapa Geologico de España, escala 1:50.000: Instituto Geológico y Minero de España, 58 p .

Riba, O., and Reguant, S., 1986, Una taula dels temps geològics: Institut d'estudis Catalans, Arxius de la secció de ciències LXXXI, 127 p.

Riba, O., Reguant, S., and Villena, J., 1983, Ensayo de síntesis estratigráfica y evolutiva de la cuenca terciaria del Ebro: Libro Jubilar J.M. Ríos: Geología de España, v. 7, p. 131–159.

Roca, E., Anadón, P., Utrilla, R., and Vázquez, A., 1996, Rise, closure and reactivation of the Bicorb-Quesa evaporite diapir, eastern Prebetics, Spain: Geological Society of London Journal, v. 153, p. 311–321.

Roca, E., Sans, M., and Koyi, H., 1999, The role of polyphase deformation on the growth of evaporitic diapirs: Experimental modeling of the Bicorp-Quesa diapir (Eastern Betics, Spain): London, Royal Holloway, Thrust Tectonics, Abstracts with Programs. p. 312.

Rosell, L., and Pueyo, J.J., 1997, Second marine evaporitic phase in the South Pyrenean foredeep: The Priabonian Potash Basin, in Busson, G., and Schreiber, B.C., eds., Sedimentary deposition in rift and foreland basins in France and Spain (Paleogene and lower Neogene): New York, Columbia University Press, p. 358–387.

Sáez, A., 1987, Estratigrafía y sedimentología de las formaciones lacustres del tránsito Eoceno Oligoceno del NE de la cuenca del Ebro [Ph.D. thesis]: Barcelona, Universitat de Barcelona, 245 p.

Sans, M., 1999, From thrust tectonics to diapirism. The role of evaportites in the kinematic evolution of the eastern south-Pyrenean front. [Ph.D. thesis]: Barcelona, Universitat de Barcelona, 197 p.

Sans, M., and Vergés, J., 1995, Fold development related to contractional salt tectonics: Southeastern Pyrenean thrust front, Spain, in Jackson, M.P.A., et al., eds., Salt tectonics: A global perspective: American Association of Petroleum Geologists Memoir 65, p. 369–378.

Sans, M., Muñoz, J.A., and Vergés, J., 1996, Triangle zone and thrust wedge geometries related to evaporitic horizons (southern Pyrenees): Canadian Petroleum Geology Bulletin, v. 44, p. 375–384.

Sans, M., Vergés, J., Gomis, E., Parés, J. M., Schiatarella, M., Travé, A., Calvet, F., Santanach, P., and Doulcet, A., 1999, Layer parallel shortening in salt-detached folds: Constraints on cross-section restoration: London, Royal Holloway, Thrust Tectonics, Abstracts with Programs, p. 321–324.

Schultz-Ela, D.D., Vendeville, B., and Jackson, M.P.A., 1993, Mechanics of active salt diapirism: Tectonophysics, v. 228, p. 275–312.

Serrano, A., and Martínez del Olmo, W., 1990, Tectónica salina en el Dominio Cántabro-Navarro: evolución, edad y origen de las estructuras salina, in Ortí, F., and Salvany, J. M., eds., Formaciones evaporíticas de la Cuenca del Ebro y cadenas periféricas, y de la zona de Levante: Barcelona, Universitat de Barcelona, p. 39–53.

Serrano, A., Hernáiz, P., Malagón, J., and Rodríguez Cañas, C., 1994, Tectónica distensiva y halocinesis en el margen SO de la cuenca Vasco-Cantábrica: Geogaceta, p. 131–134.

Solé-Sabarís, L., and Masachs, V., 1940, Edad de las terazas del río Cardoner en Manresa: VI Estudios Geomorfológicos de la Peninsula Ibérica, v. 2, p. 3–6.

Talbot, C.J., and Alavi, M., 1996, The past of a future syntaxis across the Zagros, in Alsop, G.I., et al., eds., Salt tectonics: Geological Society [London] Special Publication 100, p. 89–110.

Talbot, C.J., and Jackson, M.P.A., 1987, Internal kinematics of salt diapirs: American Association of Petroleum Geologists Bulletin, v. 71, p. 1068–1093.

Talbot, C.J., Koyi, H., Sokoutis, D., and Mulugeta, G., 1988, Identification of evaporite diapirs formed under the influence of horizontal compression: A discussion: Canadian Petroleum Geology Bulletin, v. 36, p. 91–95.

Twiss, R.J., and Moores, E.M., 1992, Structural geology: New York, W.H. Freeman and Company, 532 p.

Underhill, J.R., 1988, Triassic evaporites and Plio-Quaternary diapirism in western Greece: Geological Society of London Journal, v. 145, p. 269–282.

Vendeville, B., 1991, Thin-skinned compressional structures above frictional-plastic and viscous decollement layers: Geological Society of America Abstracts with Programs, v. 23, no. 5, p. A423.

Vendeville, B., and Jackson, M.P.A., 1992, The rise of diapirs during thin-skinned extension: Marine and Petroleum Geology, v. 9, p. 331–353.

Vendeville, B., and Nilsen, K., 1995, Episodic growth of salt diapirs driven by horizontal shortening: Gulf Coast Section Society of Economic Paleontologist and Mineralogist Foundation 16th Annual Research Conference, Salt, Sediment and Hydrocarbons: Houston, Texas, Society of Economic Paleontologist and Mineralogist Foundation, p. 285–295.

Vergés, J., 1993, Estudi geològic del vessant sud del Pirineu Oriental i Central. Evolució cinemàtica en 3D [Ph.D. thesis]: Barcelona, Universitat de Barcelona, 200 p.

Vergés, J., Muñoz, J.A., and Martínez, A., 1992, South Pyrenean fold-and-thrust belt: Role of foreland evaporitic levels in thrust geometry, in McClay, K.R., ed., Thrust tectonics: London, Chapman and Hall, p. 255–264.

Vergés, J., Marzo, M., Santaularia, T., Serra-Kiel, J., Burbank, D.W., Muñoz, J.A., and Gimenez-Montsant, J., 1998, Quantified vertical motions and tectonic evolution of the SE Pyrenean foreland basin, in Mascle, A., et al., eds., Cenozoic foreland basins of western Europe: Geological Society [London] Special Publication 134, p. 107–134.

Vially, R., Letouzey, J., Bernard, F., Haddadi, N., Deforges, G., Askri, H., and Boudjema, A., 1994, Basin inversion along the North African margin. The Sahara Atlas (Algeria), in Roure, F., ed., Peri-tethyan platforms: Paris, Editions Technip, p. 79–118.

Wagner, G., Mauthe, F., and Mensik, H., 1971, Der Salztock von Carona in Nordostpanien: Geologische Rundschau, v. 60, p. 970–996.

Waltham, D., 1997, Why does salt start to move?: Tectonophysics, v. 282, p. 117–128.

Wiltschko, D.V., and Chapple, W.M., 1977, Flow of weak rocks in Appalachian Plateau folds: American Association of Petroleum Geologists Bulletin, v. 61, p. 653–670.

MANUSCRIPT ACCEPTED BY THE SOCIETY APRIL 12, 2000

Printed in the U.S.A.

Diapirism in convergent settings triggered by hinterland pinch-out of viscous decollement: A hypothesis from modeling

Elisabetta Costa
Dipartimento di Scienze della Terra, Università di Parma, Italy
Bruno Vendeville
Bureau of Economic Geology, University of Texas at Austin, Texas 78713-8924 USA

Abstract

We use a series of systematic experiments to illustrate the effect of slow shortening of a brittle cover overlying a viscous shaly or evaporitic decollement and to test the parameters controlling diapirism in contractional settings. All experiments were run under identical conditions except for two varying parameters, (1) the cover/decollement thickness ratio and (2) the presence or absence of a frictional strip in the hinterland of the model. Models with no frictional strip comprised a continuous basal layer of viscous polymer and were deformed by moving a rigid plexiglass wall. In others models, a narrow strip, made of dry sand overlying glass microbeads, was intercalated between the moving wall and the model. The base of the glass bead layer acted as a low-angle detachment plane, in contrast with the polymer layer that deformed as a decollement layer. During shortening this strip deformed as a frictional wedge.

In models having no frictional wedges, the brittle layer was entirely detached from its base and deformed by box folds bounded by pairs of conjugate kink bands. The underlying viscous decollement layer thickened but never rose diapirically.

In models with a frictional wedge, a fault-propagation fold formed in front of the wedge. The viscous layer detached at or near its base and was incorporated into the hanging wall of the fold and passively carried upward to shallower levels. As the fold grew and tightened, its crest was thinned by normal faulting and slumping, which allowed the viscous material to pierce and emerge. Experiments also indicate that a minimum thickness is required for the viscous layer to rise diapirically.

We compare our model results with data from the eastern Mediterranean, where shale diapirs occur south of Crete in the crest of the Mediterranean Ridge. Diapirs there form elongate trends parallel to structures within the ridge, in front of the continental backstop, where the shaly decollement probably pinches out northward. This setting is comparable to that of the inner part of the models that compose a frictional wedge in the hinterland. Conversely, diapirs are absent in the more external part of the Mediterranean Ridge, where deformation is dominated by box folding. This situation is comparable to that of (1) models with no frictional hinterland wedge and (2) the outer part of the other models far away from the frictional hinterland wedges.

INTRODUCTION

Although diapirism has long been recognized and studied in extensional basins and passive continental margins, diapirism in convergent settings has received much less attention. Salt diapirism, in particular, has been widely associated with thin-skinned extension, to which it seems closely linked in space and time (Bishop, 1978; Woodbury et al., 1980; Vendeville and

Costa, E., and Vendeville, B., 2001, Diapirism in convergent settings triggered by hinterland pinch-out of viscous decollement: A hypothesis from modeling, *in* Koyi, H.A., and Mancktelow, N.S., eds., Tectonic Modeling: A Volume in Honor of Hans Ramberg: Boulder, Colorado, Geological Society of America Memoir 193, p. 123–130.

Jackson, 1992). Salt diapirs are typically initiated during the rift or drift phase of divergent continental margins, where thick-skinned or thin-skinned extension is caused by crustal tectonics or by gravity gliding or spreading down the continental slope.

Unlike salt diapirs, mud diapirs are most commonly found in modern and ancient accretionary wedges and the mechanism by which they form and rise in convergent settings is still unclear. To investigate these problems we used analogue experiments of shortening of a brittle cover overlying a viscous decollement and compared the experimental results with data from the Mediterranean Ridge to help better understand the mechanics and kinematics of mud diapirism in collisional and accretionary contexts.

ACCRETIONARY WEDGES

General characteristics

Since the late 1980s, accretionary wedges have been the subject of intensive investigations. There is a broad consensus that fluids in accretionary wedges play a major role in the dynamics and the deformation style. This is because rapid tectonic loading of wet sediments causes fluid overpressure, which decreases the effective strength of the sediments and thereby creates zones of weakness (Langseth and Casey Moore, 1990). Although it is generally agreed that fluid overpressure due to the accretion of wet, undercompacted sediments is the most important factor controlling mud behavior in modern accretionary prisms, many questions remain unanswered. What is the mechanism driving mud rise and extrusion? What are the major conduits for mud expulsion? How much does diapirism vary between accretionary prisms having different sediment characteristics and thicknesses, different stress and strain distribution within the wedge, and varying convergence rates and angles?

Mud diapirism

Brown (1990), following Ramberg (1967, 1968a, 1968b), inferred that mud diapirism was primarily driven by buoyancy forces related to the density contrast between the overpressured muddy mass and the denser overburden, both of which were assumed to behave viscously. However, if the sediment overburden is assumed to behave as a brittle, frictional-plastic material that obeys a Mohr-Coulomb criterion of failure, as most sedimentary rocks do in the upper continental crust (Weijermars et al., 1993), buoyancy alone is unlikely to allow for spontaneous diapir rise. Vendeville and Jackson (1992) and Schultz-Ela et al. (1993) showed that spontaneous rise of active diapirs driven by buoyancy alone is effectively prevented by the strength of overburden roofs thicker than one-third of the regional overburden thickness. Spontaneous diapiric rise requires differential loading of the source layer (i.e., the overburden must be locally thinner), which can result from thin-skinned extension and normal faulting, erosion, or differential depositional loading. Furthermore, if mud rises by vertically hydrofracturing the brittle sediments, the minimum fluid pressure necessary to keep vertical fractures open must exceed the minimum compressive stress, assumed here to be horizontal. In an accretionary setting, regional tectonics impose that the maximum compressive stress in accretionary wedges is often, at least episodically, horizontal (Byrne and Fisher, 1990), which would close vertical fractures and thus further oppose diapir rise and extrusion.

It is therefore important to determine whether sediments in an accretionary prism obey brittle or viscous behavior. Seimic-reflection data from accretionary wedges show widespread occurrence of brittle structures such as faults and fault-related folds, often rooted at depth into a main basal decollement. These observations clearly suggest that rocks in accretionary prisms behave as brittle, rather than viscous, materials (Karig and Lundberg, 1990).

Basal decollement zone

Although the mechanical properties of accretionary wedges are still debated, the properties of decollement zones are well documented and indicate that the decollement is almost always an overpressured horizon located at the top of the underthrust sediments (Langseth and Casey Moore, 1990; Byrne and Fisher, 1990). The forward decrease in normal-incidence traveltimes that is commonly observed in the underthrust section of the deformation front, is regarded as symptomatic of high pore pressure (Shipley at al., 1990; Moore et al., 1990). Estimated pressures generally exceed 70%, and are often >90% ($\lambda = 0.7 - 0.9$), of the lithostatic pressure (Davis et al., 1983). What seems to dramatically influence the overpressure buildup in the decollement zone is the presence of an impermeable unit in the underthrust section. Where the prism comprises only permeable rocks, fluids can readily migrate vertically from the deepest parts of the prism or from the underthrust section, thus preventing overpressure (Western Nankai trough; Moore et al., 1990). By contrast, where an impermeable unit is present, as in the Barbados Ridge complex, fluids appear to migrate laterally seaward, rather than vertically through the overlying wedge (Bangs et al., 1990). As shown by Davis et al. (1983), the low taper angle typical of accretionary wedges is symptomatic of the presence of a weak basal decollement, the weakness of which is caused by excess pore pressure. Where the fluid-pressure ratio is high (i.e., $\lambda = 1$), the strength of the overpressured rock is determined mostly by rock cohesion, rather than by the angle of internal friction or shape of the Mohr-Coulomb envelope. Therefore, the overpressured rocks at the base of accretionary wedges effectively behave as weak, pressure independent, materials. The two latter properties—low strength and pressure independence—are also characteristic of low-viscosity fluids deformed under low strain rates. There, viscous forces become negligible compared with pressure forces in the decollement layer and surface forces in the overlying brittle cover.

ANALOGUE MODELS

Model setup

We took into account the above properties of sediments in accretionary wedges and of their basal decollement layers to design a series of dynamically scaled physical experiments. Our models are simplified replicas of their more complex natural examples in accretion-subduction settings. We assume that sediments are brittle and deform according to a Mohr-Coulomb criterion of failure, whereas the overpressured decollement layer behaves as a weak, pressure-indipendent, material. Such an assumption seems appropriate for most modern accretionary wedges.

Although the physical process of fluid overpressure could not be reproduced in experiments, we modeled its mechanical impact, i.e., that the wedge overlies a considerably weaker decollement layer, by using a basal layer made of low-viscosity silicone polymer (EL Polymer NA, manufactured by Waker Silicone, USA) and deforming it under low strain rates. At low strain rates, the surface forces in a low-viscosity material are much lower than pressure forces. Thus the material behaves like a weak, incompressible, pressurized fluid.

Our models comprised a layer of dry sand, representing the brittle sediment overburden, overlying a viscous decollement of silicone polymer, simulating overpressured shale. The mechanical and physical properties of the materials used in the models are listed in Table 1. The geometric and kinematic characteristics of the models are listed in Figure 1 and Table 2. In all experiments, shortening was applied by slowly moving the left end wall.

Experiments were run under identical conditions except for varying two parameters. We varied the cover/decollement thickness ratio by changing the thickness of the viscous layer. We also used two different geometries for the initial source layer: in the first set of models (type I models: experiments 368a and 368b), the layer of viscous silicone extended across the entire length of the model and was in contact with the moving wall (Fig. 1).

By contrast, in the second set of models (type II models: experiments 369 and 371), the viscous decollement did not extend across the entire model but stopped about 10 cm from the moving wall (Fig. 1). Within this 10-cm-wide strip, the decollement layer was made of glass microbeads, a frictional-plastic material weaker than the overlying sand but considerably stronger than the viscous silicone. During shortening, the strip underlain by the bead layer behaved as a hinterland thrust sheet that detached at the base of the bead layer, thickened as a wedge, and eventually transmitted the shortening forward to the brittle cover underlying the viscous decollement. This frictional wedge behaved as a backstop, which (Byrne et al., 1993, p. 123) "is a region within a forearc that has significantly greater shear strenght than the sediments lying farther trenchward."

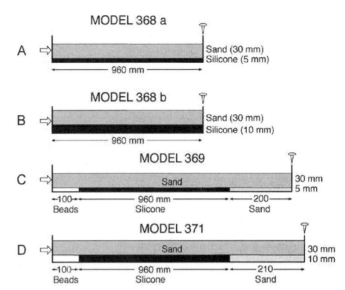

Figure 1. Initial setup for experiments. A, B: Type I models (368a and 368b), having continuous viscous decollement. C, D: Type II models (369 and 371) having frictional strip located between model and moving wall.

Model results

In type I models (no frictional wedge) and in the external part of type II models (far away from the frictional wedge: see comments below) the overburden deformed as box folds bounded by pairs of conjugate kink bands, the axial planes of which converged near the top of the decollement. During this stage, corresponding to about 20% shortening, the brittle layer was detached from its base along the entire model length. Later the kink bands evolved into genuine thrust planes, and additional shortening was accommodated by slip along these planes. Shortening was often accommodated by preferential slip along one of each pair of conjugate thrusts, causing the folds to become asymmetric and their roofs to tilt. During this late deformation stage the detachment level deepened and was located within the silicone layer. There was no significant stretching of the hinges of the box folds. The underlying viscous layer thickened regionally

TABLE 1. MECHANICAL AND PHYSICAL PROPERTIES OF THE MATERIALS USED IN THE MODELS

Material	Density (g/cm³)	Grain size (μm)	Angle of internal friction	Dynamic shear viscosity η (Pa s)
Sand	1.75	300	30°	- - - -
Silicone	0.976	- - - -	- - - -	2×10^4

TABLE 2. GEOMETRIC AND KINEMATIC CHARACTERISTICS OF THE MODELS

Models	Silicone thickness (mm)	Number of structures	Total shortening	Shortening accomplished by growth and tightening of early folds
368 a	5	6/7	31%	11%
368 b	10	5/6	32%	2%
369	5	8	35%	13%
371	10	5	36%	24%

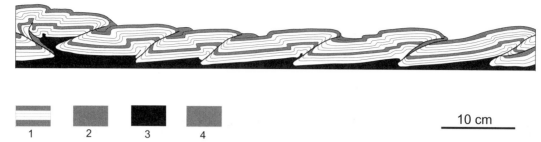

Figure 2. Regional vertical section through model 368a (type I) cut at end of experiment and located at center of model, to avoid edge effects from model sidewalls. This model showed initially symmetric box fold anticlines. Initially symmetric box folds became asymmetric during further shortening. 1, sand layers; 2, glass microbeads; 3, silicone; 4, slumped sand and/or fault gouge.

but never rose diapirically (Figs. 2 and 3). We infer that the high horizontal compressive stress causing box folding prevented the silicone from piercing the box fold. Instead, the silicone was expelled downward from the core of the anticlines. Locally, pairs of downward-diverging kink planes intersected within the brittle layer, thus defining narrow synclines that were forcefully pushed downward, below the regional datum, during further shortening.

In type II models (with frictional wedge: experiments 369 and 371), a fault-related fold formed in front of the wedge. Unlike box folds, fault-related folds appear to have detached at or near the base of the viscous layer. The viscous silicone was incorporated into the hanging wall and thus passively carried on top of the footwall (Figs. 4 and 5). During additional shortening, the fault-related folds grew continuously and the viscous silicone moved to higher levels. In type II models, all structures formed early (before 10% shortening), whereas additional shortening is merely accommodated by growth and amplification of these early structures (Table 2). As the fault-related folds grew and tightened, normal faulting and slumping at the fold front stretched and thinned the fold hinges, which eventually allowed the viscous silicone to pierce and emerge (Fig. 6). Model results also indicate that, although the mechanism of nucleation of the most internal fold above the viscous layer was controlled by the presence of a frictional hinterland wedge, a minimum thickness was also required for the viscous layer to emerge. Only type II models in which the source layer was initially thick had emergent diapirs (cf. Figs. 4 and 5).

The differences between the type I and type II models can be explained using the analytical models of stress balance by Davis et al. (1983), Platt (1988, 1990), and Byrne et al. (1993). Analytical models indicate that the main parameters controlling the location of the detachment within the source layer are (1) the existence of a transition from weak decollement to a hinterland wedge of stronger material (backstop) and (2) the presence of surface slope in the fold belt. In type I models (no frictional wedges) shortening was distributed throughout the entire model length; therefore no regional topography was created, except locally above the anticlines and fault footwalls. Therefore, stresses remained evenly distributed throughout the entire model (e.g., models 368a and 368b) and the main detachment was located at or near the top of the viscous decollement layer, at least during the early stage of fold formation. The viscous layer deformed passively in response to regionally uniform shortening, folding, and thrusting of the overlying sediments.

By contrast, in type II models, the frictional strip in front of the moving wall thickened and became an asymmetric forethrust wedge detaching at the base of the glass beads layer (experiment 369 and 371). Following Byrne et al. (1993), this wedge represents a type 2 backstop, i.e., a backstop whose contact with the accretionary wedge dips landward. Further shortening thickened the thrust sheet and steepened the topographic slope created by the wedge. Results from type II models (with frictional wedges) suggest that, as the thrust wedge moved forward the basal detachment propagated into the viscous silicone layer, forcing this part of the model to detach at or near the base of the silicone. The volume of silicone located above this detachment is effectively incorporated into the hanging wall of the structures and passively carried upward to shallower depths during further shortening.

ANALOGY TO NATURAL EXAMPLES

Type II models (with frictional strips) are analogous to the most internal basin part being incorporated in the front of an ac-

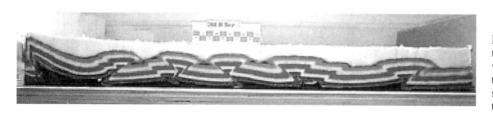

Figure 3. Regional vertical section through model 368b (type I) where viscous/brittle thickness ratio was twice that of model 368a. Although initially symmetric box fold became asymmetric, there is no regionally preferential sense of asymmetry in this model.

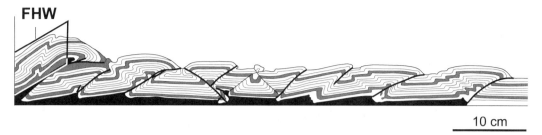

Figure 4. Regional vertical section through model 369 (type II). Note difference in geometry between fault-related fold in front of hinterland wedge (FHW, left side), and box folds on right side of model, away from frictional hinterland wedge. Symbols as in Figure 2.

Figure 5. Regional vertical section through model 371 (type II). Viscous layer was twice as thick as in model 369. More fault-related folds formed (arrows). However, only leftmost fault-related fold, located in front of hinterland wedge (FHW, left side), kept growing until end of experiment. Symbols as in Figure 2.

cretionary prism, and where shaly or evaporitic decollements pinch out against the advancing thrust sheets. Type I models (no frictional strips) are analogous to the external part of the same basins, far away from the hinterland wedges, and hence not sensitive to their influence.

We compare our model results to the Mediterranean Ridge, one of the most studied accretionary complex whose rich and updated literature is available.

Mediterranean Ridge

The Mediterranean Ridge formed in response to slow convergence of the African and Eurasian plates starting 40–46 Ma (Spakman et al., 1988). Shortening along the accretionary front was

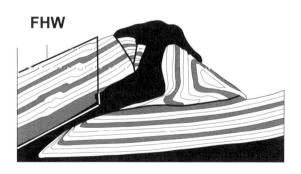

Figure 6. Detail from model 371 showing diapir piercing core of fold that formed in front of frictional hinterland wedge (FHW). Normal faulting and slumping thinned fold hinge and crest, unroofing viscous silicone and allowing it to emerge. Symbols as in Figure 2.

controlled not only by the motion of Africa and Eurasia, but also by the motion of the Aegean microplate. The mean rate of convergence is about 40 mm/yr (Le Pichon et al., 1995). The plate boundary at the front of the wedge (south) corresponds to the base of the Messinian evaporitic section, whereas the plate boundary deepens northward and cuts down to the top of the Mesozoic carbonate section (Fig. 7; Mascle and Chaumillon, 1998; Reston et al., 2000). New seismic reflection and refraction data (IMERSE project, Reston et al., in press) show that the western Mediterranean Ridge comprises two wedges: a pre-Messinian wedge that has been underplated beneath a Messianian accretionary wedge (Fig. 8). The Mediterranean Ridge is bounded landward (north) by a continental backstop, believed to consist of indurated sediments of the Hellenic nappes (Fig. 9; Robertson and Kopf, 1998; Lallemant et al., 1994).

Some questions remain unanswered about mud diapirism in the Mediterranean Ridge. Why is mud diapirism restricted along dip to the inner part of the ridge, and why did diapirs emerge only 1–2 Ma, whereas the accretionary prism has been active since the Oligocene? Answers to these questions can provide further insights about the geometry, kinematics, and dynamics of the Mediterranean Ridge.

Analogy between models and the Mediterranean Ridge

Although our models are greatly simplified analogues of natural accretionary prisms, their geometry and kinematics show some striking similarities with the Mediterranean Ridge. In both models and the Mediterranean Ridge, diapirs are located in the innermost part of the accretionary wedge, just in front of the frictional wedge, represented in nature by the Hellenic

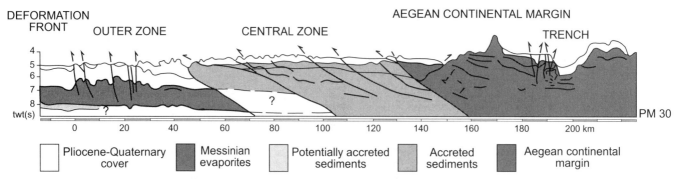

Figure 7. Simplified interpretation of cross section of eastern part of Mediterranean Ridge (modified from Mascle and Chaumillon, 1998); twt is two-way travel time.

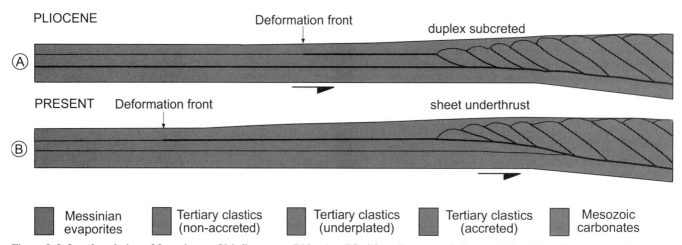

Figure 8. Inferred evolution of frontal part of Mediterranean Ridge (modified from Reston et al., in press). Basal decollement on Aptian shales that took place in Tertiary was later replaced by new detachment at base of Messinian evaporitic section (A). Farther north new detachment is believed to cut down to old decollement, causing underplating of Tertiary clastic sequence (B). Not to scale.

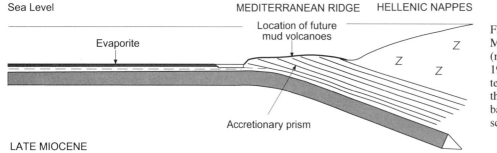

Figure 9. Schematic cross section across Mediterranean Ridge in late Miocene (modified from Robertson and Kopf, 1998). As in models 396 and 371, hinterland wedge (Hellenic nappes) overthrust accretionary prism (type 2 backstop of Byrne et al., 1993). Not to scale.

nappes. Diapirs in both nature and models emerge only during the latest stages of deformation (i.e., 1–1.7 Ma in the 40 Ma Mediterranean Ridge, and during the last 2 hrs, after 60–70 hours of deformation in experiments). The structure style, geometry, and sizes are similar in both models and the Mediterranean Ridge, at least in their central and outer parts (Figs. 10 and 11). In the inner part of the ridge it is impossible to confidently determine the actual geometry of the structures because of the poor quality of acoustic imaging. However, subbottom profiling intersecting some of the diapirs there (Robertson et al., 1996) indicates that diapirs are laterally bounded by almost vertical, shallow normal faults, as are the diapirs in our models.

CONCLUSIONS

Our experimental results show that, in contractional settings, diapirs are located where the viscous decollement pinches out against the frictional hinterland wedge (backstop). There, the stress gradient caused by the wedge topographic slope forces the system to detach at the base of the viscous layer. Some of the

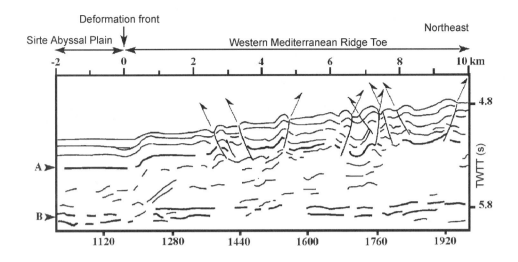

Figure 10. Similarities in geometry and size between structures in models and structures in western part of Mediterranean Ridge. Line drawings from Chaumillon et al. (1996). A: Base of Pliocene-Quaternary sediments. B: Base of Messinian evaporites. Length ratio of scaling = 10^{-5} (1 cm = 1 km). TWTT is two-way traveltime in seconds.

source layer becomes incorporated into the hanging wall of fault-related folds and is carried passively to shallower levels as the thrust moves. As additional shortening tightens the folds and stretches their hinges (Fig. 6), the viscous material in the hanging wall can pierce the thinned cover and eventually emerge. Because contractional structures in basins underlain by viscous decollements propagate out of sequence, the most internal stuctures can continue to grow late in the overall evolution of the accretionary wedge, even after more external structures have already formed.

We infer that the model results can answer some of the questions on both the location and timing of diapir emergence in the Mediterranean Ridge.

In the Mediterranean Ridge area, emergent diapirs are mainly located in the most internal part of the ridge and are geographically associated with the wedge formed by the lithified sediments of the Hellenic nappes, which are analogous to the hinterland thrust wedge in the type II models. We suggest that, as the Hellenic nappes advanced and reached the cover overlying the overpressured mud decollement, it detached along its base and deformed into fault-related folds that trapped and passively carried some of the mud to shallower depth. Eventually, combined stretching and thinning of the fold hinges as the fold tightened, and the progressive decrease in the depth of burial of the volume of overpressure mud, allowed mud to spontaneously rise and emerge at the seafloor.

In conclusion, folding above the pinch out of the viscous decollement appears to be related to the presence of a frictional

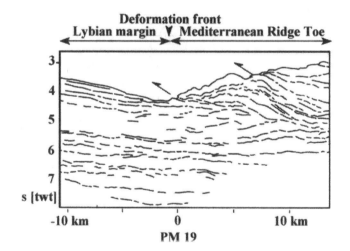

Figure 11. Similarities in geometry and size between structures in models and structures located at front of central part of Mediterranean Ridge, facing Lybian continental margin. Line drawings are from Mascle and Chaumillon (1998). In this seismic line these authors were unable to establish nature and age of detachment. Length ratio of scaling = 10^{-5} (1 cm = 1 km). s[twt] is seconds, two-way traveltime.

hinterland wedge (backstop). The rise and emergence of mud diapirs are driven by tectonics, rather than by buoyancy. Deep-seated mud masses rise passively only, as a part of the hanging wall of the main thrust planes. Then, mud-diapir piercement and extrusion are enhanced by tectonic thinning of the fold hinges. Only later, as the overpressured mud rises closer to the seafloor, can decompression cause rapid methane expansion, thereby promoting active rise and emergence of the mud.

ACKNOWLEDGMENTS

We thank Calvin Cooper and Alina Polonia for their helpful comments, the two reviewers, M.P. Coward and A. Teixell, whose suggestions greatly improved the chapter. We also thank Edvige Masini and Paolo Murelli for drawing the figures.

REFERENCES CITED

Bangs, N.L.B., Westbrook, G.K., Ladd, J.W., and Buhl, P., 1990, Seismic velocities from the Barbados Ridge Complex: Indicators of high pore fluid pressures in an accretionary complex: Journal of Geophysical Research, v. 95, p. 8767–8782.

Bishop, R.S., 1978, Mechanism for emplacement of piercement diapirs: American Association of Petroleum Geologists Bulletin, v. 62, p. 1561–1583

Brown, K.M., 1990, The nature and hydrogeologic significance of mud diapirs and diatremes for accretionary systems: Journal of Geophysical Research, v. 95, p. 8969–8982.

Byrne, T., and Fisher, D., 1990, Evidence for a weak and overpressured decollement beneath sediment-dominated accretionary prisms: Journal of Geophysical Research, v. 95, p. 9081–9098.

Byrne, D.E., Wang, W., and Davis, D.M., 1993, Mechanical role of backstops in the growth of forearcs: Tectonics, v. 12, p. 123–144.

Chaumillon, E., Mascle, J., and Hoffmann, H.J., 1996, Deformation of the western Mediterranean Ridge: Importance of Messinian evaporitic formation: Tectonophysics, v. 263, p. 163–190.

Davis, D., Suppe, J., and Dahlen, F.A., 1983, Mechanics of fold-and-thrust belts and accretionary wedges: Journal of Geophysical Research, v. 88, p. 1153–1172.

Jackson, M.P.A., and Vendeville, B.C., 1994, Regional extension as a geologic trigger for diapirism: Geological Society of America Bulletin, v. 106, p. 57–73.

Karig, D.E., and Lundberg, N., 1990, Deformation bands from the toe of the Nankai accretionary prism: Journal of Geophysical Research, v. 95, p. 9099–9110.

Lallemant, S., Truffert, C., Jolivet, L., Henry, P., Chamot-Rooke, N., and de Voogd, B., 1994, Spatial transition from compression to extension in the Western Mediterranean Rridge accretionary complex: Tectonophysics, v. 234, p. 33–52.

Langseth, M.G., and Casey Moore, J., 1990, Introduction to special section on the role of fluids in sediment accretion, deformation, diagenesis and metamorphism in subduction zones: Journal of Geophysical Research, v. 95, p. 8737–8741.

Le Pichon, X., Chamot-Rooke, N., Lallemnant, S., Noomen, R., and Veis, G., 1995, Geodetic determination of the kinematic of central Geece with respect to Europe: Implications for Eastern Mediterranean tectonics: Journal of Geophysical Research, v. 100, p. 12675–12690.

Mascle, J., and Chaumillon, E., 1998, An overview of Mediterranean Ridge collisional accretionary complex as deduced from multichannel seismic data: Geo-Marine Letters, v. 18, p. 81–89.

Moore, G.F., Shipley, T.H., Stoffa, P.L., Karig, D.E., Taira, A., Kuramoto, S., Tokuyama, H., and Suyehiro, K., 1990, Structure of Nankai trough accretionary zone from multichannel seismic reflection data: Journal of Geophysical Research, v. 95, p. 8753–8766.

Platt, J.P., 1988, The mechanics of frontal imbrication. A first order analysis: Geologische Rundschau, v. 77, p. 577–589.

Platt, J.P., 1990, Thrust mechanics in highly overpressured accretionary wedge: Journal of Geophysical Research, v. 95, p. 9025–9034.

Ramberg, H., 1967, Gravity, deformation and the Earth's crust: San Diego, California, Academic Press, 214 p.

Ramberg, H., 1968a, Fluid dynamics of layered systems in the field of gravity, a theoretical basis for certain global structures and their isostatic adjustement: Physics of the Earth and Planetary Interiors, v. 1, p. 63–87.

Ramberg, H., 1968b, Instability of layered systems in the field of gravity, I and II: Physics of the Earth and Planetary Interiors, v. 1, p. 427–474.

Reston, T.J., von Huene, R., Babassilas, D., Camerlenghi, A., Cernobori, L., Chamot-Rooke, N., Dickmann, T., Fruehn, J., Jones, K., Lallemant, S., Le Meur, S.D., Le Pichon, X., Lonergan, L., Loucoyannakis, M., Nicolich, R., Pascal, G., Tay, P.L., Warner, M., and Westbrook, G., The IMERSE project. An overview of the main results: Marine Geology (in press).

Robertson, A.H.F., and Kopf, A., 1998, Tectonic setting and processes of mud volcanism on the Mediterranean Ridge accretionary complex: evidence from LEG 160, in Robertson, A.H.F., et al., Proceedings of the Ocean Drilling Program, Scientific results, Volume 60: College Station, Texas, Ocean Drilling Program, p. 655–680.

Robertson, A.H.F., Emeis, K.C., Richter, C., Blanc-Valleron, M.M., Buloubassi, I., Brumsack, H.J., Cramp, A., De Lange, C.J., Di Stefano, E., Flecker, R., Frankel, E., Howell, M.W., Janecek, T.R., Jurado-Rodriguez, M.J., Kemp, A.E.S., Koisumi, I., Kopf, A., Major, C.O., Mart, Y., Pribnow, D.F.C., Rabaute, A., Roberts, A., Rullkotter, J.H., Sakamoto, T., Spezzaferri S., Staerker, T.S., Stoner, J.S., Whiting, B.M., and Woodside, J.M., 1996, Mud volcanism on the Mediterranean Ridge: Initial results of Ocean Drilling Program Leg 160: Geology, v. 24, p. 239–242.

Shipley, T.H., Stoffa, P.L., and Dean, D.F., 1990, Underthrust sediments, fluid migration paths and mud volcanoes associated with the accretionary wedge off the Costa Rica: Middle America trench: Journal of Geophysical Research, v. 95, p. 8753–8766.

Shultz-Ela, D.D., Jackson, M.P.A., and Vendeville, B.C., 1993, Mechanics of active salt diapirism: Tectonophysics, v. 228, p. 275–312.

Spakman, W., Wortel, M.J.R., and Vlaar, N.J., 1988, The Hellenic subduction zone: A tomographic image and its implications: Geophysical Research Letters, v. 15, p. 60–63.

Vendeville, B.C., and Jackson, M.P.A., 1992, The rise of diapirs during thin-skinned extension: Marine and Petroleum Geology, v. 9, p. 331–353.

Weijermars, R., Jackson, M.P.A., and Vendeville, B.C., 1993, Rheological and tectonic modelling of salt provinces: Tectonophysics, v. 217, p. 143–174.

Woodbury, H.O., Murray, I.B., and Osborne, R.E., 1980, Diapirs and their relation to hydrocarbon accumulation, in Facts and principles of world petroleum occurrence: Miall, A.D., ed., Calgary, Canadian Society of Petroleum Geologists, p. 119–192.

MANUSCRIPT ACCEPTED BY THE SOCIETY APRIL 12, 2000

Salt tectonics and sedimentation along Atlantic margins: Insights from seismic interpretation and physical models

Webster Ueipass Mohriak
Petrobras, Exploration and Production—GEREX/GESIP, Rio de Janeiro, Brazil
Peter Szatmari
Petrobras, Cenpes, Rio de Janeiro, Brazil

ABSTRACT

The interpretations of authochthonous and allochthonous salt tectonics in the sedimentary basins in the South Atlantic and Gulf of Mexico have undergone major changes as a result of improved seismic imaging of complex structures in the deep water regions, with important consequences for petroleum exploration. The interpretations of these features have been greatly aided by comparative analysis with analogue structures from diverse salt basins worldwide and by sandbox physical models developed to simulate the geodynamic evolution of selected structures. This work discusses a number of elusive structures observed on seismic data of selected basins in Atlantic-type divergent margins, including collapse of previous salt diapirs forming pseudoturtlebacks, pseudodownlaps associated with massive clastic progradation, and antithetic faults related to steps in the basement (forced folds) or to postextension block rotation compensating associated synthetic faults. The seismic interpretation and the insights from physical models together with analogies from other sedimentary basins indicate that major evacuation grabens controlled by landward-dipping faults may be formed by extensional tectonics affecting the overburden and basinward salt flow. Reactive salt diapirs form that may evolve to active diapirs and eventually collapse with the continuation of the extensional processes, particularly during periods of rapid clastic progradation. Major listric antithetic normal faults detaching at the base of evaporite layers may be formed by basinward salt flow during episodes of massive clastic progradation, resulting in pseudodownlaps of basinward-younging sequences and large stratigraphic gaps between the overburden and the salt weld. Systematic arrays of antithetic normal faults may be formed by forced folding of the overburden above a step in the basement and by inversion of the rotation angle caused by touchdown of the overburden blocks on the presalt layers.

INTRODUCTION

Petroleum exploration in deep-water frontier regions increasingly relies on structural and stratigraphic plays associated with salt tectonics, particularly in the Gulf of Mexico and in the South Atlantic. Recent advances in the understanding of salt tectonics and sedimentation are associated with technological improvements in seismic reflection acquisition and processing, physical modeling using different materials to simulate salt and overburden, and mathematical modeling, including simulation of physical processes by finite elements (Jackson, 1995). The stratigraphic and structural elements associated with halokinetic features may be visualized both in two and three dimensions, using computer software to represent the surfaces of selected horizons or stratigraphic sequences interpreted on seismic data or sandbox experiments. Halokinetic structures as visualized in high-quality seismic reflection profiles in the South Atlantic salt basins may have different alternative interpretations to explain

Mohriak, W.U., and Szatmari, P., 2001, Salt tectonics and sedimentation along Atlantic margins: Insights from seismic interpretation and physical models, *in* Koyi, H.A., and Mancktelow, N.S., eds., Tectonic Modeling: A Volume in Honor of Hans Ramberg: Boulder, Colorado, Geological Society of America Memoir 193, p. 131–151.

their development through time. This chapter presents some seismic profiles illustrating these structures, discusses possible genetic mechanisms, and analyzes feasible conceptual models based on the integration of seismic interpretation and sandbox physical models. The following structures are discussed: salt evacuation grabens, major antithetic faults associated with massive clastic progradation, and systematic arrays of antithetic faults associated either with basement steps (forced folds or drape monoclines) or to postextension block rotation compensating synthetic faults.

These structures are interpreted using the paradigm of regional extensional tectonics (Vendeville et al., 1987; Cobbold et al., 1989) rather than salt dissolution (Jenyon, 1986, 1988), although salt dissolution has probably taken place during the process (Ge and Jackson, 1998). The paradigm of regional extension as a geological trigger for salt tectonics (Vendeville, 1991; Vendeville and Jackson, 1991; Vendeville and Jackson, 1992a, b; Jackson and Vendeville, 1994; Jackson and Talbot, 1994) has been applied to several basins in the South Atlantic and in the Gulf of Mexico, based not only on seismic interpretation but also on comparisons with physical models that use sand to represent the overburden and silicone to represent the salt layers (Vendeville et al., 1987; Szatmari and Aires, 1987; Cobbold et al., 1989). The influence of basement-involved faults on salt tectonics has also been analyzed by centrifuged models and comparisons with seismic profiles, particularly in the Danish basin (e.g., Koyi and Petersen, 1993; Koyi et al., 1993).

A major difference between shallow and deep water domains along Atlantic-type continental margins is that in the shelf and slope the structural styles associated with salt tectonics are most often related to listric normal faults that are predominantly synthetic (dipping toward the ocean rather than toward the continent). In the shallow water province, these faults are related to collapse of the limestone platform previously deposited above the evaporites, as shown in several seismic profiles both along the eastern Brazilian and the western African margins (e.g., Ojeda, 1982; Brice et al., 1982; Guardado et al., 1989; Teisserenc and Villemin, 1989; Lundin, 1992). However, salt tectonics styles from the shelf edge toward the slope and deep basin are characterized by tall, piercing salt diapirs and by compressional features that usually indicate vergence toward the oceanic crust (Cobbold et al., 1995). The interpretation of extensional and compressional domains along divergent margins is substantiated by physical models (e.g., Letouzey et al., 1995), although some extensional features in the platform are not reproduced well by these experiments. Some centrifuge experiments (e.g., Talbot, 1992; Koyi, 1996) also indicate that the deep water regions of continental margins might be better characterized by allochthonous salt tongues, as observed in the Gulf of Mexico, rather than the authochthonous salt diapirs affected by later compression, as commonly observed in the seismic profiles in the South Atlantic.

Most faults related to salt tectonics along passive continental margins have a structural syle characterized by basinward dips, forming marginal asymmetric grabens with depocenters mainly controlled by synthetic master faults, and subsidiarily, by antithetic faults compensating for the overburden extension (Cloos, 1968). This is probably related to the thermal subsidence of the lithosphere following the breakup of the continental plates and inception of oceanic crust along divergent margins (McKenzie, 1978), which result in basinward tilting of the basement and the presalt horizons.

Physical models designed to simulate the overburden extension when the base of the model is preferentially tilted toward one side indicate that most faults have a predominant dip direction, the sense of movement pointing toward the direction of tilting (Fig. 1A). This style of extensional tectonics, i.e., most faults dipping toward the deeper portions of the basins, is pre-

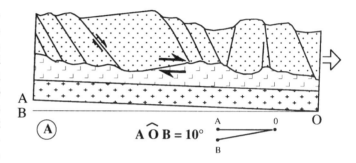

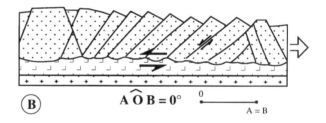

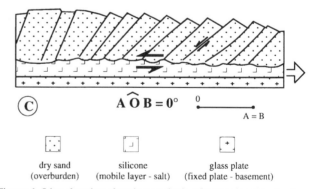

Figure 1. Line drawing showing analysis of extensional tectonics and characteristic fault patterns observed on sandbox physical models. Large arrows show sense of extension vector, and small arrows show slip sense of basal shear stress. Fault styles are associated with inclination of base of fixed plate. A corresponds to tilted base, and B and C correspond to subhorizontal bases. In B, overburden is extended, and in C, extension restricted to elastic sheet at base of overburden simulates salt flow. Based on Vendeville et al. (1987) and McClay (1990).

ponderant along divergent continental margins (Vendeville and Cobbold, 1988). The down-to-the-basin faults are often formed first during the extensional process, but the extension may be subsequently compensated by antithetic faults that dip toward the opposite side, forming a more symmetrical graben (McClay, 1989, 1990).

Physical models indicate that regional extension applied to an overburden above a lubricating layer that is more or less horizontal (Fig. 1B) is often characterized by a set of faults that dip opposite the direction of extension (Vendeville et al., 1987; McClay, 1990). In some cases, extension applied only to the subhorizontal lubricating layer is reflected on the overburden as a set of antithetic faults typically developed in a domino style (Fig. 1C). This style of salt tectonics seems to be not so well developed along divergent continental margins, but the existence of large structures associated with major antithetic faults has important consequences for understanding the basin evolution and for the characterization of petroleum systems. We discuss conceptual mechanisms for the formation of these and other related features along the South Atlantic (eastern Brazilian and western African continental margins) and in the Gulf of Mexico.

DATA ANALYSIS AND RESULTS

This section presents seismic data along divergent Atlantic-type continental margins and discusses possible interpretations based on salt tectonics principles and insights from physical models.

Figure 2 (modified from Jackson and Talbot, 1994) shows the worldwide distribution of salt basins, and the inset numbers highlight the locations of some basins from which selected seismic examples are shown in subsequent figures, particularly the Campos and Santos basins in the southeastern Brazilian region, the Cabo Frio province between these basins, and the Kwanza basin offshore Angola in the western African margin.

The main features discussed in this work correspond to salt evacuation grabens, major stratigraphic gaps associated with listric normal faults (characterized by antithetic basal shear related to salt mobilization concomitant with massive clastic progradation), and systematic arrays of antithetic normal faults associated with overburden extension above steps at the base of the salt or to block rotation reversals.

Salt evacuation grabens

Figure 3 shows a regional seismic profile shot along a dip direction along the northern portion of the Campos basin (Fig. 2). The seismic section is located in the central portion of the platform, in bathymetries ranging from 100 to 2000 m, and the bottom part of the figure shows the schematic line interpretation of the seismic data. The interpretation of the profile suggests a major collapse or salt evacuation graben related to a major listric normal fault that dips landward. Several reflectors dipping seaward are characterized in the postsalt successions, suggesting pseudodownlaps for the Albian to Upper Cretaceous sequences. One possible interpretation for these features assumes upslope gravity gliding of the overburden. Alternative interpretations include salt dissolution by meteoric waters, erosion of the Aptian evaporites and of the Albian platform by deep-water currents, thick-skinned extensional tectonics (basement-involved reactivation of rift-phase normal faults), thin-skinned extensional tectonics (involving only the overburden), or relay zones between different extensional domains. Figure 4 shows a depth-converted seismic profile with a schematic interpretation (bottom), which indicates that the rift-phase normal faults do not appear to be involved in the extensional features observed in the overburden, although there are some antithetic faults in the basement at the northwest extremity of the profile. These faults might have been reactivated either before or after some degree of tilting of the basement during the thermal phase of subsidence. Although the main faults dip landward, there are indications that the salt mass moved basinward, and possibly also perpendicular to the plane of the section during progradational episodes.

The Albian blocks seem to be detached by extension during an early phase of halokinesis, and subsequently, the structure observed near the northwestern extremity of the profile (between shotpoints 100 and 300) seems to be very young relative to the extensional episodes. The seismic data (Fig. 3) indicate that a present-day canyon with an axial trough near shotpoint 250 may be controlled by the deeper halokinetic structures that favored the capture of this channel in recent times.

There are some similarities between this structure in the Campos basin and analogue structures interpreted in the Kwanza basin, the conjugate basin along the West African margin. One of these (Fig. 5, A and B) has been interpreted as a collapse structure related to extensional salt tectonics rather than to salt dissolution (Duval et al., 1992). There is a marked growth of sedimentary thickness in the hanging wall of this major salt withdrawal trough. The growth of the upper Tertiary strata within the trough indicates that the master fault dips westward, and the Albian blocks moved basinward as rafts during the extensional processes, forming a pseudoturtleback. The resultant structure is called a pseudoturtleback because of the anticlinal shape and the stratigraphic age of the core (located between 10 and 14 km along the distance axis of Fig. 5). This depocenter is constituted by sedimentary rocks much younger than the adjacent strata, from which it is separated by two inward-dipping faults (Vendeville and Jackson, 1992b). The Albian blocks are offset from each other by translation along a detachment plane or shear zone at the base of the salt layer (between 2 and 3 s two-way traveltime [TWTT], from east to west in Fig. 5).

Figure 6 (based on Schultz-Ella, 1992) suggests that salt evacuation grabens along Atlantic-type continental margins may be formed by three phases of evolution: (1) a reactive stage associated with thin-skinned overburden extension (Fig. 6, A and B); (2) an active stage (Fig. 6C) characterized by forceful intrusion of a salt diapir into the topmost overburden layers (mainly Albian carbonates, and possibly reaching the depositional surface); and (3) a passive stage characterized by downbuilding, widening of the diapir stem, and final diapir collapse

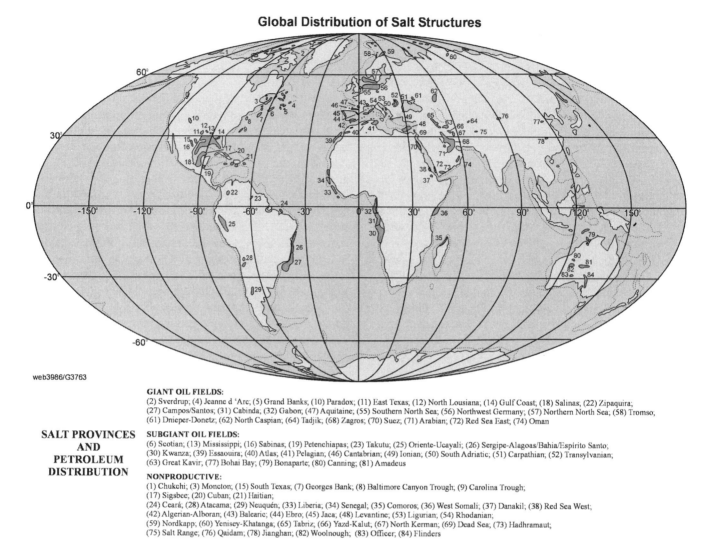

Figure 2. Global distribution of salt basins, highlighting location of discussed regions in Campos and Santos basins (27) in eastern Brazilian margin and Kwanza basin (30) in western African margin. Based on Jackson and Talbot, 1994.

by salt withdrawal, resulting in thick sedimentary wedges subsequently rotated during the continuation of the thin-skinned extensional processes (Fig. 6, D–F).

Results of physical models and computer restoration of the seismic profile in the Kwanza basin may also be applied to the interpretation of the structure imaged in the Campos basin, which may be considered a mirror image of equivalent structures along conjugate margins, with a major difference related to the direction of dip of the normal faults. Near the center of these salt evacuation structures, drilling and physical models indicate that salt diapirism may have played a major role in creating the space necessary for the sediments in the core. However, alternative hypotheses based on pure extensional tectonics with no salt diapirs reaching the surface, and relays of transfer zones, might also provide feasible explanations. Salt dissolution and salt evacuation with negligible overburden extension might also be invoked as alternative possibilities to the mechanisms discussed above.

Antithetic faults related to massive clastic progradation

Figure 7 shows a paradigm or an extreme example of the antithetic fault style in the South Atlantic, illustrating phenomenal halokinetic structures in the Santos basin, offshore Brazil (Fig. 2).

The seismic data (upper part of the figure) indicate that there are several sets of basinward-dipping reflectors extending from the platform toward the the deep water salt diapir province.

The genetic mechanisms for this structure have been recently discussed in several works (e.g., Mohriak et al., 1993, 1995; Demercian et al., 1993; Mohriak, 1995; Szatmari et al., 1996; Demercian, 1996; Ge et al., 1997). Some of these hypotheses include: (1) reativation of basement-involved landward-dipping normal faults; (2) erosion of the Albian platform and the salt layers by deep-water currents; (3) condensation of the Albian sequence and pinch out of the salt layers near the shelf edge;

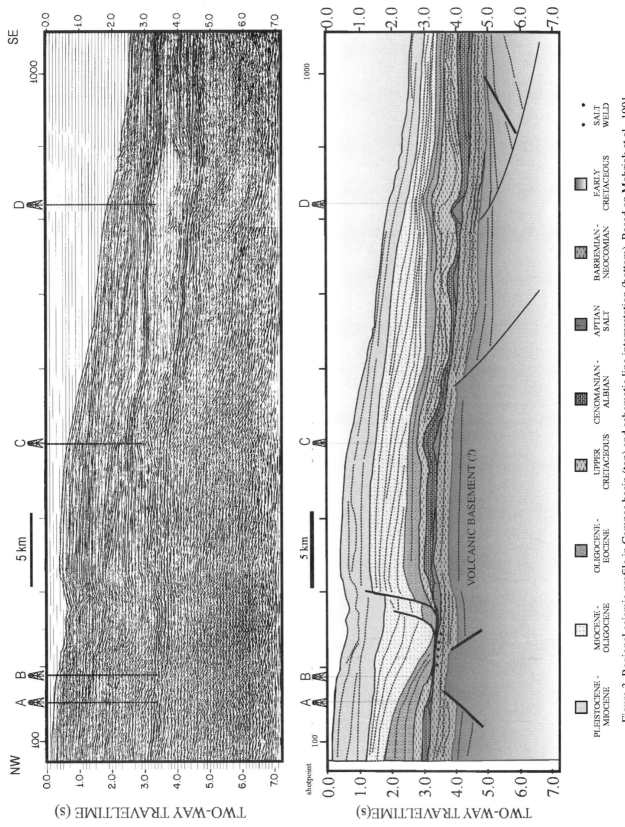

Figure 3. Regional seismic profile in Campos basin (top) and schematic line interpretation (bottom). Based on Mohriak et al., 1991.

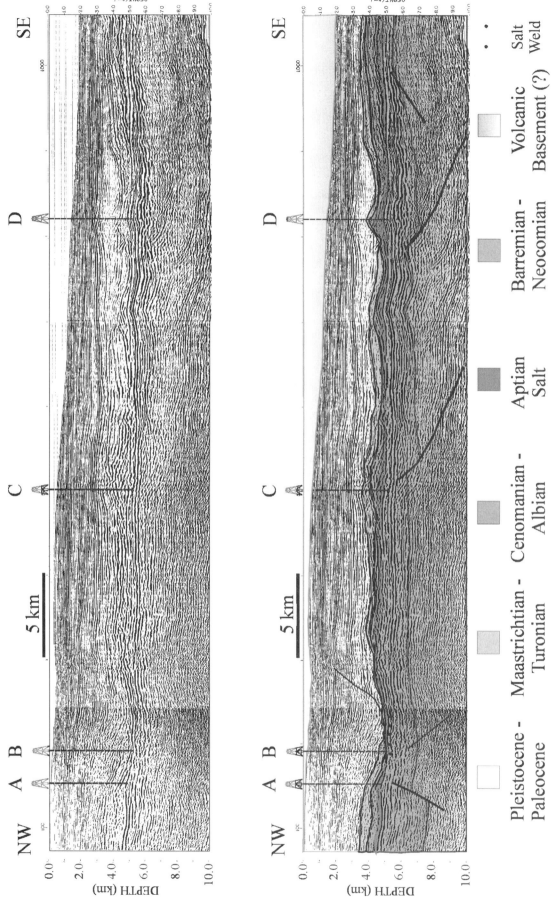

Figure 4. Depth-converted seismic profile in Campos basin (top) and intepretation (bottom). Based on Mohriak et al., 1991.

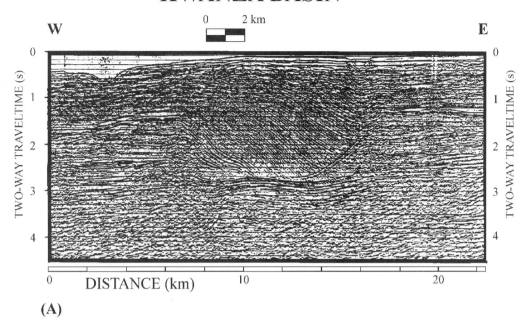

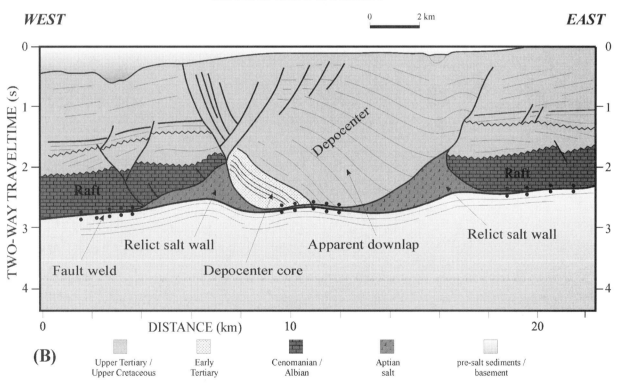

Figure 5. A, Seismic profile in Kwanza basin, offshore western Africa. B, Line drawing interpretation of seismic profile in Kwanza basin. Based on Duval et al., 1992.

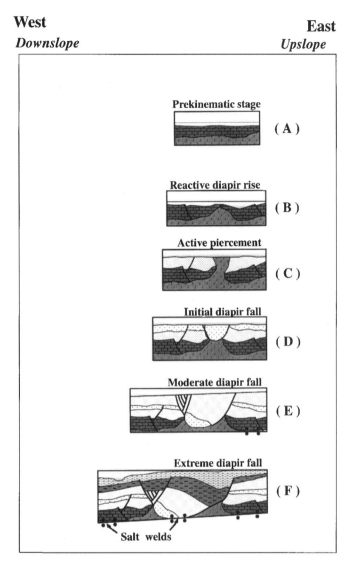

Figure 6. Schematic restoration of seismic profile in Kwanza basin, illustrating model for sequence of events responsible for observed structures. Based on Schultz-Ella, 1992.

(4) apparent downlaps of overburden strata on the base of the residual evaporites left by salt dissolution; (5) apparent downlaps of overburden strata associated with prograding wedges of siliciclastic rocks filling a starved basin; (6) apparent downlaps caused by a major antithetic, low-angle listric normal fault that was forced to advance basinward during episodes of massive clastic progradation; (7) apparent downlaps caused by vertical collapse of overburden rocks while salt was expelled basinward by the sedimentary loading.

Hypothesis 7 implies that only minimal extension of the overburden is involved in the process (Ge et al., 1997), whereas hypothesis 6 implies that major extensional processes resulted in raft blocks in the footwall of the antithetic fault moving basinward in a piggy-back style on the inflated salt layer (Mohriak et al., 1995).

Figure 7 also shows a line-drawing interpretation of the seismic data, indicating thickening of post-Aptian to early Tertiary strata from the platform toward deep water, the apparent stratigraphic downlap of reflectors that dip toward the deep water province of the Santos basin, and the younging of the stratigraphic successions in the same direction (Mohriak et al., 1993, 1995). These reflectors correspond to stratigraphic layers that prograde and become younger basinward, from Albian (10 km from the northwest end of the seismic section) to Paleocene (60 km to the southeast). The prograding wedges are separated from the rift-phase rocks by a major horizon corresponding to the interface between the base of the salt layer and the presalt strata. This interface, extending from about 4.5 to 5.5 TWTT (Fig. 6, from platform to slope), corresponds to a salt weld that marks a major detachment zone characterized by absence of significant Aptian salt and Albian platform carbonate rocks on the presalt layers, resulting in stratigraphic gaps as great as 50 m.y. at the easternmost (younger) prograding events near the present-day salt wall (Mohriak et al., 1995).

The Cabo Frio fault zone (Fig. 7, located between shotpoints 3900 and 2200) corresponds to a large area (ranging from 20 to 50 km in width, and extending for more than 100 km along the central-northern portions of the Santos basin) marked by an apparent absence of Albian to Maastrichtian stratigraphic sequences. This gap might be explained by several hypotheses, including sediment progradation in a starved basin, erosion of the carbonate platform and younger siliciclastics by deep water currents (Van der Ven, 1983), overburden extension (Demercian, 1996), salt expulsion with no extension of overburden and absence of stratigraphic sequences by nondeposition (Ge et al., 1997), overburden spreading and extension controlled by synthetic and antithetic faults (Szatmari et al., 1996), or episodes of massive clastic progradation and salt expulsion controlled by a major extensional detachment (Mohriak et al., 1995).

Based on physical models that bear intriguing similarities with structures in the Gulf of Mexico (e.g., Fig. 8, which was originally prepared to model the formation of salt tongues in the deep-water Sigsbee Escarpment), Mohriak et al. (1995) suggested that massive clastic progradation episodes on a very thick and restricted salt basin (limited basinwards by a regional high or affected by basement-involved antithetic faults) might reproduce the structures observed on seismic profiles in the deep water region of the Campos and Santos basins. Because of the clastic loading, the previous salt basin was almost completely evacuated along a salt weld underlying the prograding wedges. These inclined reflectors in the seismic profiles of the South Atlantic salt basin simulate sigmoidal progradation and also resemble the pseudoclinoforms observed in the Gulf of Mexico (Wu et al., 1990; Mohriak, 1995).

The seismic data in the Santos basin suggest that underneath the prograding wedges, only residual salt may occur in the platform and along the slope. Wide and tall deep water salt walls are characterized tens of kilometers basinward of the

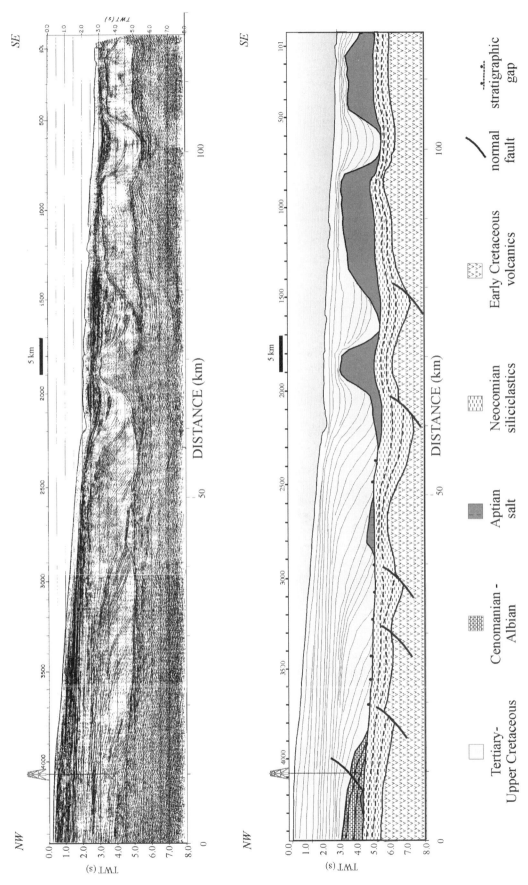

Figure 7. Seismic profile in Santos basin (top) and schematic line-drawing interpretation (bottom). TWT—two-way traveltime. Based on Mohriak et al., 1993, and Mohriak et al., 1995.

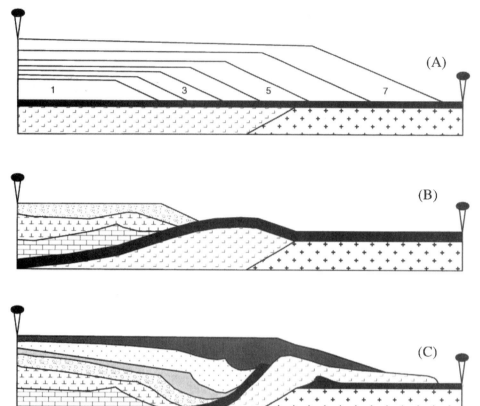

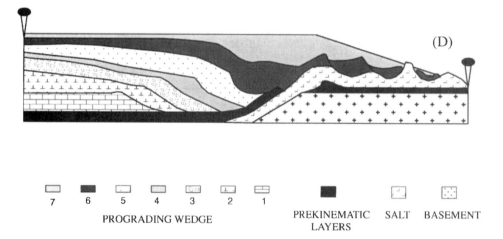

Figure 8. Sequence of events simulated by physical model designed to reproduce structures observed in Gulf of Mexico. A: Sequence of prograding episodes. B: Progradation of sedimentary sequences 1–3. C: Progradation of sedimentary sequences 4–6. D: Progradation of sedimentary sequence 7, and second generation of salt diapirs. Based on Vendeville et al., 1992, and Mohriak et al., 1995.

present-day shelf break. For example, southeast of shotpoint 2000, large salt diapirs are located at 70 km, 80–95 km, and 105–115 km from the northwest end of the profile in Figure 7. These diapirs formed because the salt masses apparently were unable to move basinward due to space problems in the latest stages of evolution, and the salt walls have grown by downbuilding and have started to deform by compression and collapse of the overburden by salt evacuation, resulting in peripheral synclines and small basins that will eventually ground onto the salt weld.

Szatmari et al. (1996) suggested that large listric normal faults with some similarities to the Santos Basin structure (Figs. 7 and 9) might be modeled by overburden extension on a passive, almost subhorizontal basement, and that major listric antithetic faults might be formed to compensatate cogenetic synthetic faults. Ge et al. (1997), however, restored the Santos profile without incorporating any regional extension of the overburden. The pseudodownlaps were interpreted as being caused by collapse of the overburden on a restricted salt basin that was totally evacuated with no overburden extension involved. According to their mod-

interplay of several geological process, including the effects of regional extensional caused by episodes of massive clastic progradation on a restricted salt basin. Figure 10 shows the conceptual mechanisms involved in the geodynamic evolution of the structure, responsible for the formation of major stratigraphic gaps related to antithetic basal shear during the salt mobilization.

The distinction between pseudodownlaps related to salt tectonics from normal siliciclastic progradation is not trivial, and there are elusive features in several sedimentary basins, the interpretation of which as a conformable weld might be related either to pseudodownlaps due to thin-skinned extension or to salt evacuation due to rolling out of the salt layer by massive clastic progradation episodes and the consequent sedimentary loading. Some of these distinctive features can only be ascertained if the prekinematic rafts are drilled in the hanging wall that moved basinward as piggybacks.

This model may also be applied to some structures observed in other sedimentary basins both in the South Atlantic and in the Gulf of Mexico. Mohriak (1995) suggested that pseudodownlaps in the northeastern Gulf of Mexico may correspond to authochthonous salt mobilization and the stratigraphic gaps might be related to antithetic faults formed during clastic progradation episodes in the Late Jurassic and Early Cretaceous. Alternatively, Get et al. (1997) interpreted the formation of pseudodownlaps in the northeastern Gulf of Mexico by salt expulsion due to clastic progradation that mobilized a thickened Louann salt layer, above which there was no Smackover carbonate sequence to be extended.

Figure 11 illustrates the restoration of a regional seismic profile in the Gulf of Mexico (modified from Diegel et al., 1995), indicating that the pseudodownlaps observed in the seismic data may be associated with clastic progradation and collapse of the salt mass that flowed both basinward and laterally, concomitantly with overburden extension and formation of allochthonous salt tongues.

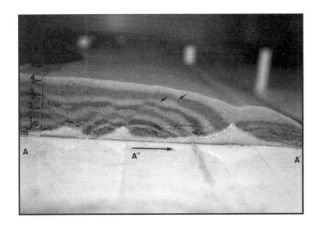

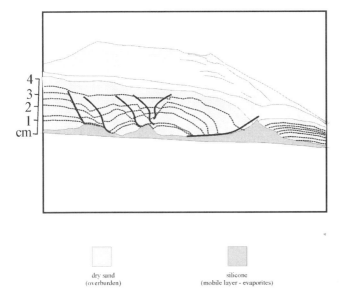

Figure 9. Physical model for Santos basin structure showing pseudodownlaps and associated salt tectonics characterized by major landward-dipping faults. Based on Szatmari et al., 1996.

els, the extension, as proposed by others (e.g., Demercian et al., 1993; Mohriak et al., 1995; Szatmari et al., 1996), would violate basin mechanics (e.g., requiring upslope gravity gliding), generate unrealistic strains, and require an initial salt layer to be much too thick, assuming conserved area.

Ge et al. (1997) suggested that balancing of the Albian gap alone in the Cabo Frio fault zone (Fig. 7) would require at least 25 km of downdip shortening, which is not observed in the data. Although some contractional structures have been reported in the deep water region of the Santos basin (Cobbold et al., 1995), their magnitude seems to be far too low to compensate for the suggested extension.

Taking into consideration the results of regional seismic interpretation and the physical models discussed here, we suggest that the structure in the Cabo Frio region may be formed by the

Antithetic fault systems associated with block rotation reversals

The Cabo Frio province at the southern part of the Campos basin is characterized by a peculiar set of antithetic faults located from the shelf break toward the deep water region (Fig. 12; see Mohriak et al., 1990). This enigmatic style of salt tectonics also seems to occur in some regions along the western African continental margin, particularly along the south Gabon basin (Liro and Coen, 1995).

Overburden extension above steps in the basement or at the base of the salt layer has been suggested as a mechanism to form antithetic faults related to forced folds (Withjack et al., 1989) or to drape monoclines (Jackson and Vendeville, 1992). The influence of basement-involved normal faulting on the development of salt structures has been recognized in a number of sedimentary basins, particularly in the North Sea (e.g., Stewart et al., 1996).

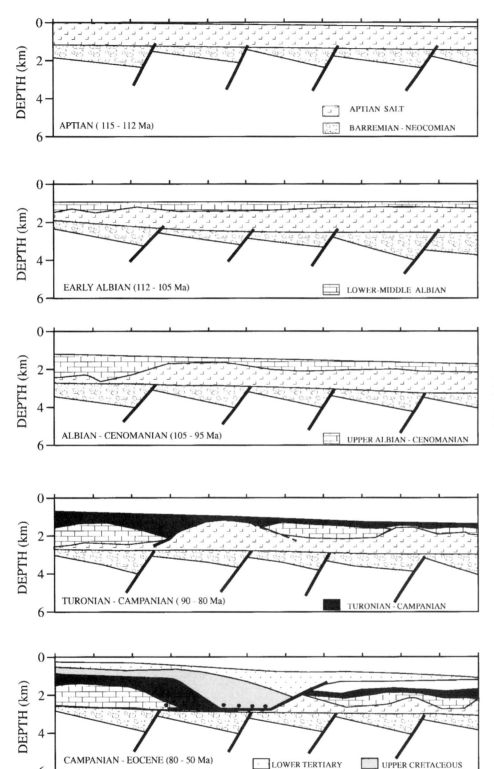

Figure 10. Conceptual model for salt tectonics structures in Cabo Frio region (Campos and Santos basins). Based on Mohriak et al., 1995.

GULF OF MEXICO

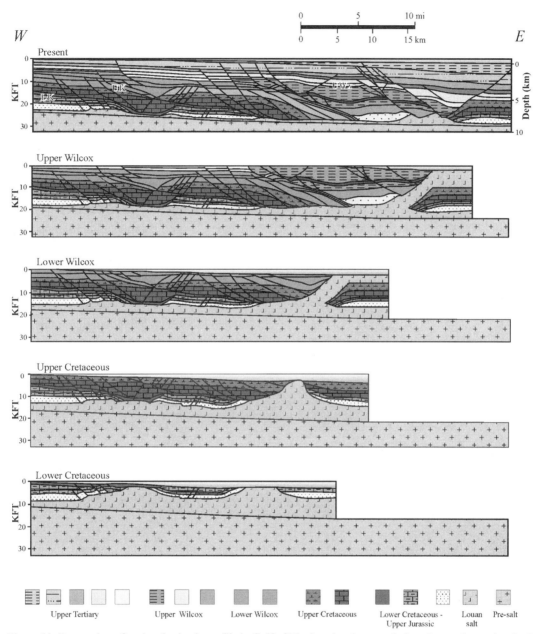

Figure 11. Restoration of regional seismic profile in Gulf of Mexico showing pseudodownlaps and associated salt tectonics characterized by salt evacuation and formation of allochthonous salt tongues. Based on Diegel et al., 1995.

This possibility is indicated in Figure 13, which indicates that the antithetic fault system may be controlled by a hinge line at the base of the salt reflector. This deepening of the presalt reflectors near the shelf break cannot be explained solely by velocity pull-down caused by the increased bathymetry. Apparently, the step at the base of the salt was controlled by subtle reactivations of the normal fault systems that controlled the rift-phase depocenters in the deep water region, and resulted in mobilization of a thicker salt layer that existed above the rift depocenter. The salt mobilization might be triggered either by reactivation of the normal faults or by episodes of clastic progradation.

Different mechanisms have been proposed to explain this style of salt tectonics. Figure 14 shows conceptual models tested by physical modeling experiments aiming at reproducing the structures observed in the seismic profiles of the Cabo Frio region (Rizzo et al., 1990; Mohriak et al., 1991). Figure 14A shows a sedimentary basin with grabens in the synrift sequence, and gliding during the thermal phase of subsidence might be

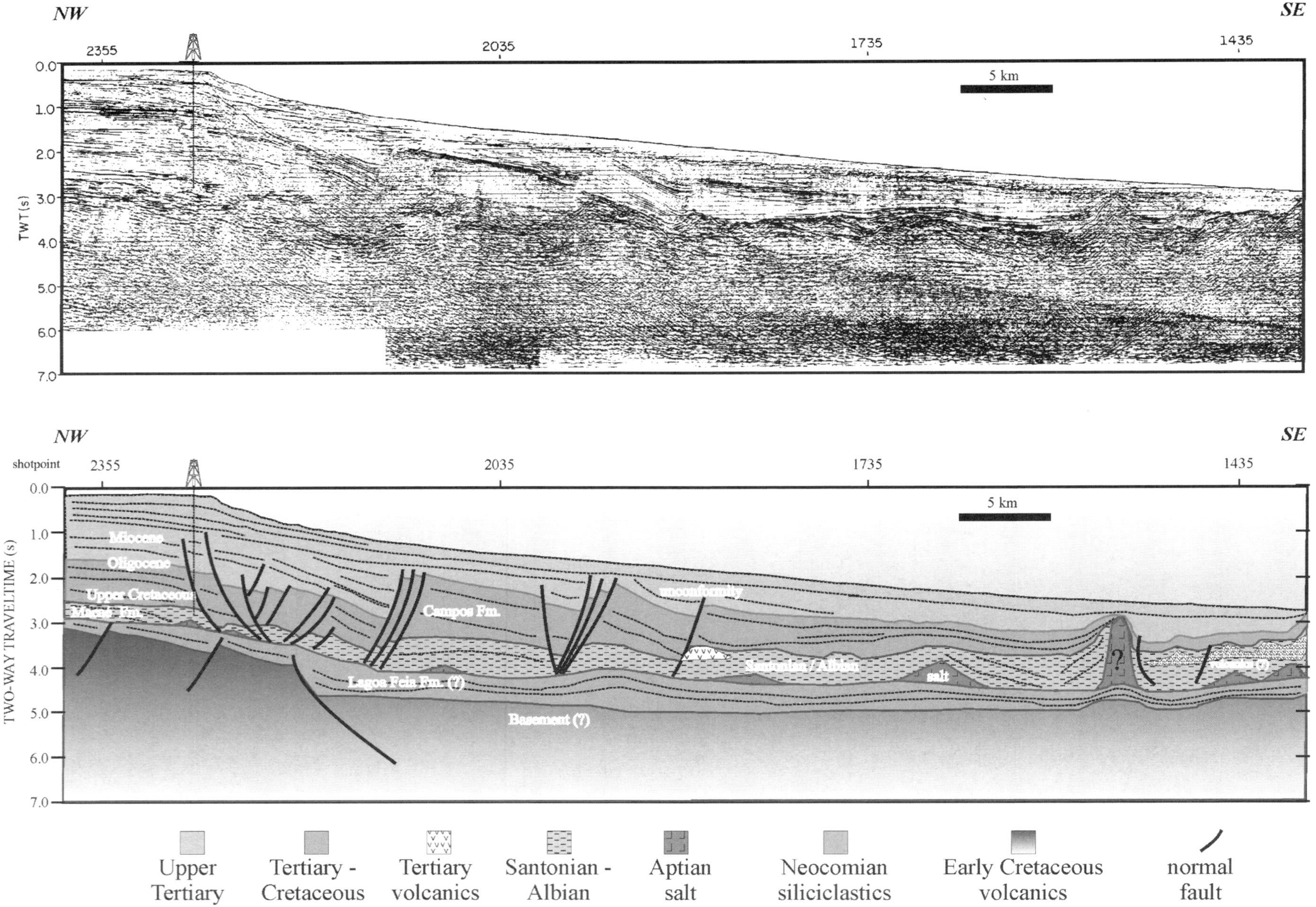

Figure 12. Regional profile in Cabo Frio region of Campos basin showing antithetic faults from shelf edge toward deep water region. Halokinetic features are imaged between shotpoints 2300 and 1800, and probable volcanic features are identified near shotpoint 1500. TWT—two-way traveltime. Based on Mohriak et al., 1990 and Mohriak et al., 1991.

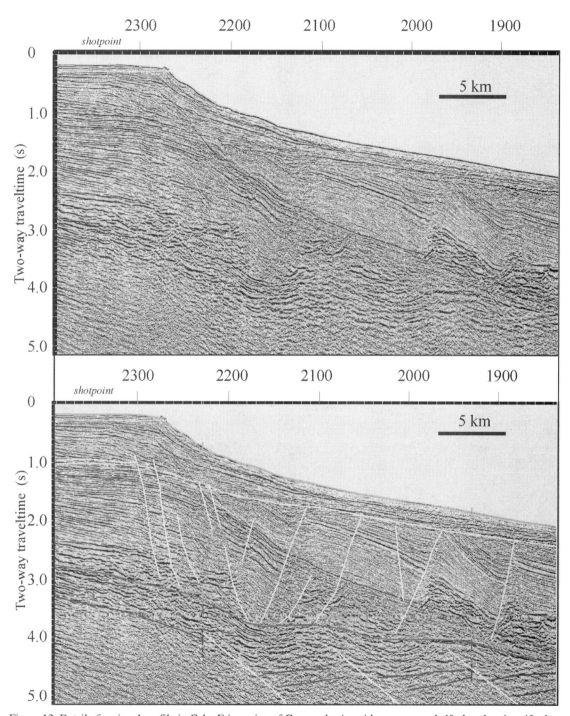

Figure 13. Detail of regional profile in Cabo Frio region of Campos basin, with zoom near shelf edge showing rift-phase structures and antithetic fault system in overburden. Based on Mohriak et al., 1991.

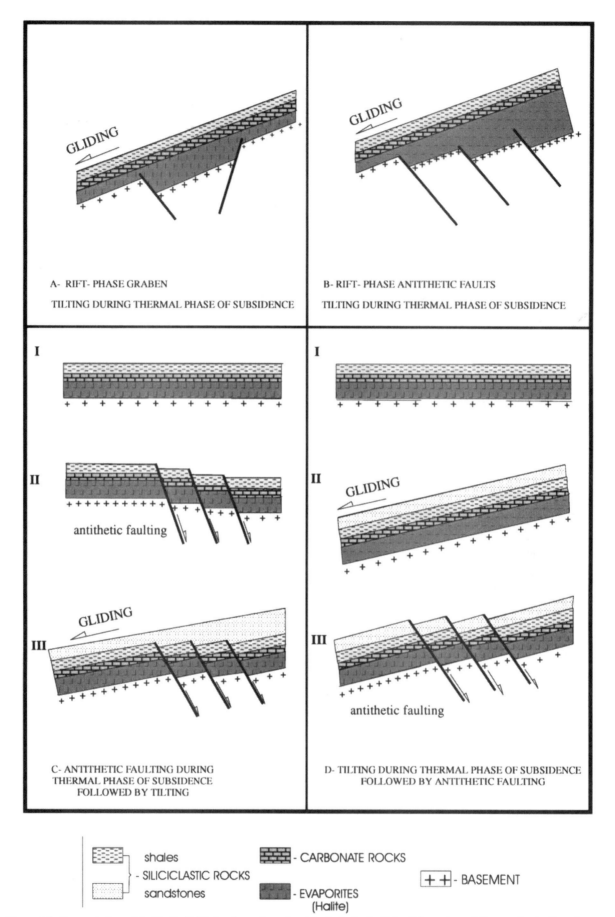

Figure 14. Conceptual models (A, B, C, D) for formation of antithetic faults in Cabo Frio region of Campos basin. Based on Rizzo et al., 1990 and Mohriak et al., 1991.

responsible for the overburden extension. Figure 14B shows a structural style characterized by antithetic faults in the basement and in the rift-phase sequences, as identified in the seismic profiles across the southern segment of the Campos basin (Lobo et al., 1983; Ojeda et al., 1983; Mohriak et al., 1991). Figure 14C illustrates a more complex sequence of events in the geologic evolution of the basin, starting with reactivation of rift-phase normal faults during the early phase of thermal subsidence, followed by gravity gliding during subsequent phases. Figure 14D illustrates a model based on early tilting during the thermal phase of subsidence, and followed by antithetic faults involving the basement during very late times of the geological evolution (such as tectonic activity during the Tertiary). This possibility is indicated in some regions of the Campos basin, as evidenced by faults affecting the basement and the overburden, and by volcanic intrusions dated from Late Cretaceous to early Tertiary (Mohriak et al., 1991).

The results of these experiments are shown in Figure 15 (A–D). Although not conclusive, the experiments indicated some resemblance of the structural styles observed in the seismic profiles with the faults identified in the conceptual model suggested in Figure 14D (and indicated in Fig. 15D). However, they did not reproduce well important features that are observed in the data, such as the marked reduction in salt thickness under the prograding wedges and the occurrence of major stratigraphic discontinuities separating structural styles in the overburden. It should also be noted that variations in siliciclastic input and sea-level fluctuations might also be important factors, as indicated by the occurrence of a major unconformity separating the rotated blocks (with marked growth sequences of Oligocene to Miocene sediments) from the normal progradational sequences dated as Pliocene to Holocene.

Another feasible mechanism may be proposed based on sandbox physical models designed to simulate the postextension rotation of blocks affected by an earlier phase of synthetic extensional faults (Vendeville et al., 1992; Vendeville and Jackson, 1992b). This model assumes that the earlier phase of overburden extension was controlled by synthetic faults, and the thinning of the salt layer resulted in touchdown of the postsalt blocks on the base of the salt. This resulted in the impossibility of further extending the overburden by anticlockwise rotation of the blocks, whereas clockwise rotation was still feasible and would account for the second phase of extension during later episodes of clastic progradation (Fig. 16), forming antithetic faults with local depocenters filled by younger stratigraphic successions. This model would imply that basinward of the shelf edge in Figure 13, the Albian blocks should be controlled by synthetic normal faults during an early phase of extension, and

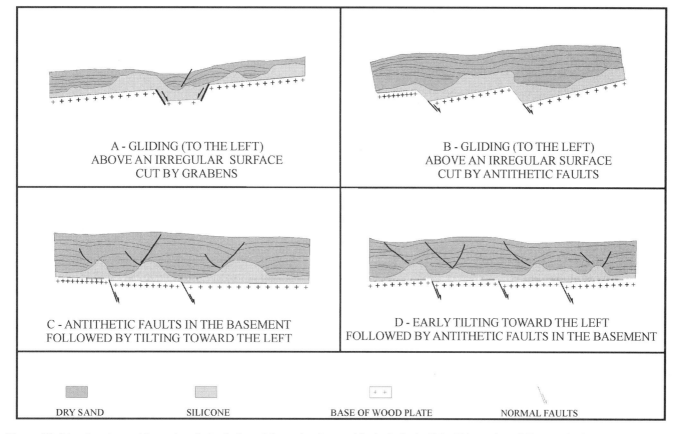

Figure 15. Line drawings with results of physical models to simulate antithetic faults in Cabo Frio region of Campos basin. Based on Rizzo et al., 1990 and Mohriak et al., 1991.

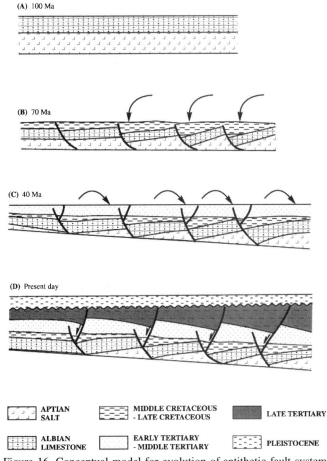

Figure 16. Conceptual model for evolution of antithetic fault system from Albian (diagram A) to present-day (diagram D). This mechanism is related to postextension block rotation after touchdown of overburden blocks on base of evaporite layers. Based on Vendeville and Jackson, 1992b.

that the synkinematic growth should point landward rather than basinward, as indicated in the upper Tertiary sequences. The seismic data shown in Figure 13 suggest a subtle thickening of the earliest postsalt strata toward the landward direction, but the predicted synthetic faults are not imaged well because of severe multiples and because of the strong effect of the antithetic faults affecting the younger sequences.

DISCUSSION

The integration of different tools (seismic interpretation and physical models) in the analysis of halokinetic structures within the salt basins provides valuable constraints on competing models of formation. The genetic mechanisms and the temporal development of salt structures may be simulated by physical experiments and almost instantaneously reproduced by computer animations (Guglielmo et al., 1997), which provide helpful insights into the geodynamic evolution of complex features. These techniques have been applied to the interpretation of selected salt tectonics structures in the South Atlantic.

Viable alternative hypotheses to explain genetic mechanisms of some halokinetic structures are discussed in this work, including salt dissolution, salt evacuation, thin-skinned overburden extension, basement-involved (thick skinned) extension, or some combination of these processes. The examples and models discussed here are derived from regional interpretation of South Atlantic basins, particularly in the Campos, Santos, and Espirito Santo basins (Brazilian margin), and in the Kwanza and Gabon basins (West Africa), but may also occur in the Gulf of Mexico and other sedimentary basins worldwide.

The structural style of faults affecting the overburden during extensional processes along continental margins is most often characterized by synthetic (basinward) dips. However, some segments of these margins seem to be characterized by predominantly antithetic (landward dipping) normal faults.

Salt evacuation structures such as observed onshore and offshore Angola might be interpreted as the result of sudden dissolution of salt diapirs, which would create space to accommodate the rapid sedimentation that is interpreted in the nucleus of these troughs. The seismic interpretation and the insights from physical models indicate, however, that these structures probably formed by extensional processes related to salt tectonics with a complex development history that resulted in a pseudoturtleback structure. These pseudoturtlebacks in the South Atlantic are characterized by a stratigraphically young nucleus, and they may have been started by a reactive diapir rise of Aptian salt as consequence of overburden extension, followed by an active piercement stage and by diapir fall during much later episodes of clastic progradation.

Several radically different interpretations have been suggested for the major antithetic fault system and the Cabo Frio stratigraphic gap zone in the Santos basin. Previous hypotheses included the pinch out of the limestone platform as a consequence of erosion by deep-water currents or by dissolution of the previously deposited salt layer followed by platform collapse. The present interpretation assumes that extension is associated with early halokinesis and faulting of the Albian overburden rocks during Late Cretaceous episodes of massive clastic progradation. The genetic mechanism proposed for the style of antithetic faults observed in the Cabo Frio region (Fig. 10) involves a combination of overburden extension, massive clastic progradation, and basinward and lateral salt flow (Mohriak et al., 1995). It may also be applied to major structures in the Gulf of Mexico (e.g., Mohriak, 1995) and to other sedimentary basins. In contrast, Ge et al. (1997) assumed almost no extension in the process and the pseudodownlaps are interpreted as caused by lateral salt mobilization or by rolling the salt basinward by prograding sediments.

Szatmari et al. (1996), on the basis of physical modeling of the Santos basin structure, corroborated the interpretation that extension of the overburden plays an important role in creating antithetic faults during episodes of progradation. Furthermore, the effects of massive clastic progradation on salt structures characterized by antithetic fault systems are clearly related to processes involving overburden extension, as observed in both the African margin (Liro and Coen, 1995) and in the Gulf of Mexico (Diegel et al., 1995).

The extremely important constraints of an outer basin barrier and massive clastic progradation are easily recognized in the physical models designed to replicate these structures. They also indicate that extensional processes may play a subsidiary rather than vital role in forming structures that have a similar geometry and may occur at various stages of the basin evolution, from the early salt mobilization phases to very late in the geological history of the basin.

The major difference between the Gulf of Mexico model and the Santos structure is related to timing of the massive clastic progradation episodes. In the Santos basin these events occurred by the end of restricted marine megasequence responsible for the Albian carbonate platform, and formed authochthonous salt structures that advanced toward the recently created oceanic crust. In the Gulf of Mexico, these events climaxed only in the late Tertiary, tens of millions of years after the deposition of the equivalent Jurassic carbonate platform (Smackover Formation), and resulted in formation of allochthonous salt structures, probably because of space problems in accommodating great extension in a basin that was not diverging, and may be regionally characterized by convergent salt flow (Cobbold and Szatmari, 1991).

Systematic arrays of antithetic normal faults may be related to collapse of sediments above a thickened salt layer on steps in the basement, or by a very low angle of the presalt layers during episodes of clastic progradation, particularly if the extension is applied only to the overburden. These structures might also be formed by reversal of the rotation angle caused by depletion of the salt mass and touchdown of the overburden blocks on the presalt layers, which would imply a two-phase extensional process, the first characterized by synthetic normal faults, and the later by antithetic normal faults. In the example discussed in the Campos basin, all three factors seem to control the array of faults observed from the shelf edge toward the deep water region, and other factors, such as the presence of volcanic intrusions within the sedimentary sequences, might also be important elements influencing the development of this style. The development of the antithetic faults in the southeastern Brazilian region coincides with the shift of deltaic depocenters, with maximum sedimentary thickness above the salt layers dated as Cretaceous in the Santos basin, and as upper Tertiary in the Campos basin.

There are important consequences of these models on the hydrocarbon exploration along continental margins. First, the presence of a mobile substratum plays a fundamental role in creating structures in the overburden, such as anticlines, turtle backs, roll-overs, and peripheral synclines. Second, the faults related to halokinesis may constitute hydrocarbon migration pathways from the presalt source rocks to feed postsalt reservoirs. Third, salt tectonics also influence sedimentation and the distribution of porosity in postsalt reservoirs. Fourth, the principles governing salt tectonics may be important elements in elucidating the evolution of some structures, and contribute to a better understanding of the geological history of the basin and to the prediction of exploratory plays.

For example, the interpretation of a raft block completely offset from its original position may affect the migration of fluids from the source rocks through carrier beds below the salt layer and upward across salt windows. Another consequence might be the formation of large stratigraphic gaps due to massive clastic progradation and salt mobilization, resulting in the absence of particular sequences of postsalt reservoirs within evacuation grabens. The timing of formation of the halokinetic structures may also influence heat flow both below and above the salt layer, thus affecting the hydrocarbon generation potential.

CONCLUSIONS

The seismic interpretation and the physical models indicate that the salt evacuation grabens located in the Campos and Kwanza basins were probably formed by extensional tectonics and were related to different stages of salt diapirism. The stages include an early phase of thin-skinned overburden extension (reactive stage), which resulted in the rise of salt masses during late Albian and Cenomanian time, a subsequent phase of forceful intrusion of salt through the ruptured depositional surface of the overburden (active stage), and a final passive stage of diapir fall concomitant with later phases of overburden extension. The final stages resulted in widening of the diapir, rotation of sedimentary layers during salt evacuation, and an extremely rapid filling of a major trough by younger sediments. These troughs are not classic halokinetic turtlebacks because of the stratigraphic relationships between adjacent beds and the nucleus of the structure.

In the Campos profile, the main extensional fault in the evacuation graben is antithetic (landward dipping), whereas in the Kwanza, the main fault is synthetic (basinward dipping); however, the genetic mechanism proposed for this structure implies that there were times when the main fault was antithetic and other times when the main fault was synthetic. This flip flop of block rotation and subsidence associated with salt tectonics seems to have occurred in many equivalent features along the Atlantic continental margins.

Physical models and restoration of seismic profiles suggest that the major stratigraphic gaps and related antithetic fault structures described in the Santos basin and possibly other analogs (e.g., in the Gulf of Mexico or in the West African margin) were probably formed by overloading a salt basin with thick wedges of sediments during episodes of massive clastic progradation, following a short-lived starved basin interval. This resulted in rising of an anticlinal salt mass that was forced to migrate basinward, overriding a previous salt barrier. The salt mass collapsed along an antithetic fault that was pushed farther basinward toward the deep-water region; older layers were continually overstepped by younger layers, and some blocks may have been carried basinward as piggy-backs on the mobile salt.

Peculiar arrays of antithetic faults as observed in some basins along Atlantic-type continental margins may be associated with overburden extension above steps in the basement (forced folding), a complex combination of thermal subsidence, reactivation of basement-involved normal faults and clastic progradation, or diverse phases of postextension block rotation affecting overburden blocks, following collapse of the salt mass and the touchdown of the overburden on the presalt layers. This latter mechanism might favor the inversion of the rotation angle of the postsalt sequences, resulting in antithetic faults formed at later phases of clastic progradation. Physical models designed to simulate this style of salt tectonics were not conclusive about which mechanisms might be responsible for the faults in the Campos basin, but indicated that steps in the basement, reactivated during the postrift phase of subsidence, might play an important role.

ACKNOWLEDGMENTS

Most of the physical models discussed in this work were executed either at Petrobras Research Center or at the Applied Geodynamics Laboratory of the University of Texas, as part of a multidisciplinary salt tectonics research project sponsored by several industrial associates. We appreciate dozens of enlightening discussions with P. Cobbold, M.P.A. Jackson, B. Vendeville, L.S. Demercian, M. Guerra, and M. Pequeno, which greatly contributed to our present understanding of the complex history of the SoutAtlantic salt basin. We are grateful for the many constructive remarks by C.J. Talbot, M. Hudec, and J.C. Harrison, who reviewed the first draft of the manuscript and provided helpful suggestions that improved this chapter. We thank H. Koyi for the invitation to prepare this contribution and for his constructive editorial comments and insightful suggestions.

REFERENCES CITED

Brice, S.E., Cochran, M.D., Pardos, G., and Edwards, A.D., 1982, Tectonics and sedimentation of the South Atlantic rift sequence: Cabinda, Angola, *in* Watkins, J.S., and Drake, C.L., eds., Studies in continental margin geology: American Association of Petroleum Geologists Memoir 34, p. 5–18.

Cloos, E., 1968, Experimental analysis of Gulf coast fracture patterns: American Association of Petroleum Geologists Bulletin, v. 52, p. 420–444.

Cobbold, P.R., and Szatmari, P., 1991, Radial gravitational gliding on passive margins: Tectonophysics, v. 188, p. 249–289.

Cobbold, P.R., Rossello, E., and Vendeville, B., 1989, Some experiments on interacting sedimentation and deformation above salt horizons: Bulletin de la Société Géologique de France, v. 8, p. 453–460.

Cobbold, P.R., Szatmari, P., Demercian, L.S., Coelho, D., and Rossello, E.A., 1995, Seismic experimental evidence for thin-skinned horizontal shortening by convergent radial gliding on evaporites, deepwater Santos Basin, *in* Jackson, M.P.A., et al., eds., Salt tectonics: A global perspective: American Association of Petroleum Geologists Memoir 65, p. 305–321.

Demercian, S., 1996, A halocinese na evolução do sul da Bacia de Santos do Aptiano ao Cretáceo Superior [MS thesis]: Universidade Federal do Rio Grande do Sul, 201 p.

Demercian, S., Szatmari, P., and Cobbold, P.R., 1993, Style and pattern of salt diapirs due to thin-skinned gravitational gliding, Campos and Santos basins, offshore Brazil: Tectonophysics, v. 228, p. 393–433.

Diegel, F.A., Karlo, J.F., Schuster, D.C., Shoup, R.C., and Tauvers, P.R., 1995, Cenozoic structural evolution and tectono-stratigraphic framework of the northern Gulf Coast continental margin, *in* Jackson, M.P.A., et al., eds., Salt tectonics: A global perspective: American Association of Petroleum Geologists Memoir 65, p. 109–151.

Duval, B., Cramez, C., and Jackson, M.P.A., 1992, Raft tectonics in the Kwanza Basin, Angola: Marine and Petroleum Geology, v. 9, p. 389–404.

Ge, H., and Jackson, M.P.A., 1998, Physical modeling of structures caused by salt withdrawal: Implications for deformation caused by salt dissolution: American Association of Petroleum Geologists Bulletin, v. 82, p. 228–250.

Ge, H., Jackson, M.P.A., and Vendeville, B.C., 1997, Kinematics and dynamics of salt tectonics driven by progradation: American Association of Petroleum Geologists Bulletin, v. 81, p. 393–423.

Guardado, L.R., Gamboa, L.A.P., and Luchesi, C.F., 1989, Petroleum geology of the Campos Basin, a model for a producing Atlantic-type basin, *in* Edwards, J.D., and Santogrossi, P.A., eds., Divergent/passive margin basins: American Association of Petroleum Geologists Memoir 48, p. 3–79.

Guglielmo Jr., G., Jackson, M.P.A., and Vendeville, B.C., 1997, Three-dimensional visualization of salt walls and associated fault systems: American Association of Petroleum Geologists Bulletin, v. 81, p. 46–61.

Jackson, M.P.A., 1995, Retrospective salt tectonics, *in* Jackson, M.P.A., et al., eds., Salt tectonics: A global perspective: American Association of Petroleum Geologists Memoir 65, p. 1–28.

Jackson, M.P.A., and Vendeville, B.C., 1994, Regional extension as a geologic trigger for diapirism: Geological Society of America Bulletin, v. 106, p. 57–73.

Jackson, M.P.A., and Talbot, C.J., 1994, Advances in salt tectonics, *in* Hancock, P.L., ed., Continental deformation: Pergamon Press, p. 159–179.

Jenyon, M.K., 1986, Salt tectonics: Elsevier Applied Science Publishers Ltd., 191 p.

Jenyon, M.K., 1988, Seismic expression of salt dissolution-related features in the North Sea: Bulletin of Canadian Petroleum Geology, v. 36, p. 274–283.

Koyi, H., 1996, Salt flow by aggrading and prograding overburdens, *in* Alsop, G.I., et al., eds., Salt tectonics: Geological Society of London Special Publication 100, p. 243–285.

Koyi, H., and Petersen, K., 1993, Influence of basement faults on the development of salt structures in the Danish Basin: Marine and Petroleum Geology, v. 10, p. 82–94.

Koyi, H., Jenyon, M.K., and Petersen, K., 1993, The effect of basement faulting on diapirism: Journal of Petroleum Geology, v. 16, p. 285–312.

Letouzey, J., Colletta, B., Vially, R., and Chermette, 1995, Evolution of salt-related structures in compressional settings, *in* Jackson, M.P.A., et al., eds., Salt tectonics: A global perspective: American Association of Petroleum Geologists Memoir 65, p. 41–61.

Liro, L.M., and Coen, R., 1995, Salt deformation history and postsalt structural trends, offshore southern Gabon, West Africa, *in* Jackson, M.P.A., et al., eds., Salt tectonics: A global perspective: American Association of Petroleum Geologists Memoir 65, p. 323–331.

Lundin, E., 1992, Thin-skinned extensional tectonics on a salt detachment, northen Kwanza Basin, Angola: Marine and Petroleum Geology, v. 9, p. 405–411.

McClay, K.R., 1989, Physical models of structural styles during extension, *in* Tankard, A.J., and Balkwill, H.R., eds., Extensional tectonics and stratigraphy of the North Atlantic margins: American Association of Petroleum Geologists Memoir 46, p. 95–110.

McClay, K.R., 1990, Extensional fault systems in sedimentary basins: A review of analogue model studies: Marine and Petroleum Geology, v. 7, p. 206–233.

McKenzie, D., 1978, Some remarks on the development of sedimentary basins: Earth and Planetary Science Letters, v. 40, p. 25–32.

Mohriak, W.U., 1995, Salt tectonics structural styles: Contrasts and similarities between the South Atlantic and the Gulf of Mexico: Houston, Texas, Gulf Coast Section SEPM, 16th Annual Research Conference, p. 177–191.

Mohriak, W.U. (coord.), 1991, Evolução tectono-sedimentar da área "offshore" de Cabo Frio, Rio de Janeiro: Petrobrás internal report: Rio de Janeiro, Brazil, Depex/Dirsul/Serab.

Mohriak, W.U., Barros, A.Z., and Fujita, A., 1990, Magmatismo e tectonismo cenozóicos na região de Cabo Frio, Rio de Janeiro: Congresso Brasileiro de Geologia, 37, Natal, Brazil: Anais, Sociedade Brasileira de Geologia, v. 6, p. 2873–2885.

Mohriak, W.U., Macedo, J.M., and Tarabini, R.T., 1993, Estilos estruturais e tectônica de sal na região de Cabo Frio, Rio de Janeiro: Sociedade Brasileira de Geologia, Rio de Janeiro, Atas do III Simpósio de Geologia do Sudeste, p. 64–70.

Mohriak, W.U., Macedo, J.M., Castellani, R.T., Rangel, H.D., Barros, A.Z.N., Latgé, M.A.L., Ricci, J.A., Mizusaki, A.M.P., Szatmari, P., Demercian, L.S., Rizzo, J.G., and Aires, J.R., 1995, Salt tectonics and structural styles in the deep-water province of the Cabo Frio region, Rio de Janeiro, Brazil, in Jackson, M.P.A., et al., eds., Salt tectonics: A global perspective: American Association of Petroleum Geologists Memoir 65, p. 273–304.

Ojeda, H.A.O., 1982, Structural framework, stratigraphy, and evolution of Brazilian marginal basins: American Association of Petroleum Geologists Bulletin, v. 66, p. 732–749.

Ojeda, H.A.O., Carminatti, M., and Rodarte, J.B., 1983, Bacia de Campos: Arcabouço estrutural regional e interpretação genética preliminar: Rio de Janeiro, Brazil, Petrobrás internal report, Depex, 50 p.

Rizzo, J.G., Mohriak, W.U., Aires, J.R., and Barros, A.Z.N., 1990, Modelagem física de falhamentos antitéticos em águas profundas da região de Cabo Frio na Bacia de Campos, Rio de Janeiro: Congresso Brasileiro de Geologia, 37, Natal, Sociedade Brasileira de Geologia, v. 5, p. 2228–2249.

Schultz-Ela, D.D., 1992, Restoration of cross sections to constrain deformation processes of extensional terranes: Marine and Petroleum Geology, v. 9, p. 372–388.

Stewart, S.A., Harvey, M.J., Otto, S.C., and Weston, P.J., 1996, Influence of salt on fault geometry: Examples from the UK salt basins, in Alsop, G.I., et al., eds., Salt tectonics: Geological Society of London Special Publication 100, p. 175–202.

Szatmari, P., and Aires, J.R., 1987, Experimentos com modelagem física de processos tectônicos no Centro de Pesquisa da Petrobrás: Boletim de Geociências da Petrobrás, v. 1, p. 13–24.

Szatmari, P., Guerra, M.C., and Pequeno, M.A., 1996, Genesis of large counter regional normal fault by flow of Cretaceous salt in the South Atlantic, Santos Basin, Brazil, in Alsop, G.I., et al., eds., Salt tectonics: Geological Society of London Special Publication 100, p. 259–264.

Talbot, C.J., 1992, Centrifuged models of Gulf of Mexico profiles: Marine and Petroleum Geology, v. 9, p. 412–432.

Teisserenc, P., and Villermin, J., 1989, Sedimentary basin of Gabon: Geology and oil systems, in Edwards, J.D., and Santogrossi, P.A., eds., Divergent/passive margin basins: American Association of Petroleum Geologists Memoir 48, p. 117–199.

van der Ven, P.H., 1983, Seismic stratigraphy and depositional systems of northeastern Santos Basin, offshore southeastern Brazil [MS thesis]: The University of Texas at Austin, 151 p.

Vendeville, B., 1991, Mechanisms generating normal fault curvature: A review illustrated by physical models, in Roberts, A.M., et al., eds., The geometry of normal faults: Geological Society of London Special Publication 56, p. 241–250.

Vendeville, B., and Cobbold, P.R., 1988, How normal faulting and sedimentation interact to produce listric fault profiles and stratigraphic wedges: Journal of Structural Geology, v. 10, p. 649–659.

Vendeville, B.C., and Jackson, M.P.A., 1991, Deposition, extension, and the shape of downbuilding diapirs: American Association of Petroleum Geologists Bulletin, v. 75, p. 687–688.

Vendeville, B.C., and Jackson, M.P.A., 1992a, The rise of diapirs during thin-skinned extension: Marine and Petroleum Geology, v. 9, p. 331–353.

Vendeville, B.C., and Jackson, M.P.A., 1992b, The fall of diapirs during thin-skinned extension: Marine and Petroleum Geology, v. 9, p. 354–371.

Vendeville, B., Cobbold, P.R., Davy, P., Brun, J.P., and Choukroune, P., 1987, Physical models of extensional tectonics at various scales, in Coward, M.P., et al., eds., Continental extensional tectonics: Geological Society of London Special Publication 28, p. 95–107.

Vendeville, B.C., Jackson, M.P.A., and Schultz-Ela, D.D., 1992, Applied geodynamics laboratory—Second semi-annual progress report to industrial associates for 1991: Bureau of Economic Geology, University of Texas at Austin, 48 p.

Withjack, M.O., Meisling, K.E., and Russel, L.R., 1989, Forced folding and basement-detached normal faulting in the Haltenbanken area, offshore Norway, in Tankard, A.J., and Balkwill, H.R., eds., Extensional tectonics and stratigraphy of the North Atlantic margins: American Association of Petroleum Geologists Memoir 46, p. 567–575.

Wu, S., Bally, A.W., and Cramez, C., 1990, Allochthonous salt, structure and stratigraphy of the northeastern Gulf of Mexico. Part II: Structure: Marine and Petroleum Geology, v. 7, p. 334–370.

Manuscript Accepted by the Society April 12, 2000

Geological Society of America
Memoir 193
2001

Compressional structures in a multilayered mechanical stratigraphy: Insights from sandbox modeling with three-dimensional variations in basal geometry and friction

Claudio Turrini*
TOTALFINAELF Italia, Via Campanini 6, 20124 Milano, Italy
Antonio Ravaglia*, Cesare R. Perotti*
Dipartimento di Scienze della Terra, Università di Pavia, via Ferrata 1, 27100 Pavia, Italy

ABSTRACT

To investigate the influence of an inhomogeneous stratigraphy on the deformation geometries and kinematics that develop within a thin-skinned compressional regime, a set of experiments was performed in a glass-sided sandbox apparatus. Different types of dry, granular, noncohesive material were used to simulate the undeformed initial mechanical stratigraphy and the vertical strength contrast that sediments show in nature. In order to force three-dimensional deformation partitioning within the developing structures, different model geometries were constructed, these being deformed against undeformable ramps or displaced over contrasting basal detaching surfaces.

Faulted folds, detached at different levels within the stratigraphic succession, are the main structural geometries resulting from all of the sandbox experiments. Different orders of thrusted anticlines, subthrust geometries, widespread intrastratal decoupling, and imbricated units can be recognized within the models. The observed structural evolution essentially indicates piggy-back deformation kinematics. Nevertheless, anomalies within such a generic hinterland to foreland deformation progression can be recognized in detail.

Three-dimensional reconstruction of the distribution of the structures across the final models suggests moderate homogeneity in the deformation partitioning within the model hinterland domains. Conversely, vertical and along-strike variations in the model foreland domains result.

Conclusions from the analysis of the models confirm that the mechanical stratigraphy is the key controlling parameter of the deformation geometries and kinematics obtained. Important anomalies regarding geometries and kinematics can be referred to the boundary constraints. Comparison with the structures that have been interpreted in the Southern Apennines of Italy demonstrates that the simulation of an accurate original mechanical stratigraphy and the definition of the possible associated boundary conditions can be important for the potential hydrocarbon evaluation of natural thrust belts.

---

*E-mails: Turrini, Claudio.turrini@fina.be;
Ravaglia, aravaglia@manhattan.unipv.it;
Perotti, cperotti@unipv.it

Turrini, C., et al., 2001, Compressional structures in a multilayered mechanical stratigraphy: Insights from sandbox modeling with three-dimensional variations in basal geometry and friction, *in* Koyi, H.A., and Mancktelow, N.S., eds., Tectonic Modeling: A Volume in Honor of Hans Ramberg: Boulder, Colorado, Geological Society of America Memoir 193, p. 153–178.

INTRODUCTION

The initiation and growth of thrust belts are influenced by many natural variables. Among these, the original mechanical stratigraphy of the undeformed succession (Erickson, 1996; Fermor, 1999), the frictional resistance at the base of the wedge (Davis et al., 1983; Davis and Engelder, 1985), the morphology of the rigid basement (Boyer, 1995), and the sedimentation and erosion rates (Dahlen and Suppe, 1988) can all contribute to control the formation and evolution of many of the compressional structures observed in natural orogens. The interaction of such phenomena through time and space seems to create at least some of the natural conditions that determine the kinematics of a compressional tectonic wedge.

The mechanical stratigraphy that characterizes a geological province has been demonstrated to be a key factor in the structural evolution of a thrust belt (Rich, 1934; Dahlstrom, 1970; Gretener, 1972; Mitra, 1986; Eisenstadt and De Paor, 1987). The strength and thickness of the competent layers being deformed impose the dominant fold wavelength, amplitude, and possible asymmetry of the growing compressional structures (Ramsay and Huber, 1987; Erickson, 1996). As such they control the associated fault locations and the related three-dimensional fracture distribution and orientation within the multilayered rock package (Gross et al., 1997). Conversely, the weak layers that compose the original stratigraphy can define the position of potential regional and/or local detachment levels. These layers probably control the development of disharmonic and/or polyharmonic deformation geometries. The composition of the layers and their primary rheological properties, the contrast between the mechanical properties of the juxtaposed layers (i.e., the layers are welded or detached), and the thickness of each layer control the deformation mechanics and kinematics within a mountain chain (Currie et al., 1962; Wiltschko and Chapple, 1977; Ramsay and Huber, 1987).

Because only the finite strain state of the entire deformation process can be seen, it is difficult to estimate the influence of all the parameters listed here. However, the physical models allow us to impose specific conditions and test their influence on the evolution and final configuration of the experiment. The comparison between natural and experimental fault and fold geometries is valuable for understanding the deformation structures commonly observed inside thrust belts. The observation of the deformation evolution during the experiments allows the progressive development of structure to be continuously tracked. The precise knowledge of the original boundary constraints to the deforming structures provides significant and unique insights about the relationship that ties the initial mechanical perturbation to the deformation localization and progression. The application of mechanical modeling to the analysis of compressional structures has become a common step in the structural work process. To this end, sandbox simulations under a normal gravity field (e.g., Malavieille, 1984; Mulugeta and Koyi, 1987, 1992; Mulugeta, 1988a; McClay, 1990; Colletta et al., 1991; Liu et al., 1992; Marshak and Wilkerson, 1992; Marshak et al., 1992; Calassou et al., 1993; Koyi, 1995; Verschuren et al., 1996) and plasticine modeling within a centrifuge apparatus (e.g., Dixon and Summers, 1985; Mulugeta, 1988b; Liu and Dixon, 1990, 1991, 1995; Dixon and Tirrul, 1991; Dixon and Liu, 1992) have been widely demonstrated as the most suitable physical modeling techniques to simulate the thin-skinned tectonic regime. The application of such methods to hydrocarbon exploration resulted in a better definition of buried prospects and the deformation distribution within them, especially where the complexity of the structural setting is difficult to unravel using ordinary seismic derived images (Aydemir and Dixon, 1999; Colletta et al., 1997; Turrini, 1998; Rennison and Turrini, 1999).

In this chapter we investigate the influence that mechanical stratigraphy plays on the development of a thin-skinned fold and thrust belt, using sandbox experiments that show different boundary conditions. The final three-dimensional deformation partitioning within the models is carefully analyzed and described, and suggestions about the model kinematics are provided. Results are discussed with reference to recent physical modeling and by comparison of the observed experimental geometries with the natural structures that can be interpreted within the Southern Apennines of Italy.

The methodology provided a wide range of structural solutions. The model results represent the first-stage approach to the number of arguments and problems presented, these definitively needing further and more detailed investigations.

INITIAL MECHANICAL STRATIGRAPHY

The initial mechanical stratigraphy of the sandbox models was established by sieving one or more layers of glass microbeads between alternating layers of differently colored quartz sand. Dry cohesionless quartz sand with a grain size of 100–300 μm was used to simulate the competent sedimentary rocks affected by brittle deformation in thrust wedges (Hubbert, 1951; Byerlee, 1978; McClay, 1990; Krantz, 1991). Conversely, glass microbeads were introduced to replicate the natural ductile rocks that enable detachment levels and disharmonic deformation. In the final stratigraphic assemblage, the two modeling materials create an efficient competent contrast due to their different rheological parameters. Whereas sand shows a Navier-Coulomb rheology, with an internal friction angle of about $\phi = 33°$ ($\mu_b = 0.65$) and a bulk density of 1.44 g/cm^3 (dry oxides used to color the sand did not significantly modify its rheological characteristics), the glass microbeads have a grain size of 300–400 μm, a bulk density of 1.6 g/cm^3, and an internal friction angle of $\phi = 23.9°$ ($\mu_b = 0.44$).

In the experiments, two apparatuses with different dimensions have been used: both are glass-sided sandboxes, where a rigid mobile backstop pushes the cohesionless material against a basement plate, generating a Coulomb thrust wedge. The first apparatus is 90 cm long, 30 cm wide, and 14 cm high, whereas the other has internal dimensions of 42 × 30 × 10 cm. The maximum speed of the backstop is <1 cm/min. Friction along the sidewalls was reduced by coating them with graphite powder. Models were built onto a PVC plate that was covered with two different kinds of adhesive paper and with glued sand. The friction angle between the sandbox base and the sand ranges from $\phi = 30°$ to $\phi = 33°$ (glued sand).

EXPERIMENTAL CONDITIONS

Because anisotropy, due to inhomogeneity of the mechanical stratigraphy, was chosen as a common constraint for all of the models, selective boundary conditions were imposed (Fig. 1; Table 1) in order to force three-dimensional deformation partitioning across the developing structures.

In experiments F1, F3, F4, and F6, a rigid, foreland ramp was positioned to occupy half of the basal surface near the side of the apparatus far from the advancing piston (the hinterland). Thus ramp templates were shaped to (1) a segmented frontal-oblique-frontal ramp (experiment F1); (2) a concave to the piston oblique ramp (experiment F3); (3) a segmented frontal-oblique-lateral-oblique ramp (experiment F4); and (4) a simple oblique-frontal ramp (experiment F6).

In experiment F2 two different plastic materials were used to achieve a lateral friction coefficient contrast at the base of the model. In experiments F10 and F11 glued sand was placed over part of the base of the apparatus to simulate a zone of progressive high basal angle of friction, resisting the hinterland-to-foreland deformation propagation.

For experiments F1, F2, F3, F4, and F6, a predeformation wedge-shaped initial configuration was imposed to simulate an advanced stage in the thrust system evolution. During the F1 and F2 experiments, the sandbox apparatus was gradually inclined to simulate foreland flexuring, reaching a basal angle of 6°, with respect to the horizontal. Such a configuration of the model apparatus was established prior to deformation for the F3 experiment. Syntectonic deposition was performed during the F1 and F2 experiments mainly at the front and on the back of the thrusted folds, these advancing toward the model foreland. Conversely, syntectonic erosion at the top of the hinterland structures was performed exclusively during experiment F1.

EXPERIMENTAL RESULTS: DEFORMATION GEOMETRIES

Folds and imbricate structures

Faulted folds, detached at different levels within the stratigraphic succession, are the main structural geometries resulting from all of the sandbox experiments. In vertical cross section (Fig. 2), such geometries clearly define separate structural levels, these being formed both above and below the incompetent microbead layers. As a result the dimensions of the structures can change dramatically so that a clear structural hierarchy is often observed. In these cases major, first-order anticlines detach at the base of the apparatus and deform the entire stratigraphic sequence, whereas minor, second-order folds form and imbricate on the backlimb and in the footwall of the first-order ones. In map view, the same structural hierarchy is confirmed: large distances separate the crests of the first-order anticlines (Fig. 3), whereas small distances characterize the distribution of second-order structures. Folds, of any size and structural order, are generically asymmetric in their geometrical profile and they clearly indicate unambiguous tectonic transport direction (toward the foreland). Decoupling and thickening of the deformed layers can be occasionally recognized in the core of major anticlines (Fig. 4A), which always show overturned forelimbs and thinning of the layers dragged along the fault surfaces (Fig. 4, A and B). Subthrust anticlines, not evident at the model surface, can be found within the deeper part of the models (Fig. 4B), where duplex and horse-like structures can also develop, caught between two important detachment horizons (Fig. 4, C and D).

Faults

Within all of the experiments faults are either planar or listric and discontinuities, (mainly) synthetic and (subordinately) antithetic with respect to major thrust planes, can be recognized (see Figs. 2 and 4). In cross section, high-angle and low-angle fault segments eventually link across the models to define ramp-flat staircase geometries. Ramps cut through the strong sandy horizons, whereas flats are localized within the mechanically weak, glass microbeads layers. In detail, it is remarkable that detachment of the strong sand layers seems to occur systematically at the top of the weak microbeads layer (i.e., at the contact between the two). In map view (see Fig. 3) major faults usually form single sharp traces evidenced by important tectonic scarps at the model surface. Here, minor faults arrange into anastomosed discontinuities that appear at the front of the major ones. No evidence for strike slip or tear fault geometry has been observed within any of the models resulting from the experiments.

EXPERIMENTAL RESULTS: DEFORMATION KINEMATICS

The kinematics observed during the experiments essentially indicate a piggy-back sequence in the deformation progression. As such, folds and faults generally developed from the hinterland (near the advancing piston) to the foreland (further from the advancing piston) domains of the sandbox models, irrespective of any of the boundary conditions imposed. Nevertheless, interaction among the defined constraints has selectively resulted in detailed anomalies within the recognized general deformation propagation.

Deformation kinematics versus mechanical stratigraphy

The inhomogeneous mechanical stratigraphy allowed simultaneous folding and thrusting of different structures, at deep and shallow levels, to develop during most of the experiments.

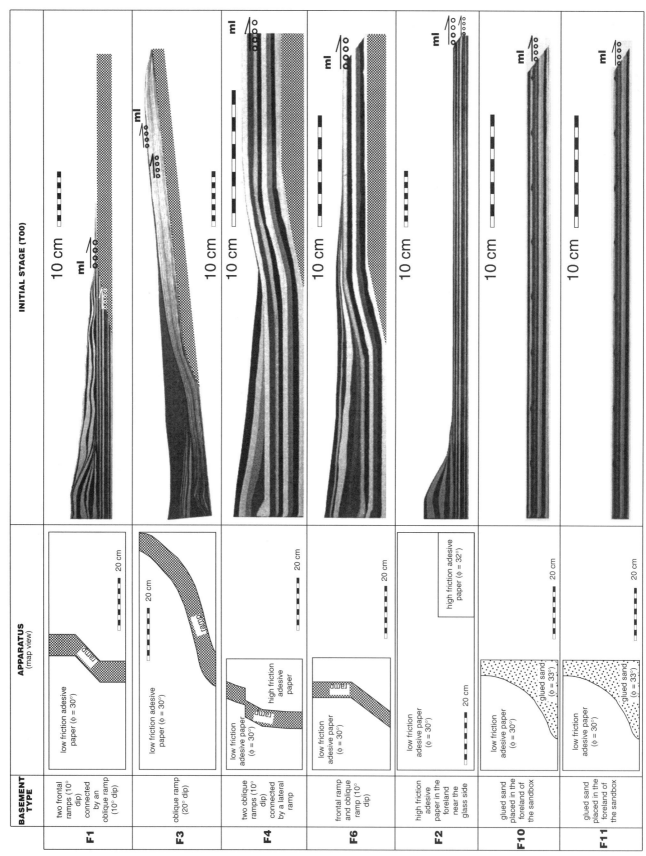

Figure 1. Experimental conditions and initial geometry of undeformed model. Direction of shortening is from left to right for any of experiments; ml: glass microbead layer.

TABLE 1. SYNTHESIS OF THE EXPERIMENTAL CONDITIONS APPLIED TO THE SANDBOX MODELS

Exp.	Initial thickness (cm)		Apparatus dimension (cm)	Deposition event	Erosion event	Base dip event	Final shortening (cm)
	max	min					
F1	8.3	5.4	90×30×14	T06	T06 and T09	T06 ($\beta = 6°$)	18.1
F3	8.3	3.0	90×30×14	-	-	T00 ($\beta = 6°$)	24.0
F4	5.4	3.0	42×30×10	-	-	-	11.5
F6	5.5	2.3	42×30×10	-	-	-	12.0
F2	8.3	2.9	90×30×14	T06 and T09	-	T09 ($\beta = 6°$)	29.2
F10	2.0		42×30×10	-	-	-	17.9
F11	1.6		42×30×10	-	-	-	22.0

T00, T01, and T03 are shortening stages. B—apparatus basal dip. S02 and S04 are final internal sections at 2 and 4 cm from the glass side.

In the F4 experiment (Fig. 5), deformation of the major, first-order anticlines, detached at the base of the model, is accomplished by the synchronous formation of minor, second-order structures at the model surface (i.e., above the microbead layer). These structures progressively develop and imbricate in the footwall of the major hinterland fault and on the back of the major, ramp-related fold.

During experiment F2 (Fig. 6), anisotropy and intrastratal decoupling within the sand-microbead package control the break-backward evolution of duplex structures within the deeper sand layers while enhancing nearly horizontal thrusting along the duplex roof. Such deformation kinematics may have occasionally and diffusely occurred within the F1 and F3 experiments, as serial slicing of those models demonstrates.

Deformation kinematics versus the presence of the foreland ramp

The presence of the rigid foreland ramp in experiments F1, F4, and F6 localized the development of early structures in the model external domain, thus controlling the progressive propagation of the active thrust front.

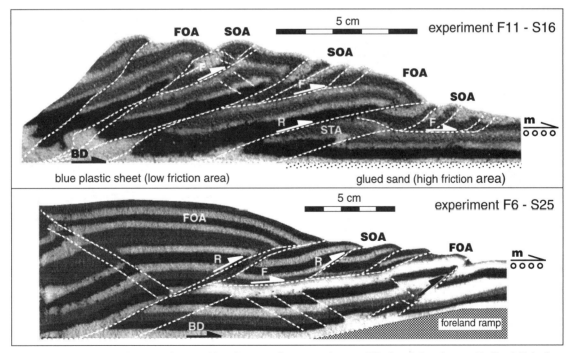

Figure 2. Main structural geometries resulting from sandbox experiments. BD: basal detachment; F: flat; FOA: first-order anticline; ml: glass microbead layer; R: footwall ramp; SOA: second-order anticline; STA: subthrust anticline. Refer to Figure 1 and Table 1 for experimental constraints.

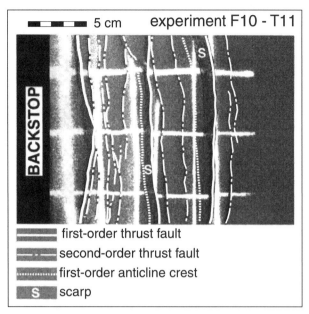

Figure 3. Map view of experiment F10 at 15.5 cm of shortening (T11). Note different distribution and spacing of first- and second-order thrust structures. Refer to Figure 1 and Table 1 for experimental constraints.

During experiment F1 (Fig. 7), the external oblique fault that develops far ahead of the major thrust front at time step T07 (fault 3 in Fig. 7A) clearly mimics the basement geometry. This fault propagates along strike in the foreland synchronously with the initiation of the hinterland fault 2. At time step T08 (Fig. 7B) deformation is stepping backward in the model as fault 2 becomes clearly active. While this fault and the fault 3 develop, a new fault (4) appears and propagates along strike across the entire model (Fig. 7, C and D). The overall deformation sequence suggests out of sequence kinematics for faults 2 and 4.

Deformation kinematics versus the basal friction contrast

The existence of a basal friction contrast in the foreland domain of models F10 and F11 has moderately influenced the basic piggy-back deformation sequence. The map view analysis of the F10 experiment (Fig. 8) shows that at time step T09 (Fig. 8B) en echelon faults develop over the high basal friction area initially defined in the model foreland domain. Here, as deformation progresses, the faults propagate laterally along strike and they converge toward the wider sector of the high basal friction area (Fig. 8, C and D). Comparative analysis between the recognized deformation history and the model's internal geometries (Fig. 8, E and F) suggests that interaction between the anisotropy of the mechanical stratigraphy and the basal friction contrast has been controlling the kinematics of the structures while enhancing disharmonic deformation within the model.

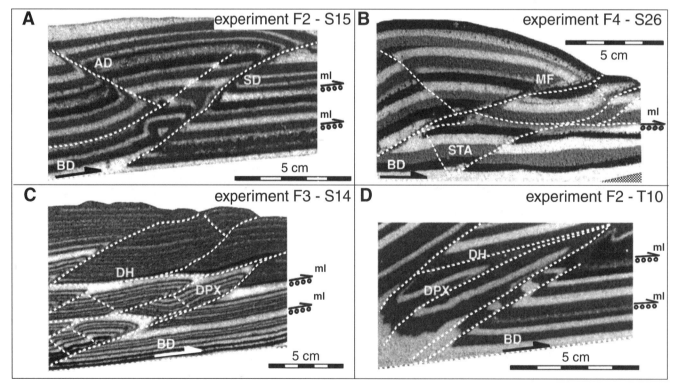

Figure 4. Detailed view of some structural features developed within inhomogeneous mechanical stratigraphy of model. AD: antithetic discontinuity; BD: basal detachment; DH: detachment horizon; DPX: duplex; MF: major fault surface; ml: glass microbead layer; SD: synthetic discontinuity; STA: subthrust anticline. Refer to Figure 1 and Table 1 for experimental constraints.

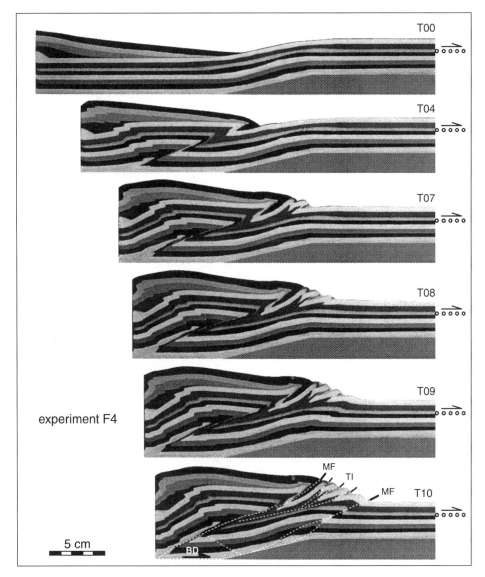

Figure 5. Experiment F4: deformation kinematics vs. mechanical stratigraphy. Progressive stages of shortening (sh) observed on sandbox glass side. Refer to Figure 1 and Table 1 for experimental constraints. T00: undeformed stage; T04: sh = 4.1 cm; T07: sh = 8.2 cm; T08: sh = 9.7 cm; T09: sh = 10.8 cm; T10: sh = 11.5 cm. BD: basal detachment; MF: major fault surface; TI: tectonic imbricates. Microbead layer is indicated.

Deformation kinematics versus syntectonic deposition

Syntectonic deposition performed during some of the experiments localized important deviations from the basic piggyback kinematics generally interpreted.

In experiment F2 (Fig. 9), the addition of syntectonic sediments at the front of the compressional wedge has repeatedly locked displacement along the external thrust. At time step T07 (Fig. 9A) syntectonic deposition forces break-backward deformation within the predeposition, thrusted anticline. As a result a subthrust structure develops below the main fault surface. Successively, at time step T10, addition of syntectonic sediments controls mechanical decoupling and disharmonic folding in the core of the external thrust front anticline (Fig. 9B).

EXPERIMENTAL RESULTS: DEFORMATION PARTITIONING ACROSS THE FINAL MODELS

The perspective view of the internal sections across some of the performed models (experiments F1, F3, F4, F6, and F11) confirms how variations in the basal geometry and friction controlled the three-dimensional partitioning of deformation. The hanging-wall cut-off interpretation derived from the serial cross-section analysis results in the fault trend reconstruction at both shallow (i.e., the model top surface) and deep (i.e., the sand layer immediately below the microbeads layers) structural levels. Integrating the map views of the fault trends with the cross-section analysis allows the distribution and the associated displacement transfer of the three-dimensional structures to be tracked and investigated within the model experiments.

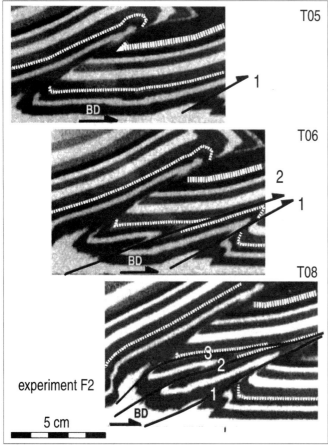

Figure 6. Experiment F2: deformation kinematics vs. mechanical stratigraphy. Detailed view of progressive development of break-backward sequence of deformation. Refer to Figure 1 and Table 1 for experimental constraints. sh—shortening. T05: sh = 14.2 cm; T06: sh = 18.5 cm; T08: sh = 23.6 cm. BD: basal detachment; dotted lines: microbead layers. Numbers indicate fault chronology.

For each of the models discussed hereafter, the faults traced in map view are named accordingly, and the fault families are indicated in the associated cross-section panels.

Three-dimensional structure distribution across the F1 model

The F1 experiment structure distribution defines two structural levels (shallow and deep) in the deformed inhomogeneous stratigraphy (Fig. 10). Within the model, mechanical decoupling generally decreases from the foreland to the hinterland and, subordinately, along strike from section S22 to section S02.

In the hinterland domain structural imbrication can be observed across the entire model and the associated faults cut through the entire sand-microbead package. Within such a domain disharmonic deformation occurs from section S12 to section S22 and exclusively below the deepest microbead layer.

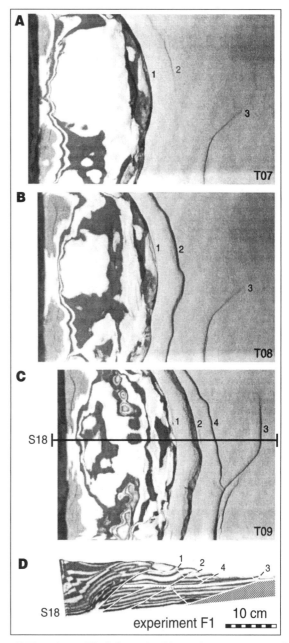

Figure 7. Experiment F1: deformation kinematics vs. presence of foreland ramp. Note out of sequence development of thrust 4 during experiment F1. Refer to Figure 1 and Table 1 for experimental constraints. A, B, C: Map views of model at different stages of shortening (sh) (T07: sh = 12.5 cm; T08: sh = 14.3 cm; T09: sh = 18.1 cm). D: Cross section at 18 cm from glass side of sandbox (see C for section location).

Here, vertical stacking in the sand layers creates horse and duplex structures that are partially to completely decoupled from the anticlines that can be recognized at the model surface.

In the foreland domain mechanical decoupling between the shallow and the deep structural levels is outstanding. The shallow faulted folds detach above the uppermost microbead layer

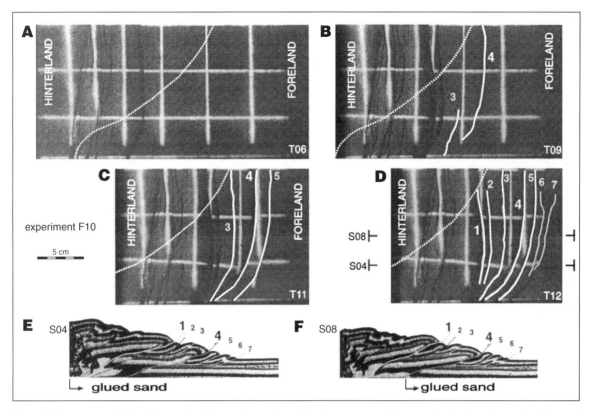

Figure 8. Experiment F10: deformation kinematics vs. basal friction contrast. Note en echelon development of thrust 4. Refer to Figure 1 and Table 1 for experimental constraints. Dotted lines in A–D represent boundary between low basal friction area (toward hinterland) and high basal friction area. A–D: Map views of model at different stages of shortening (sh) (T06: sh = 6.5 cm; T09: sh = 11.9 cm; T11: sh = 15.5 cm; T12: sh = 17.9 cm). E, F: Cross sections at 4 and 8 cm from glass side of sandbox, respectively (see D for section location). Numbers indicate fault kinematics; 1 and 4 represent major faults.

and they are displaced toward the foreland. Continuity along the strike of the derived anticline-syncline pair can be followed as well as the close correlation between the more external fold and the foreland ramp leading edge. Below the shallow structures, the deep structural level deforms as one single unit, the deepest microbead layer (see Fig. 10C) only occasionally being active (i.e., weak disharmony within the deep black and white horizons near the foreland ramp trailing edge, in sections S02 and S12). Within this unit, spacing among the structures increases across the model from section S24 to section S02. Conversely, the number of units decreases in the same direction. The structure distribution in the deep model domain is clearly constrained by the hinterland buttresses and the foreland ramp geometry. From section S02 to section S10, a major fold deformed over the rigid foreland ramp. Moving along strike across the model, from section S10 to section S22, this major fold is transferred down the oblique ramp template and hinterlandward. Simultaneously backthrusts appear as they are localized near the trailing edge of the foreland ramp oblique segment (see sections S18 and S22). Farther away from the rigid ramp and below the major fault that separates the hinterland from the foreland domains, a continuous subthrust structure can be followed.

In the F1 experiment fault maps, two main trends result (Fig. 11, A and B). The f4 and F6 faults, interpreted at shallow and deep level, respectively, define the more external trend. This seems to mimic the buried rigid barrier configuration, and comparison with the cross-section perspective view (see Fig. 10) indicates that it clearly relates to the foreland ramp leading-edge geometry. The trend is completely developed at shallow level whereas it is only partially developed at depth.

The more internal fault trend is defined by a number of arch-shaped fault traces. These are concave toward the hinterland as they elongate across the entire model. At shallow level, the derived structural pattern is simple and the faults (f1, f2, f3) repeat themselves in similar trains almost irrespective of the buried foreland ramp geometry. Nevertheless, minor backthrusts (b3, b4) are significantly distributed within a restricted domain oriented parallel to the foreland ramp oblique segment. At deep level the F4 frontal thrust is suddenly (from section S08 to section S14 in Fig. 11B) transferred across the oblique segment of the rigid ramp and toward the hinterland. Here the B1 and B2 backthrusts appear while the F1–F3 thrusts, oriented almost parallel to the hinterland wall, define the very internal domain of the model.

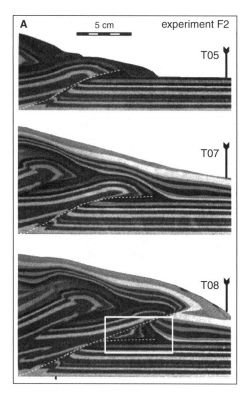
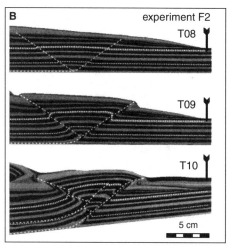

Figure 9. Experiment F2: deformation kinematics vs. syntectonic deposition within (A) internal domain and (B) external domain of the model. Note break-backward deformation in frontal thrust anticlines at time steps T08 and T10. Refer to Figure 1 and Table 1 for experimental constraints. sh—shortening. T05: sh = 14.2 cm; T07: sh = 18.5 cm; T08: sh = 23.6 cm; T09: sh = 25.7 cm; T10: sh = 29.8 cm.

Three-dimensional structure distribution across the F3 model

In the F3 experiment (Fig. 12), faulted folds detach, from top to bottom, (1) above the shallow microbead layer, (2) above the deep microbead layer, and (3) above the model basal surface.

The serial cross sections show that the deformation geometries can be moderately to highly disharmonic. In the hinterland domain mechanical decoupling between the weak and the strong layers is negligible. The faults cut through the entire model package and vertical stacking of the structures results. Toward the foreland, the mechanical decoupling increases and the shallow, middle, and deep structural levels deform by independent mechanics. Within this region, the shallow units are strongly detached from the deeper ones and the derived faults sole into the upper microbead layer. Below this layer, the middle and deep structures deform by small spaced embricated units. Forethrusts and backthrusts appear and duplex imbricated units can be interpreted as they are caught between the upper microbead layer (i.e., the roof thrust) and the model basal detachment. Farther toward the foreland a major thrust is ramping from the bottom toward the top of the model, where it branches into high-angle and low-angle faults. These account for development of the shallow frontal imbricated structures. The major structures can be followed across the serial cross sections and even the minor imbricated units seem to develop continuously within the shallow, intermediate, and deep levels.

The model fault maps (Fig. 13, A–C), reconstructed at the top of each of the defined structural levels, confirm the lateral continuity of the deformation geometries interpreted by the serial sections. The associated fault trends are oriented perpendicular to the shortening direction and they show a gentle curvature, convex toward the model foreland domain. The number of backthrust increases from shallow to deep levels, whereas the number of forethrusts increases from the deep to the shallow level. Spacing among the faults is larger at shallow level and smaller at deep level. Control from the foreland ramp on the structure distribution seems not particularly relevant, although the maximum curvature of the middle- and shallow-level external thrusts seems to be localized at the intersection of the derived fault trend with the buried oblique ramp.

Three-dimensional structure distribution across the F4 model

In the F4 experiment partitioning of the deformation can be clearly observed as along-strike variation of geometry results both vertically and horizontally in the deformed model (Fig. 14). In the hinterland domain no mechanical decoupling is evidenced. Thus a large, continuous faulted fold has been thrusted across the entire model package and displaced toward the foreland ramp. At the front of this major unit a number of imbricated minor ones deform the shallow structural level as they detached at the top of the weak microbeads layer. The number of these

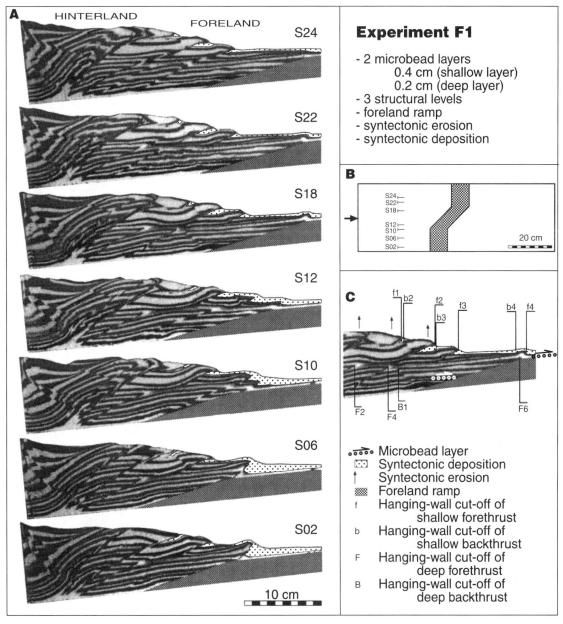

Figure 10. Experiment F1: structures developed at final stage of shortening. A: Serial cross sections. B: Map view of apparatus showing cross-section positions and basement ramp location (see also Fig. 1 and Table 1). C: Explanation of symbols used in Figure 11. Refer to Figure 1 and Table 1 for detailed experimental constraints.

minor units decreases along strike from section S02 to section S26. At a deep level (i.e., below the microbeads) a major fold developed over the rigid foreland ramp. This first-order fold structure is progressively transferred down the ramp template and into a hinterland subthrust anticline. Backthrusts appear at the footwall of the major hinterland thrust fault and at the trailing edge of the ramp template.

In map view (Fig. 15, A and B) the influence of the foreland ramp on the shallow and deep structure distribution is revealed. The shallow f1 to f4 faults (Fig. 15A) are fairly continuous across the entire model; they strike almost perpendicular to the shortening direction, yet obliquely to the foreland ramp orientation. These faults do not show any indication of the buried ramp presence. However, the arch-shaped geometry of the f5 and f6 external faults suggests that the basement ramp geometry is partially controlling the development of the shallow structures. Such a control of the foreland ramp on the localization of the deep structures is outstanding from the associated fault pattern (Fig. 15B). Here the F5 external thrust is strongly arch shaped and it completely overlaps the foreland ramp in the right side of

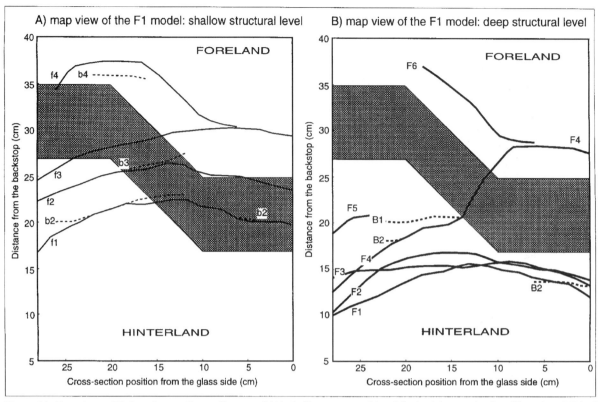

Figure 11. Experiment F1: hanging-wall cut-off maps of (A) shallow structural level faults (above microbead layer) and (B) of deep structural level faults (below microbead layer) (see Fig. 10C). Refer to Figure 1 and Table 1 for detailed experimental constraints.

the model. Transition of the F5 fault toward the hinterland is abrupt as the derived structure is transferred from the foreland side to the hinterland side of the rigid ramp.

Three-dimensional structure distribution across the F6 model

In experiment F6 (Fig. 16) the foreland ramp dips 10° toward the piston, and has consistently controlled the geometry of the external thrust front across the entire model. Toward the hinterland a major faulted anticline shows a strong continuity across the model and no decoupling can be observed between the shallow and the deep level.

In the model domain between sections S03 and section S15, the external thrust climbs up from the basal detachment at the leading edge of the rigid ramp with a dip varying from 20° to 25°. The associated anticline shows strong asymmetry and no backthrusts. Toward the opposite side of the apparatus (sections S17 to S27), the same external thrust ramps up from a position over the middle of the rigid ramp and has a higher dip that ranges from 27° to 40°. In this part of the model, one or more backthrusts develop near the trailing edge of the rigid ramp and in the backlimb of the associated anticline.

Moderate to strong imbrication of the shallow level units appears across the model foreland domain. The derived structural units are completely decoupled from the deep forethrust and backthrust mechanics as a function of the thick, incompetent microbead horizon.

In the fault map reconstruction, the thrusts show a clear arch-shaped geometry, concave to the piston both at shallow and deep levels (Fig. 17, A and B). Faults developed within the two structural levels show a weak geometrical compatibility that confirms mechanical decoupling between the derived structures (see Fig. 16).

At a shallow level (Fig. 17A) the faults show along-strike continuity as they strike across the deep foreland ramp, and the influence of the ramp geometry on the shallow structures seems negligible.

At a deep level (Fig. 17B) the fault curvature is more evident and the resulting asymmetry suggests that the foreland ramp has constrained the development of the structures. Backthrusts occur within a restricted zone of the model (i.e., hinterlandward of the foreland ramp frontal segment) and their distribution is likely due to transfer of displacement along the trailing edge of the foreland ramp.

Three-dimensional structure distribution across the F11 model

In experiment F11 (Fig. 18) a basal high friction area, concave toward the advancing piston, defines a clear lateral de-

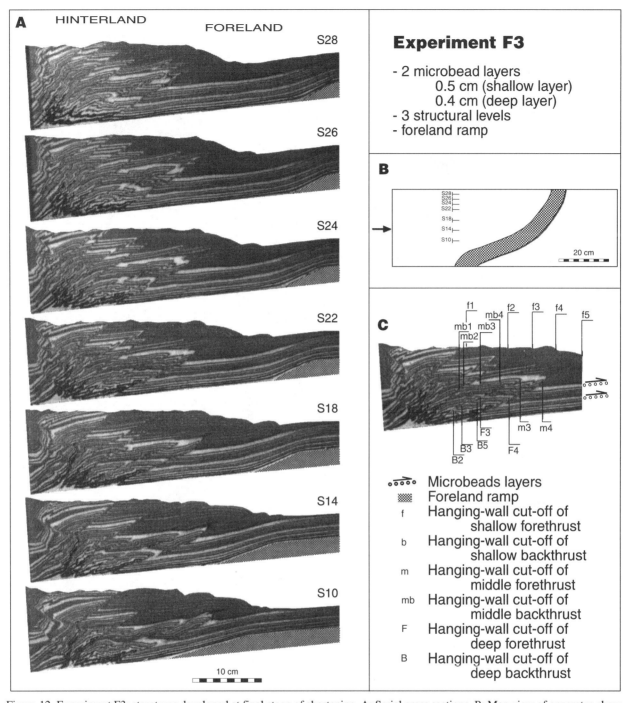

Figure 12. Experiment F3: structures developed at final stage of shortening. A: Serial cross sections. B: Map view of apparatus showing cross section positions and basement ramp location (see also Fig. 1 and Table 1). C: Explanation of symbols used in Figure 13. Refer to Figure 1 and Table 1 for detailed experimental constraints.

formation gradient. The structural units show increasing imbrication and vertical stacking as we move from the low-friction to the high-friction basal surface area. In particular, the vertical sections located in the area where the high-friction basal surface (glued sand) is near the advancing piston (sections S04 to S08) are characterized by (1) a steeper taper angle of ~19°; (2) a strong imbrication at the front of the tectonic wedge both at deep (first-order structures) and shallow (second-order structures) levels; (3) a tendency for a low dip and general concave-down profile of the major thrusts; and (4) the near absence of backthrusts.

In the area where the high-friction basal surface is far away from the piston (foreland side of the apparatus, sections

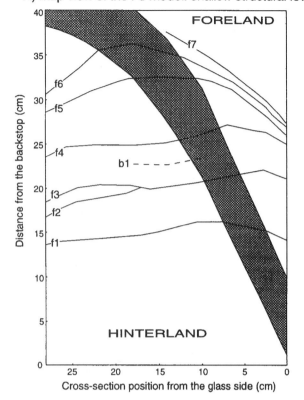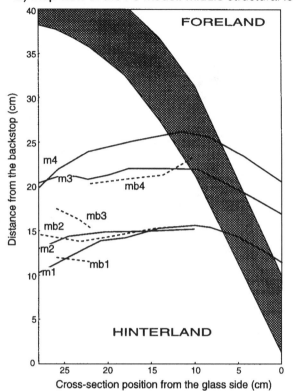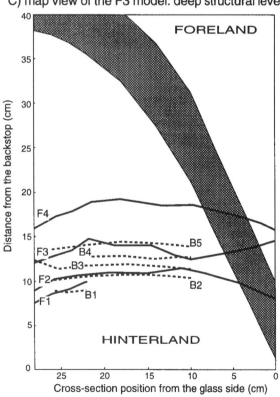

Figure 13. Experiment F3: hanging-wall cut-off maps of (A) shallow structural level faults (above upper microbead layer), (B) middle structural level faults (between upper and lower microbead layers; see Figure 12C), and (C) deep structural level faults (below lower microbead layer) (see also Fig. 12C). Refer to Figure 1 and Table 1 for detailed experimental constraints.

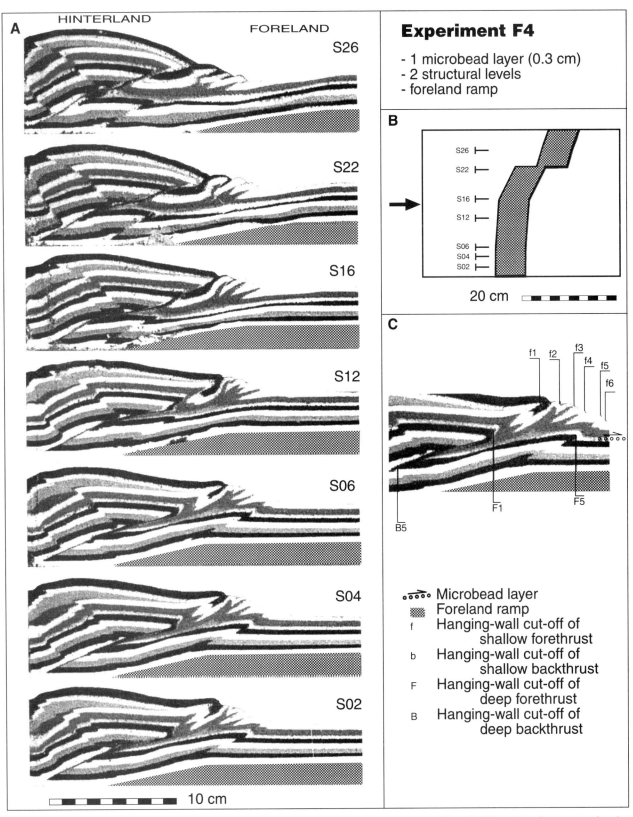

Figure 14. Experiment F4: structures developed at final stage of shortening. A: Serial cross sections. B: Map view of apparatus showing cross-section positions and basement ramp location (see also Fig. 1 and Table 1). C: Explanation of symbols used in Figure 15.

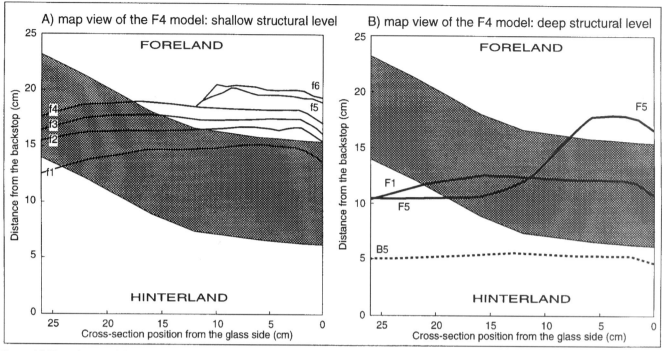

Figure 15. Experiment F4: hanging-wall cut-off maps of (A) shallow structural level faults (above microbead layer) and (B) deep structural level faults (below microbead layer) (see also Fig. 14C). Refer to Figure 1 and Table 1 for detailed experimental constraints.

S20–S26), the deformation geometries show (1) a smaller value (~15°) of the taper angle; (2) a larger number of regularly spaced first-order structures; (3) a relatively higher dip and general concave-up profile of the major thrusts; and (4) the presence of multiple backthrusts.

Analysis of the model fault maps (Fig. 19, A and B) reveals that curvature of thrusts becomes important from the hinterland toward the foreland structural domain. Here, a restricted transfer zone can be interpreted within both the shallow and deep structural levels at the boundary between the high-friction basal area and the low-friction basal area. At deep levels the forethrusts show anastomosed geometries, while farther toward the hinterland discontinuous backthrusts are localized. At shallow level the faults clearly replicate the deep fabric in general terms of geometry and forethrust-backthrust distribution. This suggests that while the mechanical stratigraphy controls the strong structural disharmony that the serial sections document (see Fig. 18), variation in the three-dimensional basal friction controls the major structural wavelength across the model.

DISCUSSION

Comparison of the model results

The sandbox experiments provided a number of results that show strong similarity as well as notable differences. A comparison among the models shows that the structural geometries and kinematics are considerably similar in general terms of shape and basic deformation progression, respectively. In detail, important differences have been observed in the model three-dimensional distribution and evolution of the structures. A major similarity is the hinterland to foreland piggy-back deformation progression that can be commonly interpreted irrespective of any of the boundary conditions selectively applied. Further similarities relate to the inhomogeneous mechanical stratigraphy that was chosen as a common constraint to the experiments and they essentially refer to the following.

1. The structural geometries, i.e., faulted folds, embricated units, staircase-shaped overthrusts, and duplexes, are commonly recognized within any of the model experiments (see Figs. 2 and 4).

2. Differential structural levels develop in the stratigraphic succession (the formation of two or more structural levels is straightforward in any of the experiments performed; see Figs. 10, 12, 14, 16, and 18).

3. The models exhibit structural disharmony (vertical partitioning of the deformation within the structural levels of the models is widely recognized, although poor to strong disharmonic tectonics can be interpreted; see Figs. 2, 4, and 10–18).

4. Synchronicity between the structures' kinematics developed within the shallow and the deep levels (simultaneous deformation of different structures within different structural levels has always been recognized, although occasionally strongly localized, within any of the sandbox experiments; see Figs. 5 and 6).

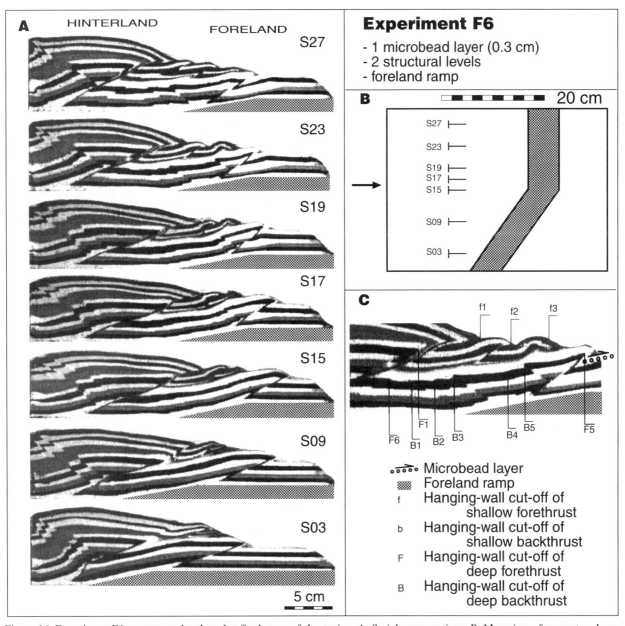

Figure 16. Experiment F6: structures developed at final stage of shortening. A: Serial cross sections. B: Map view of apparatus showing cross-section positions and basement ramp location (see also Fig. 1 and Table 1). C: Explanation of symbols used in Figure 17.

Major differences among the experimental results are mainly due to the boundary conditions initially established. Those differences develop in response to variations in the three-dimensional basal geometry and friction of the models and as a function of the syntectonic deposition occasionally performed. In detail they refer to the following.

1. Anomalous kinematics are recognized within the generic piggy-back deformation progression. In the experiments F1, F4, and F6, the presence of a rigid barrier localizes early deformation within the model foreland ramp, which is progressively hunted by the advancing thrust belt front (see Fig. 9). In the experiments F10 and F11, the presence of a basal friction contrast within the model foreland domain, localizes transversal kinematics and along-strikes propagation of the deformation between the high-friction and low-friction areas (see Fig. 8). In the experiment F2, syntectonic deposition on top of the model enhances break-backward kinematics within the frontal thrust anticline (see Fig. 9).

2. Displacement transfer mechanics are interpreted within the different models, both vertically (across the structural

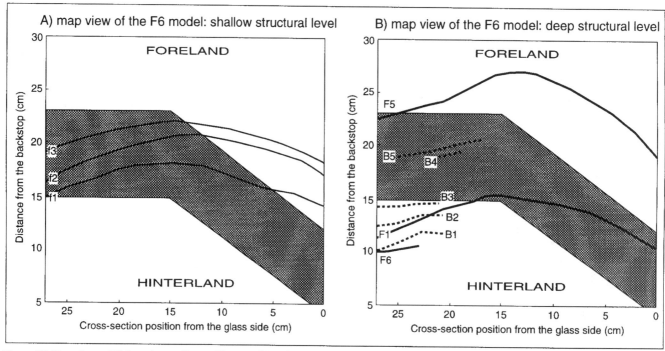

Figure 17. Experiment F6: hanging-wall cut-off maps of (A) shallow structural level faults (above microbead layer) and (B) deep structural level faults (below microbead layer) (see also Fig. 16C). Refer to Figure 1 and Table 1 for detailed experimental constraints.

levels) and horizontally (along strike). The experiments F1, F4, and F6 show that the presence of a foreland ramp causes along strike displacement transfer to occur by oblique thrusting and forethrust-backthrust structural associations (see Figs. 11, 15, and 17). The experiments F10 and F11 show that the presence of a basal friction contrast within the model foreland domain causes lateral displacement transfer by en echelon fault arrangement (see Fig. 8) and overlapping thrust geometries (Fig. 19B), respectively. Further toward the model hinterland in the experiment F11, minor backthrusts contribute to accommodate the overall deformation (see Fig. 19B).

3. Three-dimensional deformation partitioning can be established among the different (shallow, intermediate, deep) structural levels, within each of the model experiments. In the experiments F1, F4, and F6, the shallow structures deform cylindrically and almost irrespective of the rigid barrier position and form, whereas the deep structures can be discontinuous while showing important along-strike variations, closely related to the foreland ramp geometry (see Figs. 11, 15, and 17). In the experiment F3, despite the serial cross sections indicating a moderate to strong structural disharmony (see Fig. 12), the mechanical levels defined (shallow, intermediate and deep) show top to bottom geometrical compatibility of the derived structures (see Fig. 13). In the F11 experiment, although the models' internal geometries indicate strong mechanical decoupling, the shallow structure trends are definitively coupled to the deep ones (see Figs. 18 and 19).

Comparison of the model results with recent sandbox modeling literature

Comparison of our model results with recent sandbox modeling literature suggests analogies and differences in terms of deformation geometries, kinematics, and three-dimensional partitioning.

The following considerations can be ruled out.

The deformation geometries we observed closely resemble the fold-fault structures presented and discussed in recent sandbox modeling literature (e.g., Malavieille, 1984; Mulugeta and Koyi, 1987; Mulugeta, 1988a; McClay, 1990; Colletta et al, 1991; Liu et al., 1992; Storti et al., 1997). The resulting hinterland to foreland piggy-back kinematics are clearly similar to the structural evolution that different authors have described.

The inhomogeneous mechanical stratigraphy we defined allowed vertical structural disharmony, and multiple detachments to be performed within any of the models. The resultant structural hierarchy is outstanding in most of the experiments. Among the many solutions attempted in the past (Ballard et al., 1987; McClay, 1990; Colletta et al., 1991; Baby et al., 1995; Letuzey et al., 1995; Merle and Abidi, 1995; Storti and McClay, 1995; Verschuren et al., 1996; Mugnier et al., 1997; Storti et al., 1997; Corrado et al., 1998), only the utilization of Newtonian, nongranular materials (i.e., silicone, oil-water emulsion) resulted in mechanical decoupling mechanics and an overall structural framework similar to the one we present herein.

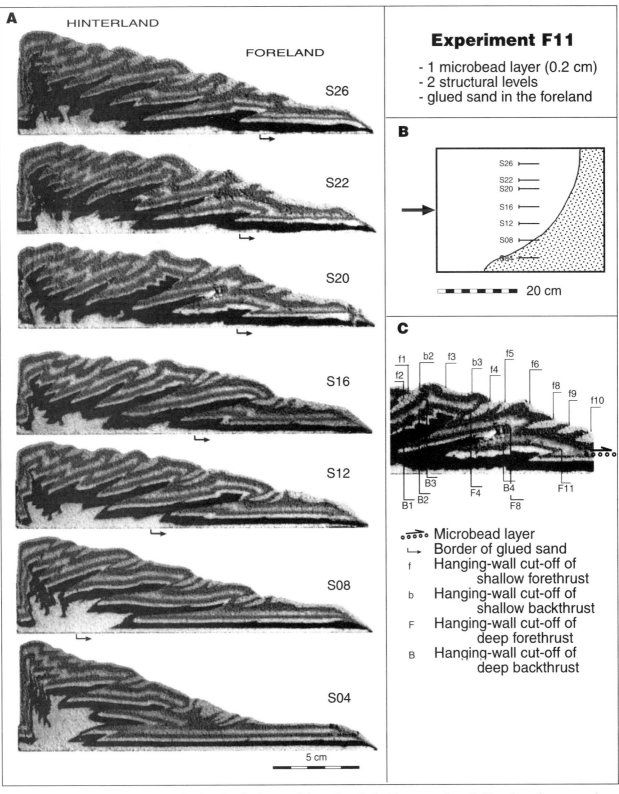

Figure 18. Experiment F11: structures developed at final stage of shortening. A: Serial cross sections. B: Map view of apparatus showing cross-section positions and basement ramp location (see also Fig. 1 and Table 1). C: Explanation of symbols used in Figure 19.

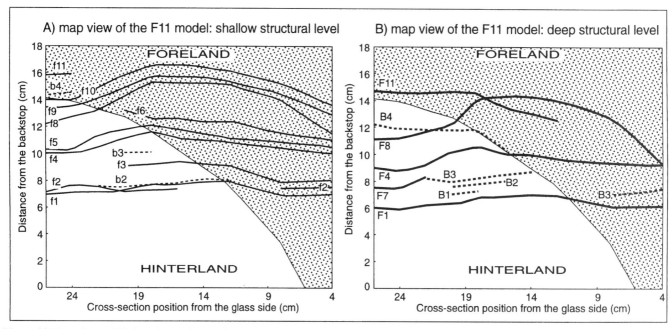

Figure 19. Experiment F11: hanging-wall cut-off maps of (A) shallow structural level faults (above microbead layer) and (B) deep structural level faults (below microbead layer) (see also Fig. 18C). Refer to Figure 1 and Table 1 for detailed experimental constraints.

The presence of a rigid foreland ramp within some of our model experiments resulted in vertical and along-strike partitioning of the deformation (see F1, F3, F4, and F6 experiments) while localizing anomalous deformation kinematics during the generic piggy-back fold and thrust propagation (see Fig. 7). Wilkerson et al. (1992) and Merle and Abidi (1995) studied the influence of a rigid ramp on the derived sandbox compressional geometries. The influence of a foreland ramp configuration on the associated structure kinematics was simulated by Philippe et al. (1998). Results from Merle and Abidi are not comparable to the conclusions we present because of the syntectonic erosion effect they considered. Wilkerson et al. indicated out of plane strains (developing in response of the oblique-ramp template configuration), results that we cannot support or disprove, because our experiments did not address such details. Philippe et al. documented deformation kinematics similar to the ones we propose, both in terms of along-strike propagation of the foreland faults and out of sequence progression of the hinterland faults (cf. Fig. 13 of Philippe et al. [1998] with Fig. 7 herein).

The presence of basal friction contrast in our F10 and F11 experiments essentially controls transfer displacement mechanics that develop over the foreland boundary between the high basal friction area and the low basal friction area (see Figs. 8 and 19B). Colletta et al. (1991) and Calassou et al. (1993) carefully analyzed deformation partitioning related to such a boundary condition. In particular, the three-dimensional reconstructions performed by Calassou et al. could show synchronous thrust deformation and/or propagation and anastomosed faults geometry similar to those we describe.

Although the simulation of syntectonic deposition and erosion during our sandbox experiments represented a secondary target, the results we obtained suggest detailed deformation kinematics (intrastratal decoupling, sindepositional breakbackward thrusting; see Fig. 9) and specific geometries (disharmonic folds, subthrust structures; see Fig. 9) that cannot be found in the current literature (Storti and McClay, 1995; Mugnier et al., 1997).

Application of the modeling results to the Southern Apennines

The comparison of the sandbox models performed with the structural setting interpreted from the Calabro-Lucano Southern Apennines chain sector indicates that the experiments can suggest deformation geometries and kinematics that reasonably replicate the thin-skinned compressional structures that develop within the region.

The Calabro-Lucano Southern Apennine chain sector is interpreted as an arch-shaped, west-southwest–east-northeast–displaced, fold and thrust belt wherein compressional and extensional events overprinted each other through time to form the current mountain chain structure (Mostardini and Merlini, 1986, and references therein; Casero et al., 1988; D'Andrea et al., 1993; Pasi et al., 1995; Roure and Sassi, 1995; Ferranti et al., 1997; Menardi Noguera and Rea, 1997; Bonini and Sani, 1998; Monaco et al., 1998). The structural style indicates mainly thin-skinned (in the external domains of the chain, to the northeast) to possibly thick-skinned (in the internal domains of the chain, to the southwest) tectonics. An upper, allochthonous wedge of carbonate and/or clastic units (Sicilide–Lagonegro–Apenninic platform–Irpinian units) has been thrusted great distances over the associated autochthonous deformed units of Apulian platform carbonates.

The mechanical stratigraphy recognized by the available data set (surface geology, wells, seismic interpretation) largely accounts for the wide range of deformation structures, mechanics, and kinematics that have been described in the region (Rennison and Turrini, 1999). Such stratigraphy emphasises the strength contrast that the sediments, belonging to the main tectonic units progressively described, demonstrate within the structural setting. With reference to the regional cross section shown in Figure 20, three principal tectono-stratigraphic domains can be recognized: internal, external, and marginal (foredeep related) with respect to the regional architecture and the related southwest-northeast displacement direction.

In the internal domain the following mechanical layering is observed (from surface to depth): allochthonous massive carbonate units (a strong layer); 3–4-km-thick, mainly limestones-dolostones (Miocene-Triassic "Apenninic platform units"). The next layering comprises allochthonous, multilayered, mainly carbonate units (a relative weak layer); 1–3-km-thick limestones, dolostones, radiolarites, and shales (Oligocene-Triassic Lagonegro, Sicilide, and Irpinian units). Underlying these are autochthonous clastic units (a weak layer); 0–800(?)-m-thick shales (lower Pliocene Pre-Bradano foredeep unit). The deepest layering comprises autochthonous massive carbonate units (a strong layer); maximum 6000-m-thick, mainly carbonates, rocks (Miocene-Triassic internal Apulian platform units).

Within the external domain the following mechanical layering is defined: allochthonous clastic units (a weak layer); 3–4-km-thick shales, sandstones, and interlayered limestones (Oligocene-Miocene Irpinian units). These are underlain by autochthonous clastic units (a weak layer); 0–1000(?)-m-thick shales (lower Pliocene Pre-Bradanic units). The lowest layering comprises autochthonous massive carbonate units (a strong layer); maximum 6000(?)-m-thick, mainly carbonate, rocks, (Miocene-Triassic external Apulian platform units).

The marginal domain shows the following mechanical layering: autochthonous clastic units (a weak layer); 0–1-km-thick sand and shale (Pliocene-Pleistocene Bradano foredeep). These are underlain by allochthonous clastic units (a weak layer); 1–2-km-thick shales, sandstones, and interlayered limestones

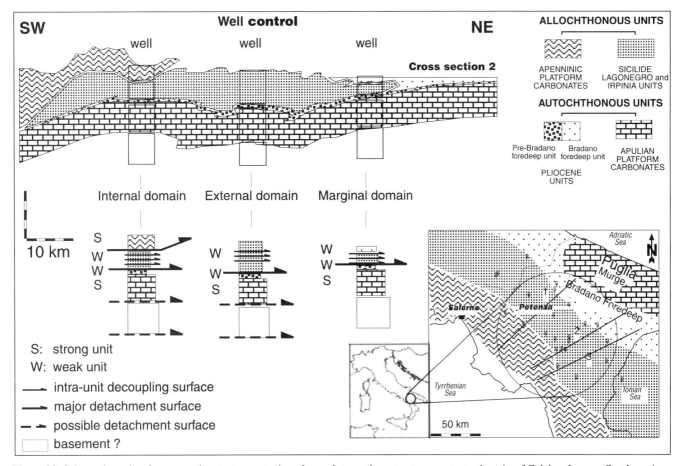

Figure 20. Schematic regional cross section, tectono-stratigraphy, and strength contrast among tectonic units of Calabro-Lucano Southern Apennines (modified from Rennison and Turrini, 1999). Location of sections of Figure 21 is shown in map.

(Oligocene-Miocene Irpinian units). The next mechanical layering comprises autochthonous clastic units (a weak layer); 0–1(?)-km-thick shales (lower Pliocene pre-Bradano foredeep units). These are underlain by autochthonous massive carbonate units (a strong layer); maximum 6000-m-thick, mainly carbonate, rocks (Miocene-Triassic Apulian platform units).

Within all of domain (internal, external, and marginal) stratigraphies, multiple weak horizons break the vertical mechanical continuity of the packages. Among these there is clear evidence for a major regional detachment surface at the top of the deepest autochthonous clastic unit (lower Pliocene) on which the allochthonous clastic units are decoupled and displaced.

The F1 sandbox experiment can be used to constrain and validate the interpretation of the structures across the Campano-Lucano Southern Apennines fold and thrust belt. The experimental wedge-shaped initial configuration (see Fig. 1) simulates an advanced stage in the Southern Apennines deformation (i.e., after the lower Pliocene emplacement of the allochthonous unit over the autochthonous ones). The mechanical stratigraphy of the models represents a possible analogue to the Southern Apennines tectono-stratigraphy mentioned here. It follows that the model shallow sandy layers (above the upper weak microbead layer) represent the far-traveled allochthonous clastic units overthrusting the Apulian external and marginal domains. Conversely, the deep model sandy layers (below the upper weak microbead layer) are intended to replicate the relatively autochthonous structural units belonging to the Apulian internal, external, and marginal domains. The model upper microbead layer represents the weak Pliocene pre-Bradano deposits that regionally decouple the allochthonous structural wedge from the autochthonous foreland domain. Within such a context, the F1 model foreland ramp, resisting the advancing synthetic thrust belt, can be reasonably compared to the undeformed Apulian foreland, which crops out to the northeast in the Murge region, and to the southwest deepens below and regionally opposes the Apennines overthrusting.

Given the analogies established between the experimental setting and the natural one, the application of the F1 model results to the interpreted Southern Apennines tectonics (Figs. 21 and 22) supports the following considerations.

1. The tectono-stratigrapy interpreted within the Calabro-Lucano region has enhanced mechanical decoupling between the derived major structural levels in a way similar to that shown by the F1 structural architecture (Fig. 21). Minor faulted folds and imbricated units develop within the allochthonous units (the model shallow structural level) while major anticlines deform the autochthonous units (the model deep structural level). The autochthonous pre-Bradano Pliocene deposits (the model shallow microbead layer in the F1 model) allows such a mechanical decoupling to take place between the allochthonous and the autochthonous unit structures.

2. As in the F1 model experiment (Figs. 21A and 10), such mechanical decoupling between the shallow and the deep structural units seems to increase from the hinterland to the foreland domain of the Southern Apennines thrust belt (Fig. 21).

3. The crustal-scale rigid ramp defined by the foreland Apulian carbonates, at the front of the Apennines thrust belt, imposes a horizontal deformation distribution similar to that indicated by the F1 model experiment (Fig. 22). The map view of the major thrust fronts in the area can be compared to the map-view fault patterns interpreted in the F1 mechanical simulation (Fig. 22, A and B).

4. The deformation events that took place during the model simulations replicate the Pliocene-Pleistocene deformation kinematics recognized within the Calabro-Lucano region. This has likely followed a basic piggy-back hinterland-to foreland structural propagation. Nevertheless, the F1 kinematics (see Fig. 7) indicate that development of the external structure trend within the Apulian carbonates might have initiated as an early-stage structure with respect to the more internal ones, this likely being reworked by late-stage, out of sequence events.

5. Such a deformation history could have enhanced early migration and accumulation of the hydrocarbons toward and within the early-stage foreland structures before the current tectonic setting was reached.

6. Due to the decoupling mechanics interpreted between the shallow allochthonous structures and the deep autochthonous ones, subthrust anticlines that are not evident at the surface may be within the deeper part of selected structural domains of the Southern Apennines chain, thus representing future targets for the hydrocarbon exploration activity.

CONCLUSIONS

The sandbox modeling demonstrated that vertical inhomogeneity within an initial mechanical stratigraphy can play a critical role in controlling initiation and development of the deformation in a thin-skinned type compressional tectonic regime. The experiments provided a number of results that show strong similarity as well as notable differences. A comparison among the models shows that the structural geometries and kinematics are similar in general terms of shape and basic deformation progression, respectively. In detail, important differences have been observed in the model three-dimensional structure distribution and evolution.

The hinterland to foreland piggy-back deformation progression represents the common structural kinematics observed during the model experiments. The inhomogeneous mechanical stratigraphy that was chosen as a common constraint to the experiments has driven the development of the major similarities observed. These similarities mainly relate to: (1) the structural geometries, (2) the development of differential structural levels in the stratigraphic succession, (3) the model structural disharmony, and (4) synchronicity between the structure kinematics developed within the shallow and the deep levels.

The major differences seen in the experiment results developed in response to variations in the model three-dimensional

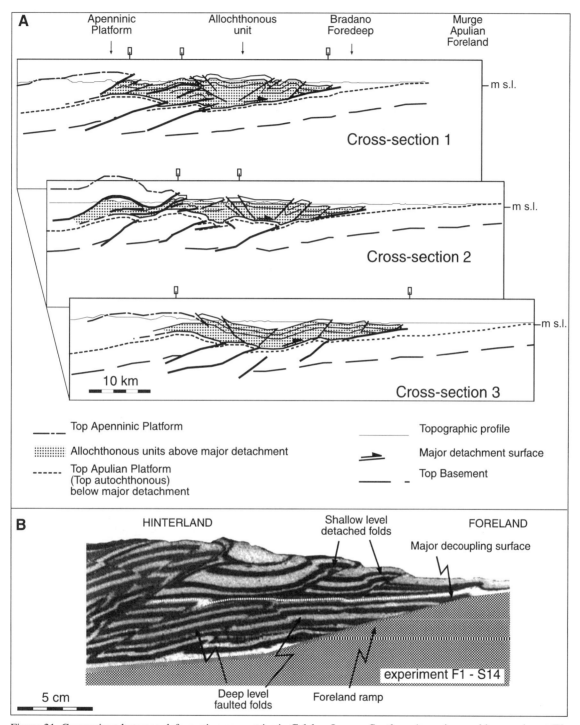

Figure 21. Comparison between deformation geometries in Calabro-Lucano Southern Apennines and in experiment F1. A: Serial structural cross sections (see Figure 20 for location; modified from Rennison and Turrini, 1999). B: Structures developed at final stage of shortening (sh = 18.1 cm) in experiment F1 at 14 cm from glass side of sandbox (see text). m s.l.—mean sea level.

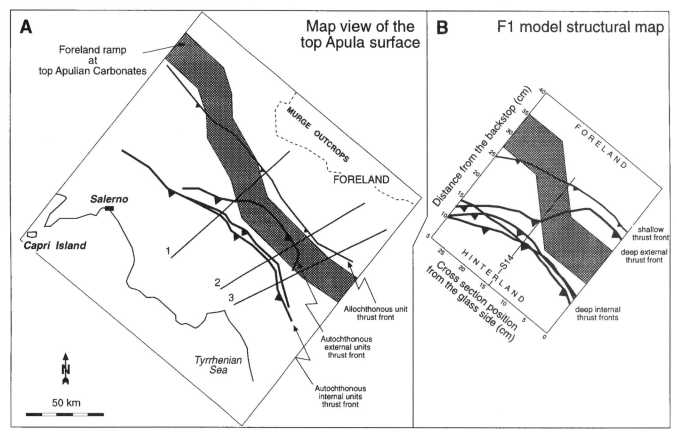

Figure 22. Comparison between schematic structural map of (A) Apula Platform top surface (modified from Rennison and Turrini, 1999) and (B) structural map of experiment F1 (see text). Figure 21 cross section position is shown.

basal geometry and friction and as a function of the syntectonic deposition occasionally performed. In detail they could be referred to: (1) the anomalous kinematics recognized within the generic piggy-back deformation progression, (2) the displacement transfer mechanics interpreted within the different models, both vertically (across the structural levels) and horizontally (along strike), and (3) the three-dimensional deformation partitioning that can be established among the different (shallow, intermediate, deep) structural levels within each experiment.

We conclude the following.

1. The structural geometries obtained are significantly similar throughout the entire set of experiments presented.

2. The structural kinematics derived are similar for all of the models discussed in terms of basic deformation progression. In detail, important differences have been demonstrated among them.

3. The three-dimensional deformation partitioning within the models is moderately homogeneous across the model hinterland domain as shallow and deep structures show minor mechanical disharmony. Conversely, three-dimensional deformation partitioning is strongly inhomogeneous across the model foreland domain, both vertically and horizontally (along strike).

4. Vertical strain distribution is mainly due to the inhomogeneous mechanical stratigraphy defined, whereas horizontal strain distribution essentially relates to the boundary conditions selectively applied (variations in three-dimensional basal geometry and friction, syntectonic sedimentation). The results obtained shed new light on the understanding of thrust belt architectures and suggest deformation kinematics that may be different from those suggested by outcrop and/or seismic data reconstruction. Consequently, they suggest new hints for comprehension of buried and/or poorly imaged highly complex structures.

Comparison of the experiment results with serial cross sections from the Southern Apennines of Italy demonstrate that the reconstruction of an accurate mechanical stratigraphy and the definition of the predeformation and syndeformation boundary constraints (barriers in the rigid basement, rates of erosion and/or deposition) can contribute to the understanding of the deformation geometries in the area while enhancing the evaluation of the regional hydrocarbon potential.

ACKNOWLEDGMENTS

TOTALFINAELF Italia S.p.A. is acknowledged for permission to present the sandbox models discussed herein. Special thanks to the reviewers, Deborah A. Spratt and Josep Poblet, for their critical comments. The experimental work for this study was carried out in the Structural Geology Laboratory of the Dipartimento di Scienze della Terra, Università di Pavia.

REFERENCES CITED

Aydemir, E.A., and Dixon, J.M., 1999, Along-strike variation of geometry and strain partitioning in fold thrust structures: 3-D seismic interpretation and physical modelling: Thrust tectonic conference, London, Abstracts with Programs: London, Royal Holloway University of London, p. 131–133.

Baby, P., Colletta, B., and Zubieta, D., 1995, Étude géométrique et expérimentale d'un bassin transporté: Exemple du synclinarium de l'Alto Beni (Andes centrales): Bulletin de la Société Géologique de France, v. 166, p. 797–813.

Ballard, J.F., Brun, J.P., Dressche, J.V., and Allemand, P., 1987, Propagation des chevauchements au-dessus des zones de décollement: modèles expérimentaux: Paris, Academie des Sciences Comptes Rendus, v. 305, ser. II, p. 1249–1253.

Bonini, M., and Sani, F., 1998, Structural evolution of the Lucanian Apennines between Potenza and the S. Arcangelo Basin, in Atti del 79° Congresso nazionale della Società Geologica Italiana: Palermo, Società Geologica Italiana.

Boyer, S.E., 1995, Sedimentary basin taper as a factor controlling the geometry and advance of thrust belts: American Journal of Science, v. 295, p. 1220–1254.

Byerlee, J., 1978, Friction of rock: Pure and Applied Geophysics, v. 116, p. 615–626.

Calassou, S., Larroque, C., and Malavieille, J., 1993, Transfer zones of deformation in thrust wedges: An experimental study: Tectonophysics, v. 221, p. 325–344.

Casero, P., Roure, F., Endignoux, L., Moretti, I., Mueller, C., Sage, L., and Vially, R., 1988, Neogene geodynamic evolution of the Southern Apennines: Memorie della Società Geologica Italiana, v. 41, p. 109–120.

Colletta, B., Letouzey, J., Pinedo, R., Ballard, J.F., and Bale, P., 1991, Computerized X-ray tomography analysis of sandbox models: examples of thin-skinned thrust systems: Geology, v. 19, p. 1063–1067.

Colletta, B., Roure, F., De Toni, B., Loureiro, D., Passalacqua, H., and Gou, Y., 1997, Tectonic inheritance, crustal architecture, and contrasting structural style in the Venezuela Andes: Tectonic, v. 16, p. 777–794.

Corrado, S., Di Bucci, D., Naso, G., and Facenna, C., 1998, Influence of palaeogeography on thrust system geometries: An analogue modelling approach for the Abruzzi-Molise (Italy) case history: Tectonophysics, v. 296, p. 437–453.

Currie, J.B., Patnode, H.W., and Trump, R.P., 1962, Development of folds in sedimentary strata: Geological Society of America Bulletin, v. 73, p. 655–674.

Dahlen, F.A., and Suppe, J., 1988, Mechanics, growth, and erosion of mountain belts, in Clark, S.P., Jr., Burchfiel, B.C., and Suppe, J., eds., Processes in continental lithospheric deformation: Geological Society of America Special Paper 218, p. 161–178.

Dahlstrom, C.D.A., 1970, Structural geology of the eastern margin of the Canadian Rocky Mountains: Bullettin of Canadian Petroleum Geology, v. 18, p. 332–406.

D'Andrea, S., Pasi, R., Bertozzi, G., and Dattilo, P., 1993, Geological model, advanced methods help unlock oil in Italy's Apennines: Oil & Gas Journal, v. 91, p. 53–56.

Davis, P., and Engelder, T., 1985, The role of salt in fold and thrust belts: Tectonophysics, v. 119, p. 67–88.

Davis, P., Suppe, J., and Dhalen, F.A., 1983, Mechanics of fold-and-thrust belts and accretionary wedges: Journal of Geophysical Resources, v. 88, p. 1153–1172.

Dixon, J.M., and Liu, S., 1992, Centrifuge modelling of the propagation of thrust faults, in McClay, K.R., ed., Thrust tectonics: London, Chapman and Hall, p. 53–69.

Dixon, J.M., and Summers, J.M., 1985, Recent developments in centrifuge modelling of tectonic processes: Equipment, model construction techniques and rheology of model materials: Journal of Structural Geology, v. 7, p. 83–102.

Dixon, J.M., and Tirrul, R., 1991, Centrifuge modelling of fold-thrust structures in a tripartite stratigraphic succession: Journal of Structural Geology, v. 13, p. 3–20.

Eisenstadt, G., and De Paor, D.G., 1987, Alternative model of thrust-fault propagation: Geology, v. 15, p. 630–633.

Erickson, S.G., 1996, Influence of mechanical stratigraphy on folding vs faulting: Journal of Structural Geology, v. 18, p. 443–450.

Fermor, P., 1999, Aspects of the three-dimensional structure of the Alberta Foothills and Front Ranges: Geological Society of America Bulletin, v. 111, p. 317–346.

Ferranti, L., D'Argenio, B., Marsella, E., Oldow, J.S., Pappone, G., and Sacchi, M., 1997, Orogen-parallel extension during foreland imbrication: Southern Apennines fold and thrust belt, Italy: Geological Society of America Abstracts with Programs, v. 29, no. 6, p. A318.

Gretener, P.E., 1972, Thoughts on overthrust faulting in a layered sequence: Bulletin of Canadian Petroleum Geology, v. 20, p. 583–607.

Gross, M.R., Gutierrez, A.G., Bai, T., Wacker, M.A., and Collinsworth, K.B., 1997, Influence of mechanical stratigraphy and kinematics on fault scaling relations: Journal of Structural Geology, v. 19, p. 171–183.

Hubbert, M.K., 1951, Mechanical basis for certain familiar geologic structures: Geological Society of America Bulletin, v. 62, p. 335–372.

Koyi, H., 1995, Mode of internal deformation in sand wedges: Journal of Structural Geology, v. 17, p. 293–300.

Krantz, R.W., 1991, Mesurements of friction coefficients and cohesion for faulting and fault reactivation in laboratory models, using sand and sand mixtures: Tectonophysics, v. 188, p. 203–207.

Letouzy, J., Colletta, B., Vially, R., and Chermette, J.C., 1995, Evolution of salt related structures in compressional settings, in Jackson, M.P.A., et al., eds., Salt tectonics: A global perspective: American Association of Petroleum Geologists Memoir 65, p. 41–60.

Liu, H., McClay, K.R., and Powell, D., 1992, Physical models of thrust wedges, in McClay, K.R., ed, Thrust tectonics: London, Chapman and Hall, p. 71–81.

Liu, S., and Dixon, J.M., 1990, Centrifuge modelling of thrust faulting: Strain partitioning and sequence of thrusting in duplex structures, in Knipe, R.J., and Rutter, E.H., eds., Deformation mechanisms, rheology and tectonics: Geological Society [London] Special Publication 54, p. 431–434.

Liu, S., and Dixon, J.M., 1991, Centrifuge modelling of thrust faulting: Structural variation along stike in fold-thrust belts: Tectonophysics, v. 188, p. 39–62.

Liu, S., and Dixon, J.M., 1995, Localization of duplex thrust-ramps by buckling: Analog and numerical modelling: Journal of Structural Geology, v. 17, p. 875–886.

Malavieille, J., 1984, Modélisation expérimentale des chevauchements imbriqués: Application aux chaînes de montagnes: Bulletin de la Société Géologique de France, v. 7, p. 129–138.

Marshak, S., and Wilkerson, M.S., 1992, Effect of overburden thickness on thrust belt geometry and development: Tectonics, v. 11, p. 560–566.

Marshak, S., Wilkerson, M.S., and Hsui, A.T., 1992, Generation of curved fold-thrust belts: Insight from simple physical and analytical models, in McClay, K.R., ed., Thrust tectonics: London, Chapman and Hall, p. 83–92.

McClay, K.R., 1990, Deformation mechanics in analogue models of extensional fault systems, in Knipe, R.J., and Rutter, E.H., eds., Deformation mechanisms, rheology and tectonics: Geological Society [London] Special Publication 54, p. 445–453.

Menardi Noguera, A., and Rea, G., 1997, The Campano-Lucano arc deep structure (Southern Apennines, Italy), Palermo, Università di Palermo, Eighth Workshop of the ILP Task Force, Origin of Sedimentary Basins, Palermo, Italy, Abstracts with Program, p. 90–92.

Merle, O., and Abidi, N., 1995, Approche expérimentale du fonctionnement des rampes émergentes: Bulletin de la Société Géologique de France, v. 166, p. 439–450.

Mitra, S., 1986, Duplex structures and imbricate thrust systems: Geometry, structural position and hydrocarbon potential: American Association of Petroleum Geologists Bulletin, v. 70, p. 1087–1112.

Monaco, C., Tortorici, L., and Paltrinieri, W., 1998, Structural evolution of the Lucanian Apennines, southern Italy: Journal of Structural Geology, v. 20, p. 617–638.

Mostardini, F., and Merlini, S., 1986, Appennino centro-meridionale: sezioni geologiche e proposta di modello strutturale: Memorie della Società Geologica Italiana, v. 35, p. 177–202.

Mugnier, J.L., Baby, P., Colletta, B., Vinour, P., Bale, P., and Leturmy, P., 1997, Thrust geometry controlled by erosion and sedimentation: A view from analogue models: Geology, v. 25, p. 427–430.

Mulugeta, G., 1988a, Modelling the geometry of Coulomb thrust wedges: Journal of Structural Geology, v. 10, p. 847–859.

Mulugeta, G., 1988b, Squeeze-box in a centrifuge: Tectonophysics, v. 148, p. 323–335.

Mulugeta, G., and Koyi, H., 1987, Three-dimensional geometry and kinematics of experimental piggyback thrusting: Geology, v. 15, p. 1052–1056.

Mulugeta, G., and Koyi, H., 1992, Episodic accretion and strain partitioning in a model sand wedge: Tectonophysics, v. 202, p. 319–333.

Pasi, R., Dattilo, P., Bertozzi, G., and Lakew, T., 1995, A novel approach to the exploration of the Southern Apennines, Italy: Geological models and oil discoveries [abs.]: American Association of Petroleum Geologists Bulletin, v. 79, p. 1241.

Philippe, Y., Deville, E., and Mascle, A., 1998, Thin-skinned inversion tectonics at oblique basin margins: Example of the western Vercors and Chartreuse Subalpine massifs (SE France), in Mascle, A., et al., eds., Cenozoic foreland basins of Western Europe: Geological Society [London] Special Publication 134, p. 239–262.

Ramsay, J.G., and Huber, M.I., 1987, The techniques of modern structural geology, Volume 2, Folds and faults: London, Academic Press, 391 p.

Rennison, P., and Turrini, C., 1999, Structural style from the Southern Apennines hydrocarbon province: An integrated view: London, Royal Holloway University of London, Thrust tectonic conference, Abstracts with Programs, p. 302–306.

Rich, J.L., 1934, Mechanics of low angle overthrust faulting as illustrated by Cumberland thrust block, Virginia, Kentucky, and Tennessee: American Association of Petroleum Geologists Bulletin, v. 34, p. 672–681.

Roure, F., and Sassi, W., 1995, Kinematics of deformation and petroleum system appraisal in Neogene foreland fold-and-thrust belt: Petroleum Geoscience, v. 1, p. 253–269.

Storti, F., and McClay, K., 1995, Influence of syntectonic sedimentation on thrust wedges in analogue models: Geology, v. 23, p. 999–1002.

Storti, F., Salvini, F., and McClay, K., 1997, Fault-related folding in sandbox analogue models of thrust wedges: Journal of Structural Geology, v. 19, p. 583–602.

Turrini, C., 1998, West Bulmoose area: The Palsson structure post well analysis: Milano, Fina Italiana internal report, 15 p.

Verschuren, M., Nieuwland, D., and Gast, J., 1996, Multiple detachment levels in thrust tectonics: Sandbox experiments and palinspastic reconstruction, in Buchanan, P.G., and Nieuwland, D.A., eds., Modern developments in structural interpretation, validation and modelling: Geological Society [London] Special Publication 99, p. 227–234.

Wilkerson, M.S., Marshak, S., and Bosworth, W., 1992, Computerized tomographic analysis of displacement trajectories and three-dimensional fold geometry above oblique thrust ramps: Geology, v. 20, p. 439–450.

Wiltschko, D.V., and Chapple, W.M., 1977, Flow of weak rocks in Appalachian Plateau folds: American Association of Petroleum Geologists Bulletin, v. 61, p. 653–670.

MANUSCRIPT ACCEPTED BY THE SOCIETY APRIL 12, 2000

Four-dimensional analysis of analog models: Experiments on transfer zones in fold and thrust belts

Guido Schreurs*, Reto Hänni
Geological Institute, University of Bern, Baltzerstrasse 1, CH-3012 Bern, Switzerland
Peter Vock
Institute of Diagnostic Radiology, Inselspital, CH-3010 Bern, Switzerland

ABSTRACT

Spiral X-ray computed tomography (CT) was applied to the analysis of brittle-viscous analog models. This technique allows volume scanning at regular time-lapse intervals, making it possible to perform a four-dimensional analysis, i.e., the three-dimensional evolution through time. The experiments simulate transfer zone development in fold and thrust belts. Models showed a marked contrast in structural style between domains that were underlain or not underlain by a thin viscous layer, representing brittle-viscous and brittle rheological domains, respectively. Closely spaced, forward-propagating thrusts formed a narrow and high fold and thrust belt over the brittle domain. A wider and lower fold and thrust belt, characterized by out of sequence thrusting, coeval fault movement, and pop-up structures formed over the brittle-viscous domain. Transfer zones developed in the transition zone between the two domains. The location and orientation of transfer zones were related to the geometry of the boundary between basal viscous layer and adjacent brittle layers. Lateral ramps formed where this boundary was parallel to the shortening direction, whereas oblique ramps formed where this boundary was oblique. Lateral and oblique ramps had shallow dips and dip angles that changed along strike. The presence of a rheologically weak layer at the base of a sedimentary sequence may favor out of sequence thrusting, coeval displacement along different thrusts, and development of forward and backward thrusts with associated folds. The location and orientation of transfer zones in nature may be controlled by rheological changes in the basal detachment. The main advantage of spiral X-ray CT analysis is that it provides the possibility to rerun the experiment on the computer screen and examine each recorded stage, using three-dimensional computer visualization techniques. Such animations are an important aid in establishing geometric and kinematic concepts necessary for understanding the development of geological structures in nature.

INTRODUCTION

Analog models have long been popular for simulating geological structures, because they offer the opportunity to determine the relation between imposed boundary conditions and resulting structures. The strength of such models is mainly in stimulating the conception of testable hypotheses about the development of geological structures in nature. Since the late 1980s, analog models have been analyzed by X-ray computed tomography (CT) originally developed for medical purposes (Hounsfield, 1973). This technique makes it possible to visualize the interior of a model without destroying it (e.g., Mandl, 1988; Colletta et al., 1991; Wilkerson et al., 1992; Schreurs, 1992, 1994; Philippe, 1995). On the basis of the varying attenuation of X-rays by

*Corresponding author. E-mail: schreurs@geo.unibe.ch

different materials, this technique generates cross-sectional slices through an analog model and allows a detailed analysis of its internal geometry and kinematics. Attenuation is mainly dependent on material density, effective atomic number, and thickness (Colletta et al., 1991). Layering can be produced by using materials with distinct attenuation. Model experiments are performed within the acquisition field of the scanner. The number of cross-sectional slices that can be obtained of a particular stage in the evolution of the model is mainly a function of the X-ray dose intensity needed for proper visualization, and the performance of the X-ray source. The time between consecutive runs is dictated by the required cooling of this source below a certain threshold value. Because of the time involved in CT data acquisition and the long cooling duration of the X-ray source, early medical scanners could only record a limited number of cross-sectional slices. Another limiting factor was the computational capacity needed to calculate images from projectional raw data profiles. Periodic acquisition of sections at similar positions during deformation made it nevertheless possible to follow the two-dimensional evolution of structures in time. However, the time-consuming recording of closely spaced sequential cross-sectional slices, necessary for a full three-dimensional analysis of the model, was generally not carried out until the end of the experiment (e.g., Mandl, 1988; Colletta et al., 1991; Philippe, 1995; Schreurs and Colletta, 1998).

Recent technological improvements of X-ray tubes, acquisition systems, and computation capacity have resulted in more powerful X-ray CT techniques. The acquisition system of spiral or helical CT scanners revolves around the object as the object moves in the longitudinal direction (orthogonal to the scanning plane) through the scanning plane. In this manner, three-dimensional volume raw data on analog models are easily acquired. The slice profile must be chosen in advance, but an unlimited number of closely spaced serial cross sections can be calculated retrospectively from the raw data. Data acquisition time for such a three-dimensional data set depends on the X-ray dose necessary to adequately penetrate the analog material (i.e., material composition and thickness), detector quality, and the performance of the entire acquisition system (e.g., rotation time, table speed, data transfer, and computation speed). Slice spacing will only affect the postprocessing time. Periodic acquisition of such volumetric data sets makes it possible to follow the three-dimensional evolution of models from the initial undeformed stage to the final deformed stage. This opens new and exciting perspectives for a complete four-dimensional analysis (three-dimensional through time) of analog models and is of special importance when investigating complex structural settings, where changes in three-dimensional geometry are common.

Three-dimensional structural changes along strike are a characteristic feature of fold and thrust belts and have been documented, e.g., in the Alberta Foothills and Front Ranges of the Canadian Rocky Mountains (Fermor and Moffat, 1992; Fermor, 1999), in the western Salt Range of Pakistan (McDougall and Khan, 1990), and in the Jura Mountains (Philippe, 1994, 1995). Faults within such belts are connected by complex systems of sub-

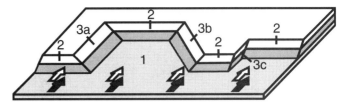

1. Subhorizontal detachment
2. Frontal ramp
3. Transfer zones: 3a. Lateral ramp, 3b. Oblique ramp, 3c. Tear fault

Figure 1. Schematic representation of three-dimensional fault system in a fold and thrust belt (modified after McClay, 1992). Transfer zones transfer displacement from one frontal ramp to next.

horizontal detachments, frontal ramps, and transfer zones (Fig. 1). Transfer zones in fold and thrust belts transfer displacement from one frontal ramp to the next. Three types of transfer zones are generally distinguished (McClay, 1992): lateral ramps, oblique ramps, and tear faults. The origins of transfer zones are diverse and can be related to, e.g., preexisting basement faults, basal rheological changes, and variations in lithological thicknesses.

Several experimental studies have investigated along-strike changes in fold and thrust belts (e.g., Colletta et al., 1991; Dixon and Liu, 1992; Liu et al., 1992; Marshak et al., 1992; Wilkerson et al., 1992; Calassou et al., 1993; Lallemand et al., 1994; Philippe, 1995; Philippe et al., 1998). Parameters such as variations in sedimentary thickness, preexisting discontinuities, basal friction conditions, stationary obstacles, and geometry of the mobile backstop have been shown to control the development of transfer zones. The effects of variations in basal friction on the resulting structures using granular materials as brittle analogs were studied by Colletta et al. (1991) and Calassou et al. (1993). These experiments were characterized by structural domains separated from one another by a transition zone. In high basal friction domains, thrusts were closely spaced and the thrust belt was narrow and high. In the low basal friction domains, thrusts formed far from the mobile backstop, and the thrust belt was wider and lower. A transfer zone formed in the transition zone between the two domains. Colletta et al. (1991) used a U-shaped basal discontinuity to induce two lateral ramps parallel to the shortening directions in the overlying sand cake. X-ray CT tomography analysis of the final stage of their experiment showed that lateral ramps root at the boundary between high- and low-friction domain and had shallow dips (~30°). Philippe (1995) studied the influence of geometrical changes of viscous layers that were placed at one or two levels within a multilayer brittle-viscous model. Transfer zones formed at the boundaries between brittle-viscous domains and purely brittle domains. In contrast to purely brittle domains, thrust propagation in brittle-viscous domains was no longer in sequence, and several thrusts were active at the same time (Philippe, 1995).

In our experiments, both brittle and viscous analog materials are used. The boundary conditions differ from those described by Philippe (1995) in that only one viscous layer at the base of the model was used and this layer did not extend over the entire width of the model. Spiral X-ray computerized tomography was applied to the study of transfer zones in the analog models. Transfer zones were induced by deforming a basal viscous layer that was placed adjacent to the mobile backstop and was overlain by brittle, granular materials. In this chapter we present a detailed four-dimensional analysis of the models and discuss the influence of the shape of the basal viscous layer on the geometric and kinematic evolution of transfer zones.

MODEL SETUP AND PROCEDURE

Models consisted of brittle and viscous analog materials and were deformed in the experimental apparatus shown in Figure 2. A 5 mm thick layer of a viscous polydimethyl-siloxane (PDMS) polymer was placed over part of the wooden base. Viscous polymers are considered a good analog material for simulating viscous flow of salt and/or evaporites in the upper crust or rocks in the lower crust (Vendeville et al., 1987). PDMS has a density of 0.965 g/cm^3 and a Newtonian viscosity of 5×10^4 Pa s at room temperatures and at strain rates below 3×10^{-3} s^{-1} (Weijermars, 1986). Dry quartz sand and corundum powder with an average grain size of about 100 μm were sprinkled in alternate layers on top of the viscous PDMS and the adjacent uncovered wooden base to produce a stratified model. These granular materials obey the Coulomb criterion of failure, and their angle of internal friction of ~30°–35° is closely similar to those determined experimentally for upper crustal rocks (40° for normal stresses <0.2 GPa and 31° for normal stresses in the range 0.2–2 GPa; Byerlee, 1978). Sand and corundum powder represent good analogs for brittle competent rocks in the upper crust (Horsfield, 1977; Byerlee, 1978). The initial model was 80 cm long, 27 cm wide, and its thickness, including the basal PDMS layer, amounted to 3 cm. Displacement of a vertical mobile wall, driven by a motor, produced shortening of the model at a constant velocity of 4.8 cm/h. At the end of the experiment the model width was reduced by 9 cm. Surface photographs were taken at 0.5 cm increments of shortening.

Our models were scaled for lengths, viscosities, and time using methods discussed by Hubbert (1937) and Ramberg (1981). At basin scale, scale ratios between models and natural examples are 2×10^{-5} for length (1 cm in the model represents 500 m in nature), 5×10^{-15} for viscosities (implying a viscosity of 10^{19} Pa s for salt) and 6.25×10^{-10} for time (1 h of experiment represents about 180 k.y. in nature). Parameters such as geothermal gradient, pore pressure, and differential compaction, which are difficult to model, were not considered in our experiments. Furthermore, the use of granular materials neglects the effect of cohesive strength in natural rocks, because it is practically zero in dry sand (Naylor et al., 1986). Scaling of grain size is also not considered, and Horsfield (1977) pointed out that fault-zone width in granular material is dependent on grain size. Nevertheless, partially scaled analog models may generate ideas about the origin and development of geological structures that should then be tested or verified by field studies.

We used a Siemens Somatom Plus 4 spiral X-ray computed tomographer (CT) for the analysis of our experiments. This scanner can acquire an entire volume of the model without interruption in a short time. In our experiments we applied a 120 kV peak voltage, 130 mA, 2 mm X-ray beam width to differentiate analog material attenuations. Rotation time was 1 s and table feed per rotation was 4.2 mm. One three-dimensional volume scan lasted about 60 s and was obtained from that part of the model where important three-dimensional geometrical changes were expected (see Fig. 3). We show the results of two experiments analyzed by spiral CT technique. From the three-dimensional volume raw data, 88 (experiment 73) and 74 (experiment 75) cross-sectional slices with 2 mm spacing were computed. Thus, the field of investigation width covered 17.6 cm and 14.8 cm, respectively (Fig. 3). From the three-dimensional data set it is possible to reconstruct images at any desired position and to postprocess them using computer visualization techniques. Periodic CT acquisition each 12.5 min during deformation allowed us to follow the three-dimensional evolution of the model structures at successive stages of deformation, and thus to examine our models in four dimensions. We obtained the three-dimensional geometry of the initial, undeformed state and subsequently after every full centimeter

Figure 2. Experimental apparatus used to simulate transfer zones.

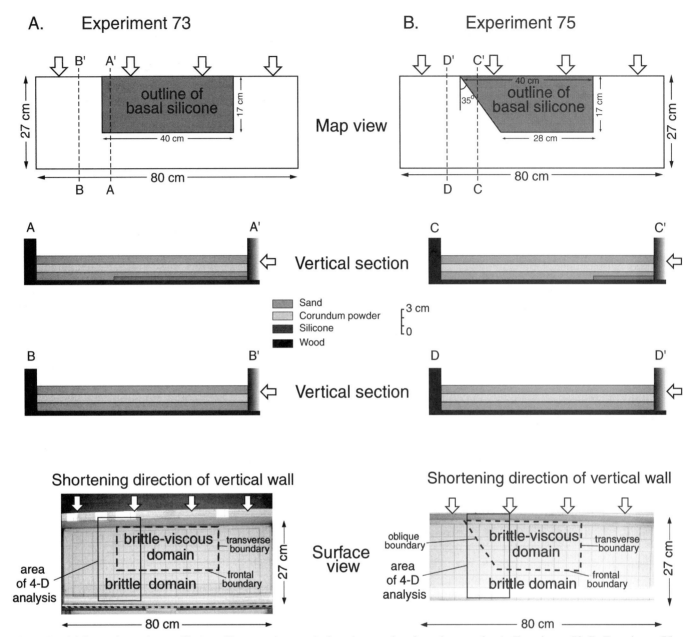

Figure 3. Initial experimental setup illustrated by map view, vertical sections, and surface photographs. A: Experiment 73. B: Experiment 75. Boundary conditions were identical except for shape of basal layer of viscous polydimethylsiloxane (PDMS).

of progressive shortening, giving us a total of 10 three-dimensional data sets for each experiment.

In our brittle-viscous analog models we tested the influence of one parameter on the evolution of transfer zones: the shape of the basal viscous layer. In one experiment (73) we placed a rectangular layer of viscous PDMS, 40 cm long, 17 cm wide, and 5 mm high on top of the wooden base next to the mobile backstop (Fig. 3A). Alternating layers of sand and corundum powder were poured on top. Thus this model had two lateral PDMS-sand boundaries parallel to, and one frontal boundary perpendicular to, the shortening direction. In another experiment (75) the shape of the basal vis-

cous layer was different: one side of the PDMS layer was oblique to the shortening direction (at an angle of 35°), while the other side was parallel to it (Fig. 3B). As a consequence, the frontal PDMS-sand boundary was now shorter and had a length of 28 cm. In each experiment a 3 mm slit at the base of the mobile wall ensured that the initial detachment horizon was located within either sand or PDMS. Each type of experiment was performed twice, with and without X-ray CT analysis, and showed very similar surface evolution as judged from photographs, indicating satisfactory experimental reproducibility. For ease of description, we refer to the sand-corundum-sand layers as the brittle domain, and to the sand-corundum-sand layers with a basal viscous PDMS layer as the brittle-viscous domain.

EXPERIMENTAL RESULTS

In all experiments there was a marked difference in the structural evolution between the brittle and the brittle-viscous domain. For this reason, the evolution is first described separately for the two domains using successive transverse CT images of the progressively deformed model of experiment 73. Subsequently, the structures in surface view and in the transition zone between brittle and brittle-viscous domain are discussed on the basis of surface photographs, and perspective and cut-out three-dimensional views.

Brittle domain (experiment 73)

After initial distributed grain flow, progressive shortening was accommodated by thrust faults with opposite vergence (Fig. 4A). Forward thrusts dipped more shallowly (26°–28°) than backthrusts (37°–43°). The downward-converging thrust faults defined a pop-up structure that rooted near the base of the lowermost sand layer. The forward thrust in the brittle domain showed a slight refraction, with a steeper dip in the upper sand layer than in the corundum powder layer, giving the thrust fault a slightly concave-upward shape along its upper part (Fig. 4A). Subsequently, a low-angle thrust fault developed in the hanging wall (Fig. 4, B and C), creating a volume of brittle material completely bounded by thrust faults, i.e., a horse.

As deformation increased, forward thrusts were predominantly active, while backthrusts formed at the bend from flat to lower ramp (Fig. 4C). The backthrusts showed little movement and were passively transported over the forward thrust. Fault-bend folds formed as the thrust sheet moved over the ramp of the forward thrust. With progressive shortening, the basal detachment was activated in front of the pop-up structure and a new thrust imbricate developed (Fig. 4, D and E). Activity along the previously formed forward thrust ceased and fault movement began to occur along the newly developed forward thrust in its footwall. With increasing deformation, another in sequence imbricate thrust formed (Fig. 4H).

Brittle-viscous domain (experiment 73)

A pop-up structure formed initially near the mobile wall (Fig. 4J), with forward thrusts dipping at 30°–34° and backthrusts at 30°–37°. Thrust faults rooted at the top of the viscous layer. With increasing strain a horse formed as another forward thrust developed in the hanging wall, dipping subparallel to the former frontal thrust (Fig. 4K). The younger hanging-wall thrust initiated directly at the boundary between the brittle and the viscous material as the latter was carried upward along the frontal ramp. The horses were progressively deformed as shortening continued (cf. Fig. 4, K–M). Fault activity was dominantly along forward thrusts and a fault-bend fold formed as material moved up the ramp. As deformation progressed, a new frontal thrust formed far away from the mobile wall, at the forward boundary between basal PDMS and sand (Fig. 4M). Displacement along the forward thrust of the pop-up structure near the mobile wall ceased. Because of continuing movement of the mobile wall, the forward thrust of the pop-up became progressively steeper and the triangular block bounded by forward thrust, backthrust, and mobile wall underwent a rotation about a horizontal axis. This boundary effect caused bulging and extension in the upper part of the fault-bend fold and small normal faults formed in its outer arc (Fig. 4M). As progressive shortening increased, backthrusts formed at the lower bend of the active frontal ramp. At a slightly more advanced stage, an out of sequence thrust developed in the region between the existing forward thrusts (Fig. 4O). Movement along this thrust occurred at the same time as displacement along the foremost forward thrust. Displacement along this out of sequence thrust was also coeval with movement along backthrusts of the frontal pop-up structure and resulted in a complex interference pattern.

Transfer zone (experiment 73)

In surface view, thrust faults developed first in the brittle domain and propagated along strike into the brittle-viscous domain. Downward-converging thrust faults bound a pop-up structure and strike parallel to the mobile wall (Fig. 5A). Bending of thrusts near the lateral walls of the deformation box resulted from friction along these boundaries. Because forward thrusts in the brittle-viscous domain rooted at the top of the viscous layer and dipped slightly steeper than in the brittle domain, the thrust front surfaced slightly closer to the mobile wall in the brittle-viscous domain (Fig. 5A). With progressive shortening, a forward thrust originated at the frontal boundary between brittle and viscous material (Fig. 5B). Deformation then took place in two completely different parts of the model: close to the mobile wall in the brittle domain, and near the frontal termination of the basal PDMS in the brittle-viscous domain (Fig. 5B).

The frontal thrusts linked up along transfer zones that formed subparallel to the shortening direction (Fig. 5C). In longitudinal sections parallel to the mobile wall, faults in the transfer zone rooted at the sand-PDMS boundary and had a listric shape (Fig. 6A). These faults are lateral ramps striking subparallel to the shortening direction. The dip of these lateral ramps decreased along strike toward the frontal ramp in the brittle-viscous domain, from about 30° to <10° (Fig. 6, A–D). The curvature of the thrust front is clearly visible in horizontal sections through the transfer zone (Fig. 7). These sections also show how the backthrust of the frontal pop-up dies out along strike as it approaches the transfer zone. In surface view, the thrust system has a broad U shape at this stage (Fig. 5C). In the brittle-viscous domain, coeval movement occurred along several thrusts, whereas fault activity in the brittle domain was

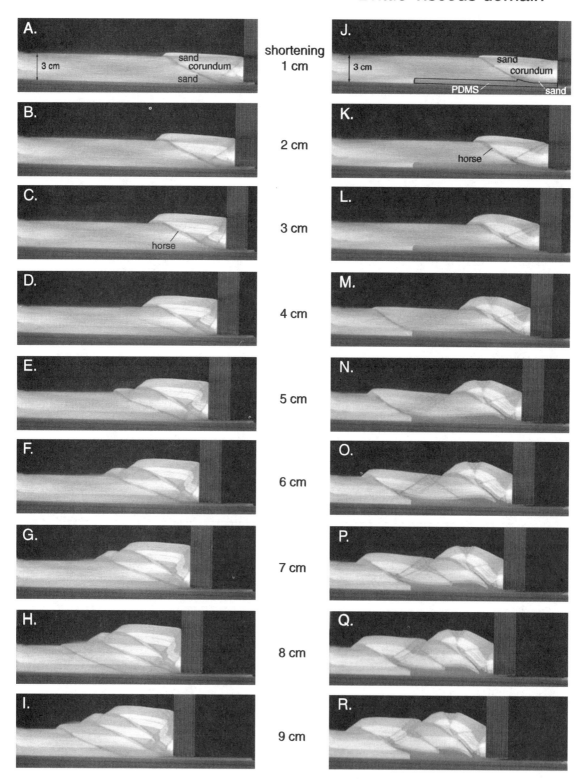

Figure 4. Comparison of structural evolution between brittle and brittle-viscous domains of experiment 73 by successive transverse computed tomography (CT) images of progressively deformed model. Shortening is indicated in centimeters. A–I: Brittle domain. J–R: Brittle-viscous domain. Algorithms used for image computation are optimized for round-shaped objects and hence CT image quality improves as length to thickness ratio of model decreases with progressive shortening.

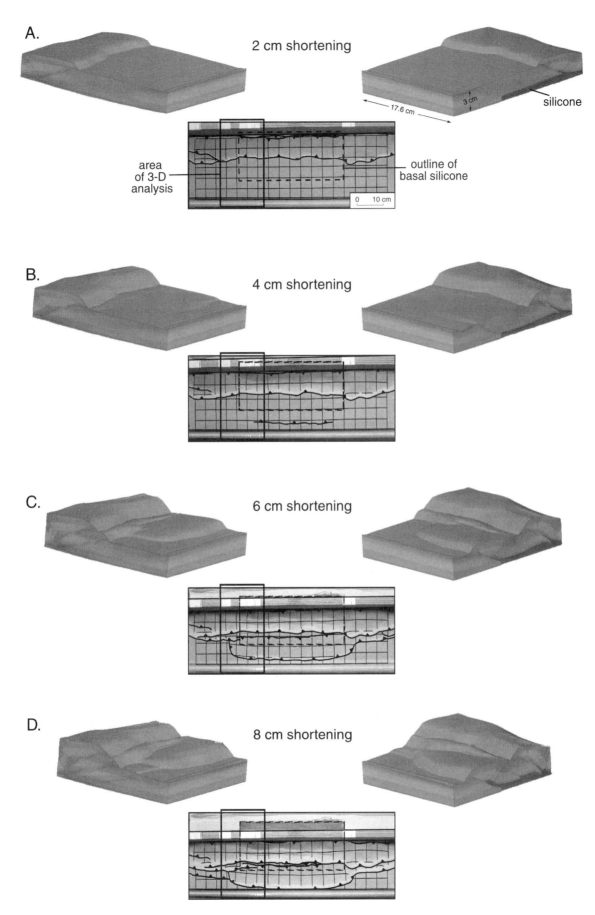

Figure 5. Four-dimensional analysis of experiment 73 showing surface photographs, interpretation, and two oblique views each for four consecutive stages in evolution of model. A: 2 cm shortening. B: 4 cm shortening. C: 6 cm shortening. D: 8 cm shortening. Rectangle indicates area of three-dimensional analysis of one particular stage; dashed rectangle indicates initial outline of basal viscous silicone.

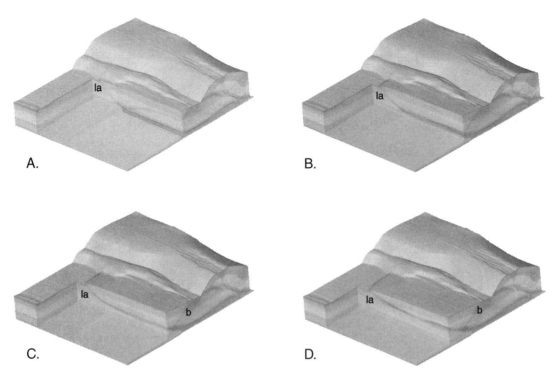

Figure 6. Four cut-out three-dimensional views of experiment 73 at 6 cm shortening including longitudinal sections through transfer zone. Diagrams show clearly shallow dip of lateral ramp and change in dip angle along strike.

restricted to the foremost forward thrust (Fig. 5, C and D). As a consequence, the initially curved geometry of the thrust front in surface view became less pronounced with progressive shortening (Fig. 5D).

Experiment 75

In this experiment one boundary of the basal viscous layer was oblique and made an angle of 35° with the shortening direction. The initial evolution of structures was similar to that in the previous experiment. A pop-up structure formed close to the mobile wall (Fig. 8A). As deformation increased, in sequence imbricate thrusts developed in the brittle domain, while a new pop-up structure formed in the brittle-viscous domain. That structure formed far away from the mobile wall at the frontal boundary between sand and PDMS (Fig. 8B). With progressive shortening, the less advanced thrusts linked up with the farther advanced thrusts by means of transfer zones (Fig. 8C). In contrast to the previous experiment, however, the transfer zone underlain by the oblique-trending sand-PDMS boundary developed a strike oblique (35°) to the shortening direction (Fig. 8, C and D). Maximum dip on this transfer zone was 30° and decreased along strike toward the frontal ramp. Cut-out three-dimensional views reveal the geometry of the oblique ramp (Fig. 9) and show how backthrusts of the frontal pop-up structure in the brittle-viscous domain disappear along strike in the brittle domain (Fig. 9, C and D). A forward thrust that formed in the brittle domain propagated along strike into the domain underlain by the viscous layer. The transfer zone development is shown for three successive stages during progressive shortening in Figure 10.

COMPARISON WITH OTHER EXPERIMENTAL STUDIES

The differences in deformation style between brittle and brittle-viscous domain in our experiments resemble to a first degree the differences between high and low basal friction domains (e.g., Colletta et al., 1991; Calassou et al., 1993; Philippe, 1995; Gutscher et al., 1998). In both our brittle domains and high basal friction domains, thrusts are closely spaced and the thrust wedge is high and steep, whereas the thrust wedge is lower and wider in the brittle-viscous and low basal friction domains. In the transition zone between the two domains transfer zones form, the orientations of which depend strongly on the geometry of the basal boundary. Lateral variations in basal friction conditions induced lateral ramps in the overlying sand cake (Colletta et al., 1991). These shallow-dipping lateral ramps rooted at the boundary between the low-friction zone (glass microbeads) and the high-friction zone (sandpaper). This is similar to experiment 73, where lateral ramps with shallow dips rooted at the lateral boundary between viscous and brittle material. However, there are also fundamental differences between our experiments and those with high and low basal friction domains. In the latter type of experiments, the sequence of imbricate thrusting in the low-friction zone involved a sys-

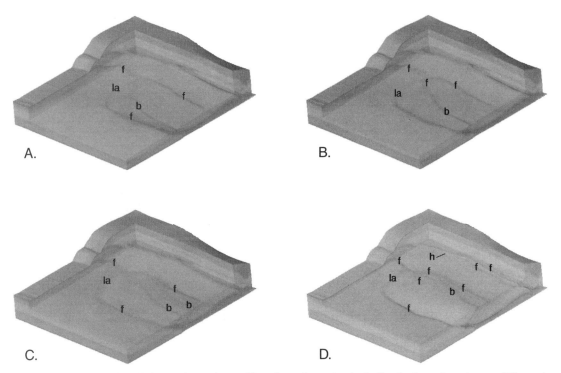

Figure 7. Four cut-out three-dimensional views of experiment 73 at 6 cm shortening including horizontal sections at different levels through transfer zone. f = forward thrust, b = backthrust, la = lateral ramp, h = horse. Orientation of lateral ramp is subparallel to shortening direction. Backthrusts in brittle-viscous domain disappear laterally or interfere with forward thrusts that originated in brittle domain and have propagated sideways into brittle-viscous domain.

tematic forward propagation, and only the frontal thrust was generally active (Colletta et al., 1991; Calassou et al., 1993). This is in contrast to the out of sequence thrusting and coeval thrusting observed in the brittle-viscous domain of our experiments. We attribute this difference to the presence of the basal viscous layer.

CONCLUSIONS RELEVANT TO THE METHOD OF ANALYSIS

This chapter shows the great potential of CT volume scanning for four-dimensional analysis of analog models. The following strengths of the method are highlighted. (1) CT volume scans are of special importance in modeling complex structural regimes in which lateral changes in three-dimensional geometry are common. Three-dimensional imaging of models can provide constraints for interpretations of complex zones based on seismic data that are often fragmentary and difficult to interpret. (2) Knowledge of three-dimensional evolution through time can help geologists and seismic interpreters in developing kinematic models, likely to be useful for petroleum geologists in determining the history of maturation, migration, and trapping of hydrocarbons. (3) Computer animations of three-dimensional perspective views can be created using computer visualization techniques. Such animations are very instructive in that the experiment can be rerun on the computer screen and thus facilitates the development of geometric and kinematic concepts relevant to the evolution of geological structures in nature. The geometry of the structures and their evolution through time can be analyzed in detail because of the flexibility of the animations. One can control the speed of the animation and stop at any point for a detailed three-dimensional study of a particular image. It is also possible to retrodeform the model and restore it to an earlier stage of deformation. (4) Sections through the model in any direction can be generated by using computer visualization software and can be compared to geological and seismic sections. (5) The digital data of analog model experiments can easily be disseminated through the internet and used for training and teaching purposes. Animations of the experiments presented in this paper can be viewed at: http://www.earthsci.unibe.ch/people/schreurs/Main.htm.

CONCLUSIONS RELEVANT TO THE EXPERIMENTS

The experiments showed a contrast in structural style between domains underlain and not underlain by a thin viscous layer. In the brittle domain thrust faults were closely spaced and a narrow and high fold and thrust belt formed. The sequence of thrusting propagated forward and the belt had a dominant vergence of thrusts and associated folds. In the brittle-viscous domain, however, spacing between faults was greater and the thrust belt was wider and lower. In contrast to results obtained in the brittle domain, there was no consistent vergence of thrusts and

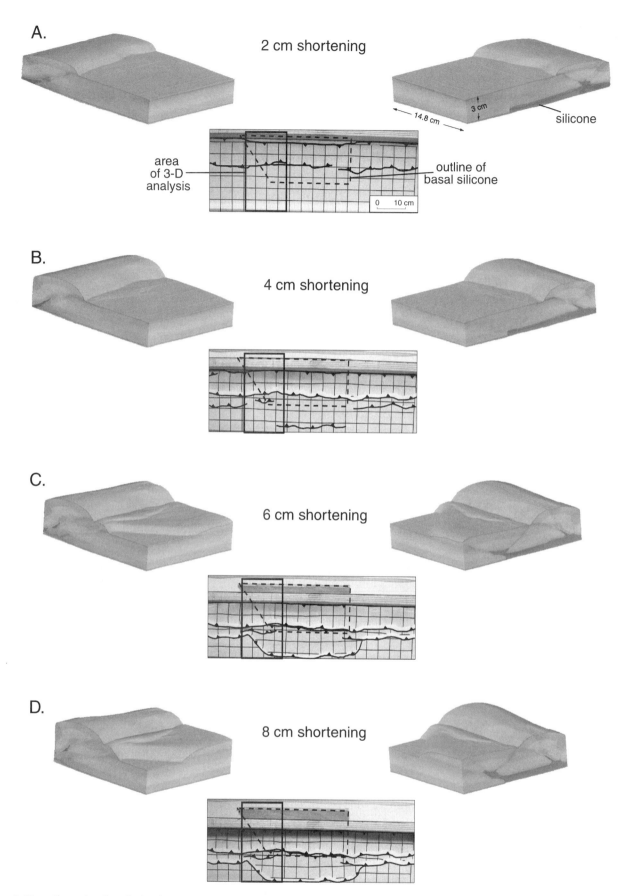

Figure 8. Four-dimensional analysis of experiment 75 showing surface photographs, interpretation, and two oblique views each for four stages in evolution of model. A: 2 cm shortening. B: 4 cm shortening. C: 6 cm shortening. D: 8 cm shortening. Note development of transfer zone striking oblique to shortening direction, but parallel to basal boundary between brittle and brittle-viscous domain. Rectangle indicates area of three-dimensional analysis of one particular stage. Area bounded by dashed line indicates initial outline of basal viscous silicone.

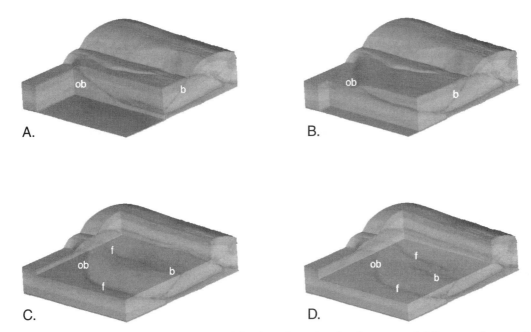

Figure 9. Four cut-out three-dimensional views of experiment 75 at 6 cm shortening showing nature of oblique transfer zone. A, B: Longitudinal sections through oblique ramp. C, D: Horizontal sections through transfer zone. Longitudinal sections (A, B) show shallow dip of transfer zone, whereas horizontal sections (C, D) show strike of oblique ramp, and interference between laterally propagating backthrust and forward thrust. f = forward thrust, b = backthrust, ob = oblique ramp.

folding. Out of sequence thrusting and coeval activity of different thrusts occurred only in the brittle-viscous domain. This indicates that out of sequence thrusting in nature may be related to the rheology of the detachment. Transfer zones developed in the transition zone between brittle and brittle-viscous domains. They linked thrusts that formed at different locations and stages in the evolution of the model. Location and orientation of the transfer zones were directly related to the basal geometry of the thin viscous layer. The transfer zones rooted in the viscous layer and their strike mimicked the orientation of the basal boundary between viscous and brittle material. A lateral ramp formed where this boundary was parallel to the shortening direction, whereas an oblique ramp formed where this boundary was oblique.

In summary, our experiments indicate that: (1) salt or other rheologically weak layers at the base of a sedimentary sequence favor out of sequence thrusting, coeval displacement along different forward thrusts, and the development of pop-up structures; (2) location and orientation of transfer zones in nature may be controlled by basal rheological changes; (3) lateral and oblique ramps in transfer zones may have shallow dips (<30°) and dip angles that vary along strike; (4) backthrusts above basal salt may interfere with forward thrusts that formed in purely brittle domains, thus contributing to the complexity of transfer zones in nature.

ACKNOWLEDGMENTS

We thank Hans-Peter Bärtschi and Robert Roth (University of Bern) for technical assistance, Andrea Schneider and Klaus-Dietz Toennies (Institute of Diagnostic Radiology, Inselspital, Bern) for computed tomography data acquisition, Ulrich Langlotz and Frank Langlotz (Maurice-Müller Institute, Bern) for data transfer, Bernard Drayer, Ulrich Linden, and Peter Strunck (Geological Institute, Bern) for help in data visualization, and Marco Herwegh (Geological Institute, Bern) for stimulating discussions. We benefited from reviews by Kevin Burke, Marc-André Gutscher, and Francis Odonne. Funding by Hochschulstiftung Bern and Swiss National Science Foundation Grant 2000–0554.11 98/1 is gratefully acknowledged.

REFERENCES

Byerlee, J., 1978, Friction of rocks: Pure and Applied Geophysics, v. 116, p. 615–626.

Calassou, S., Larroque, C., and Malavieille, J., 1993, Transfer zones of deformation in thrust wedges: An experimental study: Tectonophysics, v. 221, p. 325–344.

Colletta, B., Letouzey, J., Pinedo, R., Ballard, J.F., and Balé, P., 1991, Computerized X-ray tomography analysis of sandbox models: Examples of thin-skinned thrust systems: Geology, v. 19, p. 1063–1067.

Dixon, J.M., and Liu Shumin, 1992, Centrifuge modelling of the propagation of thrust faults, in McClay, K.R., ed., Thrust tectonics: London, Chapman & Hall, p. 53–69.

Fermor, P., 1999, Aspects of the three-dimensional structure of the Alberta Foothills and Front Ranges: Geological Society of America Bulletin, v. 111, p. 317–346.

Fermor, P.R., and Moffat, I.W., 1992, Tectonics and structure of the western Canada foreland basin, in Macqueen, R.W., and Leckie, D.A., eds., Foreland basins and fold belts: American Association of Petroleum Geologists Memoir 55, p. 81–105.

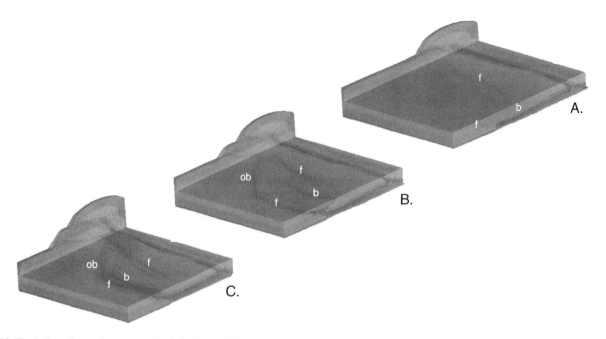

Figure 10. Evolution of transfer zone at depth for three different stages. A: 4 cm shortening. B: 6 cm shortening. C: 8 cm shortening. f = forward thrust, b = backthrust, ob = oblique ramp.

Gutscher, M.-A., Kukowski, N., Malavieille, J., and Lallemand, S., 1998, Material transfer in accretionary wedges from analysis of a systematic series of analog experiments: Journal of Structural Geology, v. 20, p. 407–416.

Horsfield, W.T., 1977, An experimental approach to basement-controlled faulting: Geologie en Mijnbouw, v. 56, p. 363–370.

Hounsfield, G.N., 1973, Computerized transverse axial scanning (tomography): British Journal of Radiology, v. 46, p. 1016–1022.

Hubbert, M.K., 1937, Theory of scale models as applied to the study of geologic structures: Geological Society of America Bulletin, v. 48, p. 1459–1520.

Lallemand, S.E., Schnürle, P., and Malavieille, J., 1994, Coulomb theory applied to accretionary and nonaccretionary wedges. Possible causes for tectonic erosion and/or frontal accretion: Journal of Geophysical Research, v. 99, p. 12033–12055.

Liu Huiqi, McClay, K.R., and Powell, D., 1992, Physical models of thrust wedges, in McClay, K.R., ed., Thrust tectonics: London, Chapman & Hall, p. 71–81.

Mandl, G., 1988, Mechanics of tectonic faulting: Amsterdam, Elsevier, 407 p.

Marshak, S., Wilkerson, M.S., and Hsui, A.T., 1992, Generation of curved fold-thrust belts: Insight from simple physical and analytical models, in McClay, K.R., ed., Thrust tectonics: London, Chapman & Hall, p. 83–92.

McClay, K.R., 1992, Glossary of thrust tectonics terms, in McClay, K.R., ed., Thrust tectonics: London, Chapman & Hall, p. 419–433.

McDougall, J.W., and Khan, S.H., 1990, Strike-slip faulting in a foreland fold-thrust belt: The Kalabagh fault and western Salt Range, Pakistan: Tectonics, v. 9, p. 1061–1075.

Naylor, M.A., Mandl, G., and Sijpesteijn, C.H.K., 1986, Fault geometries in basement-induced wrench faulting under different initial stress states: Journal of Structural Geology, v. 8, p. 737–752.

Philippe, Y., 1994, Transfer zone in the southern Jura thrust belt (eastern France): Geometry, development and comparison with analogue modelling experiments, in Mascle, A., ed., Hydrocarbon and petroleum geology of France: European Association of Petroleum Geologists, Publication 4, p. 327–346.

Philippe, Y., 1995, Rampes latérales et zones de transfert dans les chaînes plissées: géometrie, conditions de formations et pièges structuraux associés [Ph.D. thesis]: Savoie, France, Université de Savoie, 623 p.

Philippe, Y., Deville, E., and Mascle, A., 1998, Thin-skinned inversion tectonics at oblique basin margins: Example of the western Vercors and Chartreuse Subalpine massifs (SE France), in Mascle, A., et al., eds., Cenozoic foreland basins of western Europe: Geological Society [London] Special Publication 134, p. 239–262.

Ramberg, H., 1981, Gravity, deformation and the Earth's crust: New York, Academic Press, 214 p.

Schreurs, G., 1992, Analogue modelling using X-ray computed tomography analysis: Experiments on distributed strike-slip shear deformation: Institut Français du Pétrole Report 39893, 233 p.

Schreurs, G., 1994, Experiments on strike-slip faulting and block rotation: Geology, v. 22, p. 567–570.

Schreurs, G., and Colletta, B., 1998, Analogue modelling of faulting in zones of continental transpression and transtension, in Holdsworth, R.E., et al., eds., Continental transpressional and transtensional tectonics: Geological Society [London] Special Publication 135, p. 59–79.

Vendeville, B., Cobbold, P.R., Davy, P., Choukroune, P., and Brun, J.P., 1987, Physical models of extensional tectonics at various scales, in Coward, M.P., et al., eds., Continental extensional tectonics: Geological Society [London] Special Publication 28, p. 95–107.

Weijermars, R., 1986, Flow behaviour and physical chemistry of bouncing putties and related polymers in view of tectonic laboratory applications: Tectonophysics, v. 124, p. 325–258.

Wilkerson, M.S., Marshak, S., and Bosworth, W., 1992, Computerized tomographic analysis of displacement trajectories and three-dimensional fold geometry above oblique thrust ramps: Geology, v. 20, p. 439–442.

MANUSCRIPT ACCEPTED BY THE SOCIETY APRIL 12, 2000

Printed in the U.S.A.

Effective indenters and the development of double-vergent orogens: Insights from analogue sand models

Katarina S. Persson
Hans Ramberg Tectonic Laboratory, Institute of Earth Sciences, Uppsala University, Villavägen 16, S-752 36 Uppsala, Sweden

ABSTRACT

Analogue sand models have been used to investigate how changes in indenter geometry affect the rise of an orogenic wedge from the indented continent. Rigid indenters, representing cool strong continental crust, were driven laterally into hanging walls of sand, representing weaker brittle crust. The initial thickness of the sand and the angle of the face of the rigid indenter were varied. All models resulted in a two-sided wedge rising between fore-kinkbands and back-kinkbands above the advancing toe of the indenter.

The rate of kinkband formation was found to be dependent on the initial thickness of the sand hanging wall, the shape of the rigid indenter, and the rate of surficial slumping of the backslope. Changing the initial thickness of the sand hanging wall controlled the spacing and therefore the rate at which kinkbands were initiated. The greatest number of forekinks developed for rigid indenters with a frontal dip close to the internal friction angle of the sand, i.e., 30°. Effective indenters of compacted hanging wall developed if the face of the rigid ramp dipped <30° or >45°. Of those, effective indenters dipping <30° are introduced in this work. In models developing effective indenters, strain partitioning favored compaction of sand making up the effective indenter instead of building numerous closely spaced forekinks.

Surficial slumping of the backslope lowered the load on the backshear and kept it active. Surficial slumping of the backslope was most pronounced for models with rigid indenter dips of 60°–90°. The use of glass beads with a lower angle of internal friction enhanced surficial slumping on both foreslope and backslope, which kept each foreshear and backshear active longer.

Lessons learned from the models are applied to deep reflection seismic profiles of two orogens, the Middle Urals and the Svecofennian of the Gulf of Bothnia.

INTRODUCTION

Orogenesis, a geologic term of Greek roots for the process of mountain building, is one of the most complex tectonic processes known to the geosciences. To field geologists the term orogeny represents the penetrative deformation of the Earth's crust associated with phases of metamorphism and igneous activity along restricted, commonly linear, zones within a limited time interval (Dennis, 1967; Burg and Ford, 1997). However, increasing knowledge of the rheology of the lithosphere suggests that orogenesis involves the interaction of a series of large-scale geodynamic processes (Burg and Ford, 1997).

The asymmetry of subduction of one plate beneath another leads to asymmetry of continental collision and the indentation of a relatively warm soft continent by a cool stiffer continent (Tapponnier and Molnar, 1976; Tapponnier et al., 1982; Ratschbacher et al., 1991; Schmid et al., 1996; Bonini et al., 1999). Two of the most impressive characteristics of continental

Persson, K.S., 2001, Effective indenters and the development of double-vergent orogens: Insights from analogue sand models, *in* Koyi, H.A., and Mancktelow, N.S., eds., Tectonic Modeling: A Volume in Honor of Hans Ramberg: Boulder, Colorado, Geological Society of America Memoir 193, p. 191–206.

collisions are the extrusion and exhumation of deep metamorphic rocks following continental suturing and indentation. Indentation leads to vertical and lateral escape and gravitational spreading of the indented plate in patterns that depend mainly upon the lateral confinement of the indented plate, but also on the rheologies and dimensions of the indenting continent. Strong confinement of the indented plate favors crustal thickening (vertical escape), whereas weak confinement favors lateral escape (Davy and Cobbold, 1988; Ratschbacher et al., 1991).

During continental collision, the deforming brittle upper crust develops a rising wedge in profile, while the ductile rocks beneath develop a downward-growing root. The upper levels of the orogen can be interpreted on the basis of surface information, whereas the geometry of the deeper levels relies entirely on the results of deep seismic soundings (Schmid et al., 1996).

There are two extreme approaches to explain the process of vertical escape of rocks metamorphosed in orogens. One approach considers orogenic crust to be homogeneous and ductile while extruding upward as mountains and intruding downward as roots (England et al., 1985; Thompson et al., 1997, and references therein). The other approach distinguishes a brittle upper crust (or lithosphere) that extrudes upward in a fault-bounded wedge from a ductile lower crust (or lithosphere) that intrudes downward (Davy and Cobbold, 1988; Cobbold and Davy, 1988; Ratschbacher et al., 1991; Willett et al., 1993).

In this chapter we follow the second approach. Sand layers of varying thickness that represent brittle (Byerlee) upper crust or lithosphere are indented by rigid plastic blocks representing stiffer indenting upper crust or lithosphere. The models investigate how varying the geometries of the stiff indenting plate affects the geometry, rates of extrusion, and tectonic exhumation of a fault-bound wedge of metamorphosed rocks of the indented plate. All experiments have strong lateral confinement to ensure crustal thickening. Sections through these experiments give insight on deep structures where we otherwise rely on geophysical profiles. Lessons learned from the models are applied to seismic profiles of representative examples of natural orogenic wedges.

PREVIOUS MODELING

The geometry, kinematics, and progressive strain history of the accretion process and the importance of episodic rather than steady accretion of model sand wedges due to the interplay between stick-slip propagation of laterally compacted imbricate ramps and the basal decollement were discussed by Malavieille (1984), Mulugeta (1988), and Mulugeta and Koyi (1987, 1992).

An experimental study by Bonini et al. (1999) showed that the shape of the indenter plays a significant role in the development of structures in an indented sand pack. Comparing their results with the numerical model (Fig. 1A) by Beaumont et al. (1994), Bonini et al. (1999) divided their models into two categories; those where the first backthrust developed at the velocity discontinuity, S, at the toe of the rigid indenter and those that developed an effective indenter behind a backthrust that propagated from a propagation point, P, in front of the rigid indenter (Fig. 1B).

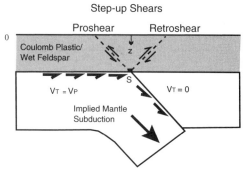

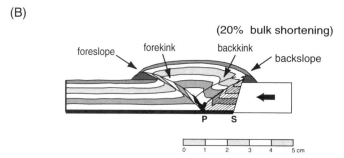

Figure 1. A: Development of step-up shear zones as mantle detaches and subducts at a velocity discontinuity (S) at the base of the crust, as described by Beaumont et al. (1994). To the right of "S" the tangent velocity (V_T) to the flexed base of the layer is equal to the rate of convergence (V_P). To the left of "S", $V_T = 0$. Model by Bonini et al. (1999), illustrating effective indenter. Backshear propagating from point P in front of S point (toe of rigid indenter) defines frontal face of effective indenter.

The plain-strain numerical model of convergent orogeny of Beaumont et al. (1994, 1996) describes a special case wherein the indenter and indented crustal layers have the same properties. This leads to the development of crustal-scale imbricate ramps as the lower ductile lithosphere detaches and subducts at a velocity discontinuity at the base of the Coulomb plastic crust, S, and begins its vertical escape as the extruding orogenic wedge (Fig. 1A). In the Beaumont et al. models (1994, 1996) it is the entire crust that remains coupled to the mantle until it reaches S (Fig. 1A), whereas in the Bonini et al. model and those presented here, the hanging wall represents only the upper crust detaching from the lower ductile crust beneath the indenter.

Bonini et al. (1999) described how effective indenters develop in front of rigid indenters with faces dipping ~60° and illustrated this with reference to the Central Alps. This work adds another category of effective indenters, i.e., those that develop above rigid indenters with faces dipping <30°, and uses the Svecofennian orogen in Sweden as a possible natural example. The Urals are used to illustrate the third category of models, those where no effective indenter develops.

ANALOGUE MODELS

Experimental material, design, and procedure

The experiments were performed at the Hans Ramberg Tectonic Laboratory of the Department of Earth Sciences, Uppsala University.

Dry quartz sand, a Mohr-Coulomb material with density of 1300 kg/m^3, and a mean angle of internal friction of 30°, was used to represent the brittle behavior of the fractured upper continental crust (Byerlee, 1978). The sand used in the experiments consists of well-sorted subrounded quartz grains that are ≤0.246 mm in diameter. An average density of 2700 kg/m^3 for the crystalline upper crust gives a density ratio ρ^* of ~0.5 between the model and nature. A thickness of 15 km of brittle upper crust is represented by 1 cm sand in the model, giving a model/nature length ratio L^* ~6 × 10^{-7}. The nature/model gravity ratio (g^*) is 1, so the stress ratio $=\rho^* g^* L^*$ is ~3 × 10^{-7}. The cohesion T_0 of rocks has the dimension of stress, so $\tau^* = \sigma^*$. Taking the measured cohesion of 90 Pa yields a value of ~300 MPa (T_0) for natural rocks. This is within the range of values listed by Goodman (1988).

Models with horizontal dimensions of 9.5 × 7 cm and heights between 0.6 and 2.2 cm were built in a Plexiglas box (Fig. 2). The hanging walls of the models consisted of sand of different colors sedimented with the same stratigraphy. The color layers were passive and inserted in order to visualize internal deformation patterns. In all models, a wedge-shaped block of rigid plastic (Fig. 2), representing crust of higher strength, was driven laterally into the sandpile at a constant displacement rate of 1.8 cm h^{-1}. To save time, a velocity of 3.2 cm h^{-1} was used for the series of models studying surface structures. However, because Coulomb materials like sand have a yield envelope essentially independent of strain rate, there is little need to scale the rate of displacement precisely (McClay and Ellis, 1987). The models were focused on the deformation of the brittle upper crust as it decoupled from its underlying ductile crust above a detachment surface represented by the Plexiglas floor of the squeeze box. A value of 0.35 for the basal friction coefficient between the sand and Plexiglas was measured.

Most models were deformed to 40% bulk lateral shortening, but some were taken to 60% bulk shortening. During the deformation, the models were measured and photographed from above at regular intervals. After shortening, the models were soaked in water, frozen, and cut to observe longitudinal cross sections. These were then photographed and photocopied for analysis and illustration. Duplicates of each model, as well as additional models deformed to 10% and 20% bulk lateral shortening, were made to assure general reproducibility.

Limitations

Increase in vertical loading due to increase in the initial thickness of the hanging wall and crustal thickening would normally result in isostatic adjustment and depression of the basal decollement on a crustal scale. This factor was not taken into account. The models were designed to focus on brittle deformation of the upper crust and took no account of any ductile strains within the extruded material. Nor were temperature variations considered. The rigid indenter could not deform as the stronger plate might deform in nature.

As the model wedge extruded, its sides collapsed under the influence of gravity, but no other effects of redistribution of material were considered. These models are therefore only applicable to areas where erosion and deposition play a minor role in orogeny.

RESULTS

Internal structures and effective indenters

Three series of experiments were used to study the structures that developed within sand packs with initial sand thicknesses of 0.6, 1.0, and 1.4 cm. For each of the three series, the experiments were repeated with dips of the front face of the rigid indenter increasing in increments of 15° from 15° to 90°.

The definition of fore-kinkbands and back-kinkbands in this work follows Malavieille (1984) and is based on the direction of movement of the rigid indenter. Structures that propagate away from the advancing rigid indenter with progressive shortening are called fore-kinkbands (or forekinks), and those that propagate backward toward the indenter are called back-kinkbands (or backkinks) (Fig. 3). Backkinks localize to form a single backshear with increasing deformation.

In each experiment, the number of forekinks and backkinks that developed are related to the amount of bulk lateral

(A)

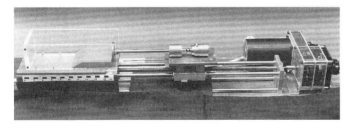

(B)

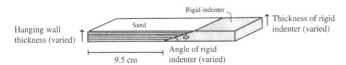

Figure 2. A: Experimental design. Rigid ramp is pushed into sand (not shown) sedimented in Plexiglas squeeze box (20 × 7 × 6 cm), at constant rate by piston driven by electric motor. B: Sketch of model design inside Plexiglas box.

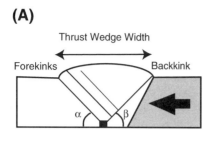

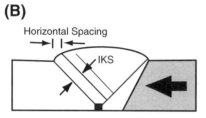

Figure 3. A: Thrust wedge width and dip of forekink (α) and backkink and/or backshear (β) as they were measured. B: Measurements of internal kink spacing (IKS; Marshak and Wilkerson, 1992) herein called perpendicular spacing, were taken between two last-formed kinks after 40% bulk shortening as distance between upper of two kinks making up kinkbands. Horizontal spacing is thrust trace width between first and second thrust. Indenter is shaded, and sand is unpatterned. Arrow indicates direction of movement of rigid indenter.

shortening. The perpendicular spacing (internal kink spacing; Fig. 3) between the two last-formed forekinks was measured as the distance between the upper of the two kinks making up the kinkbands (Fig. 3; Table 1). The dips of the last formed forekinks and backkinks were also measured (Fig. 3; Table 1).

During the first increment of displacement, the sand pack was observed through the Plexiglas wall to thicken slightly as it laterally compacted in front of the rigid indenter. With further deformation, individual forekinks and backkinks propagated from the bottom to the top surface. All models, independent of rigid ramp dip, developed triangular, wedge-shaped, pop-up structures as successive shears carried older imbricate slices upward. Such pop-up structures are typical for materials with an internal friction angle of 30° (Coletta et al., 1991). As each new imbricate foreslice developed, older forekinks became inactive and back-rotated as they were carried up an active backshear.

In models with rigid ramp dips of 30° and 45°, the first structure to appear was a forekink that nucleated at the toe of the rigid ramp and led to reverse top-to-the-left movement. In this case the toe of the rigid ramp acted as the velocity discontinuity labeled S by Beaumont et al. (1994, 1996). With further deformation, older forekinks became inactive as they were carried backward up and along a single backshear represented by the front face of the rigid indenter, confirming previous work (Malavieille, 1984; Merle and Abidi, 1995; Bonini et al., 1999, and references therein). Simultaneously, new forekinks formed at the toe of the rigid ramp. In addition, this work also found that propagation of each successive kink required an increase in the increment of bulk deformation. Thus in experiments with an initial sand thickness of 1.0 cm and a rigid ramp dip of 90°, the first kink developed after 3% bulk shortening. The second forekink developed after an additional bulk shortening of 9% (i.e., total bulk shortening of 12%), and the third forekink developed after an additional shortening increment of 10% (i.e., total bulk shortening of 22%).

Bonini et al. (1999) demonstrated that for ramp dips of ~60°, sand is transferred to the front of the rigid indenter so that the system spontaneously creates its own ramp by accreting a backshear of sand. When it was active, this new ramp was taken to be the front face of an effective indenter (Bonini et al., 1999). These workers reported that the backshear propagated from a line situated a short distance in front of the toe of the rigid indenter simultaneously with a forekink from the same line. This line, representing the toe of the effective indenter, intercepts vertical longitudinal profiles at a propagation point P. As lateral shortening progressed, the propagation of successive backkinks were shown to alternate with the propagation of successive forekinks from the P point. In other words, the area between the rigid indenter and the backshear laterally compacts until it acts as a rigid mass; i.e., the Bonini et al. effective indenter (Fig. 1). The P point is here also a separation point between slipped and unslipped zones of the basal decollement. Bonini et al. (1999) found that models having rigid ramps with dips ≤45° used the front face of the rigid indenter as a single continuous backshear, whereas models with dips ≥60° were characterized by the development of effective indenter defined by a backshear in the sand. However, in the experiments reported here, an effective indenter also appeared for rigid ramp

TABLE 1. THE INTERNAL KINK SPACING (IKS) BETWEEN THE TWO LAST-DEVELOPED KINKS WITH INITIAL OVERBURDEN THICKNESS FOR RIGID RAMP ANGLES OF 15°–75° AFTER 40% BULK SHORTENING

Initial overburden thickness (cm)	IKS for models with rigid ramp dip of 15° (mm)	IKS for models with rigid ramp dip of 30° (mm)	IKS for models with rigid ramp dip of 45° (mm)	IKS for models with rigid ramp dip of 60° (mm)	IKS for models with rigid ramp dip of 75° (mm)
0.6	3.4 (± 0.05)	2.55 (± 0.05)	4.15 (± 0.15)	3.75 (± 0.15)	3.85 (± 0.15)
1.0	3.4 (± 0.05)	4.3 (± 0.1)	4.4 (± 0.2)	5.5 (± 0.15)	5.15 (± 0.25)
1.4	6.2 (± 0.1)	5.4 (± 0.4)	4.8 (± 0.5)	6.0 (± 0.5?)	7.8 (± 0.9)

dips <30°, although, in such cases, the P point always coincided with the toe of the rigid ramp, Beaumont's S point (Figs. 4 and 5A). This allows a division of indenter geometries into three categories (<30°, 30°–45°, and >60°; Fig. 5). The phenomenon was well pronounced at rigid ramp angles of 15° for an initial sand layer thickness of 0.6 cm after 40% bulk shortening (Figs. 4 and 5A). For models presented here, the face of the effective indenter, after 40% bulk shortening, has a dip of 40° (±10°) where the rigid indenter dip is 60°. Where the rigid indenter face is <30°, the face of the effective indenter dips 30° (±4°).

In models with rigid ramp dips of 75° and thickness of 0.6 cm (Fig. 6I), kinkband nucleation started at a P point in front of the rigid toe. As the deformation proceeded, the first-formed forekinks and backkinks migrated upward along a single forekink and/or foreshear. During this upward transport, the lower parts of the backkink steepened due to lateral compaction. A new backkink formed after a bulk shortening of 10%. This new kink was also carried up the forethrust, deactivated, steepened, and rotated. The forethrust nucleated with a dip of 30° (α in Fig. 3) and the backkinks had dips of ~34° (β in Fig. 3), forming an asymmetric wedge. By 20% bulk shortening a second forekink had developed, at a new P point in front of the previous one (Fig. 6I, C). The third backkink, nucleated from this new P point, started to form a backshear with a dip of ~44° along which inactive forekinks were carried upward as new forekinks developed. By 40% bulk shortening, five forekinks have developed and a single backshear clearly defined the front face of an

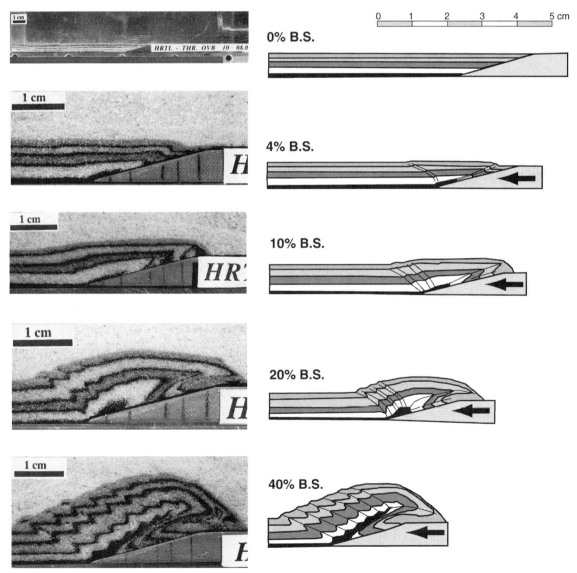

Figure 4. Photos and interpretations of progressive deformation of model with overburden thickness of 0.6 cm and rigid indenter of 15° showing stages at bulk shortening of 0%, 4%, 10%, 20%, and 40%. Arrows indicate direction of movement of rigid ramp (indenter).

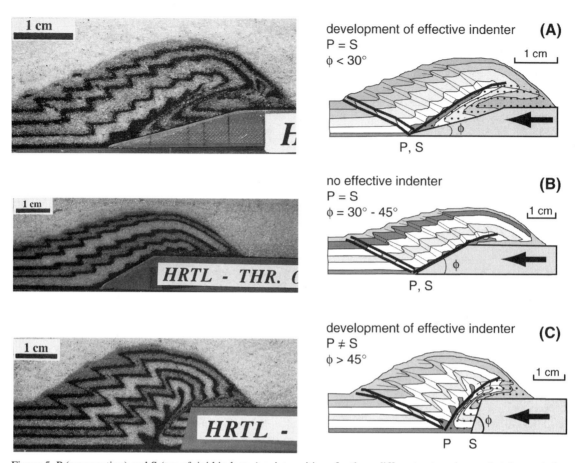

Figure 5. P (propagation) and S (toe of rigid indenter) point positions for three different categories on sketches and photos. Areas of effective indenters are shaded. Arrows indicate direction of movement of rigid ramp.

effective indenter (Fig. 6I, D). This front face was rounded with a dip of 67° near the base, and a dip of 31° at the top leading edge of the rigid ramp, as in experiments by Malavieille (1984). The backshear flattened to 0° to override the top of the rigid ramp (Fig. 6I, D).

The model with an initial sand-pack thickness of 1.0 cm and a rigid ramp dip of 75° (Fig. 6II) developed in essentially the same way to 20% bulk shortening. However, by 40% bulk shortening, the model had developed four forekinks instead of five (cf. D in Fig. 6, I and II). All forekinks traveled up a single backshear defining the face of the effective indenter with a dip of ~50° (Fig. 6II, D).

Bonini et al. (1999) used an initial sand-pack thickness of 1.4 cm. Repetitions of their experiments here confirmed their finding, that the first backkink appeared at a P point forward of the rigid indenter by 10% bulk shortening (Fig. 6III, B). This backkink deformed in the same way as in the models with lower initial thicknesses, but was steeper, having a dip of ~50°. By 20% bulk deformation, a second backkink had developed (Fig. 6III, C) and by 40% the second forekink and a third backkink had appeared (Fig. 6III, D). For rigid ramp dips of 60° and more, the importance of forekinks decreased and were replaced by brittle shear zones, i.e., the models did not develop a definite backshear up which the forekinks were transported with progressive deformation. This observation by Bonini et al. (1999) is correct for models with an initial sandpack thickness of 1.4 cm if the deformation stops after 40% bulk shortening. However, after further shortening (to 60% bulk shortening), the internal geometry of such models resembled similar models with lower initial thickness deformed to 40% bulk shortening (Fig. 6). Three forekinks developed and traveled up a backshear, defining the effective indenter as a fourth forekink was initiated (Fig. 6III, E).

Initiation and number of kinks

With increasing thickness of the hanging wall of sand, larger amounts of bulk shortening were required to initiate kinks whatever the dip of the front face of the rigid indenter. For a ramp dip of 30° and an overburden thickness of 0.6 cm, a bulk shortening of 1.3% was needed to initiate kinking of the top surface (Fig. 7A). For the same ramp angle but initial sand thickness of 1.0 cm, bulk shortening of 2.5% was needed, and for 1.4 cm, a bulk shortening of 4.1% was needed (Fig. 7A). Equiv-

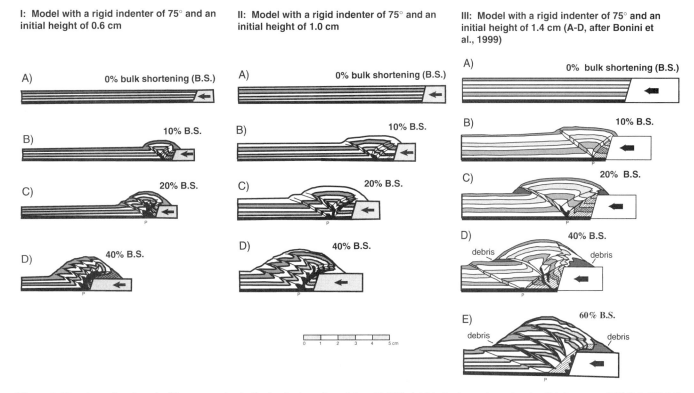

Figure 6. Structures developed with progressive bulk shortening of models with 75° rigid indenter and overburden thicknesses of (I) 0.6, (II) 1.0, and (III) 1.4 cm.

alent amounts of bulk shortening needed to initiate kinking for ramp dips of 45° for overburden thicknesses of 0.6, 1.0, and 1.4 cm are shown in Figure 7B. An increase in sand-pack thickness led to fewer kinks at a bulk shortening of 40% relative to a model with the same indenter face dip.

The greatest number of kinkbands after a particular amount of bulk shortening (here 40%) occurs with a rigid ramp dip of 30° (Fig. 8). The minimum number of kinks develop in models with rigid ramp dips of 75° and 90° (Fig. 8). These observations imply that hanging-wall thickness is not the only control on the number of kinkbands developed, but that the dip of the front face of the rigid indenter is also significant.

Relation between wedge width and kink spacing with initial sand-layer thickness and rigid ramp dip

A series of models focused on the wedge width (Fig. 3) of the first developed forekink and backkink in relation to the overburden thickness and the horizontal spacing of the subsequent kinkbands (Fig. 3A). Five different overburden thicknesses were used: 0.6, 1.0, 1.4, 1.8, and 2.2 cm. The experiments were repeated for all six previous mentioned dips of the frontal face of the rigid indenter.

Marshak and Wilkerson (1992, their Fig. 2) found a linear relationship between the wedge width of the first thrust and the initial layer thickness when using a vertical backstop. The experiments reported here found also linear relationships, even though the sand was allowed to spread down both slopes as the wedge increased in relief. These linear relations were independent of the angle of the rigid indenter (Fig. 9).

The horizontal and perpendicular spacing (Fig. 3B) between the two first kinks increased with increasing initial sand-layer thickness. In general, the spacing also increased for successive kinks in all models. Increase in spacing differs between models with different rigid ramp dips and is most pronounced for ramp dips of 75° (Table 1).

Extrusion height with bulk lateral shortening

The development of the reverse kinkbands converted lateral shortening of the sand pack into vertical escape of a thrust wedge out of the path of the rigid ramp. The extrusion of the wedge began relatively rapidly but slowed with time for all models independent of hanging-wall thickness and angle of rigid indenter (Fig. 10). Extrusion occurred at three different rates despite all the bulk shortening being at the same rates. Model wedges driven by rigid indenter dips of 45°–75° extruded at one rate and model wedges driven by rigid indenter dips of 15° and 30° extruded at two other rates (Fig. 10). The surficial extruded part of the wedge collapsed and spread sideways due to gravity. Because the growing pile was allowed to spread both backward and forward, the experiments reported

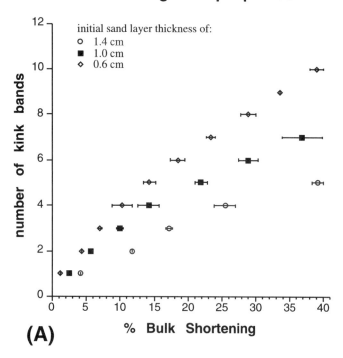

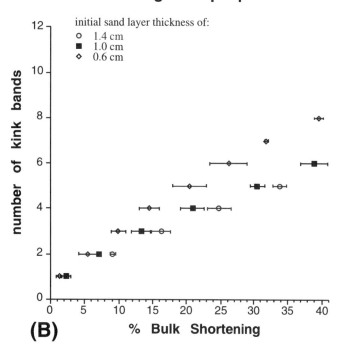

Figure 7. Percentage bulk shortening (% B.S.) needed in order to initiate kinking, with (A) rigid indenter with front face dip of 30° and (B) rigid indenter of 45°; both for initial overburden thicknesses of 0.6, 1.0, and 1.4 cm. Number of kinks initiated after 40% B.S. decreases as initial overburden thickness increases.

here are more realistic than previous models in describing orogenic wedges; the previous models used a vertical indenter much thicker than the accreted pile, applicable to accretionary wedges (Mulugeta, 1988; Mulugeta and Koyi, 1987, 1992; Huiqi et al., 1992; Marshak and Wilkerson, 1992).

The deformation mechanisms within the sand pack are lateral compaction, slip along the basal decollement, imbricate shear or kinking, and gravity-induced collapse of the extruded wedge. The wedge rises and widens stepwise and each of these steps predates the formation of a new imbricate (Fig. 11).

DISCUSSION

Initiation, number, and spacing of kinks in relation to sand-layer thickness

The thinner the sand pile, the greater the number and length of kinks and the lower the bulk shortening needed to initiate them, as Figures 7 and 8 clearly illustrate. This statement applies to models with vertical indenters higher than the rising pile (Huiqi et al., 1992; Marshak and Wilkerson, 1992), as well as to models for all ramp dips described here, and can be attributed to the different vertical load in the sand pack. The horizontal rate of advance of the rigid ramp into the sand was constant in all models reported here. As the vertical force at the base of the sand increased due to the increase of initial sand thickness and as the wedge grew, the force needed to drive movement along the shear plane had to increase as well. The properties of the sand pack changed with deformation. Initially cohesionless loose sand becomes cohesive as it laterally compacts with increasing pressure (Mulugeta, 1988). This demands larger increases in bulk shortening to initiate new kinks in thicker sand packs.

Both the horizontal and perpendicular spacing of kinkbands increased with increase in initial sand-layer thickness for models with the same rigid ramp dips and vertical load increase with the rising wedge. This increase of vertical load can also explain the increase in spacing of successive kinkbands.

The formation of an effective indenter extends the process of compaction and leads to fewer, more widely spaced kinks (Table 1). Note that Table 1 shows only the spacing of the last two kinks developed in each model. This has to be taken into account when comparing models with different rigid ramp dips, because the spacing also depends on the number of kinks initiated.

Sand-layer thickness and dips of forekinks and backkinks

Because the same sand with the same internal friction was used throughout, the initial dips of the first formed forekink did not change with initial overburden thickness. However, all forekinks steepened because of lateral and vertical compaction as they were carried from the basal decollement up the backthrust.

The dip of the backshear stayed close to the angle of internal friction (~30°) of the sand when first initiated, whatever the

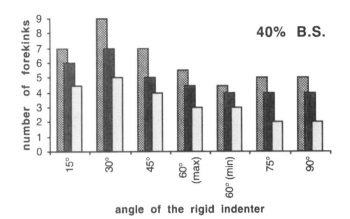

Figure 8. Histogram illustrating number of forekink bands developed after 40% lateral bulk shortening (B.S.).

angle of the rigid indenter. Any change in principal stress direction induced by changes of the rigid indenter dip was apparently insufficient to change the dip of the backshear.

Models with rigid ramp dips of 30° and 45° used the rigid ramp as the active shear plane which could not change with changes of overburden thickness (Fig. 5B).

For models building effective indenters, the dip of the active backkink relates to the backrotation, which depends on the amount of compaction in the effective indenter. This is especially clear for the model with rigid ramp dip 75° and initial overburden thickness of 1.4 cm. In this model, a new backkink with a 36° dip had just formed at a new P point by 40% bulk shortening (Fig. 6III, D). An effective indenter was clearly defined by a single backshear dipping 38° by 60% bulk shortening.

Effective indenter and its relation to number of kinks

The dip of the backshear that defines the front of the indenter depends on the coefficient of friction of the sand compacting both laterally and vertically beneath the growing sand wedge. As the wedge of compacted sand rises in response to lateral shortening, the dip of the active backkink that defines the front of the effective indenter decreases.

Models with rigid ramp dips of 30° were most mechanically efficient in the sense that no effective indenter developed. This is attributed to the dip of the rigid ramp corresponding to the angle of initial internal friction of the sand. By using the rigid indenter as the active backshear, the system was using the minimum energy to accommodate the bulk shortening. This energy efficiency is also illustrated in Figure 10, where models with rigid indenter face dips of 30° extrude the wedge least.

The 45° rigid ramp dip appears to have been sufficiently close to the 30° internal friction angle of the sand for the face of the rigid indenter to act as the active shear plane. Any backshear that did develop was so inconspicuous that it was missed in the experiments. These models did not develop as many kinks as did models with a rigid indenter dip of 30°, implying that strain partitioning along a backshear along a face of the rigid indenter dipping 45° is not as efficient as along rigid indenter faces dipping 30°. The wedges in models with rigid indenter dips of 45° extruded at the same rate (increase of wedge height with percent bulk shortening) as models developing effective indenters (Fig. 10), with P points in front of the toe of the rigid indenter.

Models with rigid ramp dips of <30° not only developed an effective indenter but also more kinks than those with rigid ramp dips of ~60° (Figs. 4 and 5). In models with rigid ramp dips of 15°, the P point coincided with the S point so that no basal slip occurred forward of the advancing indenter. The vertical load as well as the volume of laterally compacted sand transferred to the effective indenter is also less in models with rigid ramp dips of <30° compared with those with rigid ramp dips of ~60°. This suggests that the lower the volume of sand to laterally compact and the lower the vertical load, the greater the number of forekinks.

The backslopes that developed on the extruded wedge (Fig. 12) in models with rigid indenter dips of 15° were more stable than in other models (30°–90°). They underwent the least gravity collapse because the model wedge rose so slowly (Fig. 10) that its backslope only became unstable at ~35%–40% bulk shortening, depending on the initial sand thicknesses.

The deformation histories of models with 60°, 75°, and 90° rigid ramp dips were very similar. Models with rigid ramp dips of 75° and 90° developed the same number of kinks after equivalent bulk shortenings for all initial overburden thicknesses (Fig. 8). To shear along a plane close to the angle of internal friction of the sand (30°), models with rigid ramp dips of 60° developed their P point closer to the toe of the rigid ramp than models with rigid indenter dips of 75° and 90°. As a result these models had decreasing volumes of sand to laterally compact between the rigid and effective indenter. Models with rigid indenter dips of 60° develop more kinks than models of 75° and 90° because increasing strains partition along imbricate shears as the volume of sand to compact decreases (Fig. 8). However, two additional models of each thickness were made in the 60° series, as the number of kinks that had developed by 40% bulk shortening varied (see maximum and minimum values for 60° in Fig. 8). It was then appreciated how sensitive all models are to gravitation stability of the backslope. Gravity collapse of the

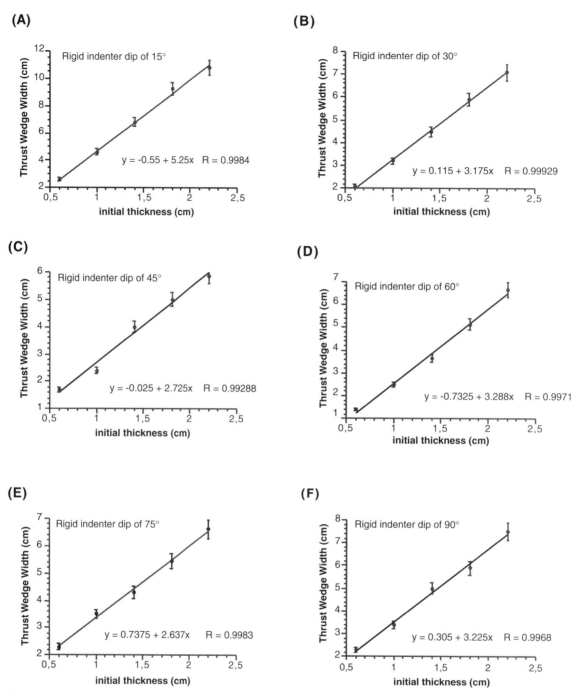

Figure 9. Wedge width for different initial overburden thicknesses and rigid indenter dips of (A) 15°, (B) 30°, (C) 45°, (D) 60°, (E) 75°, and (F) 90°. Diagrams show linear relationship for all indenter dips independent of rigid indenter. R is the deviation value from a straight line.

backslope decreases the load on the backshear. The more the gravity collapse, the easier it is for the model to accommodate strain along the existing backshear rather than initiating new forekinks. The difference in sand volume to be compacted to form the effective indenter for models with rigid indenter dips of 75° and 90° was not enough to show any differences in the number of kinks (Fig. 8).

Change of wedge height and stability of foreslope and backslope with progressive bulk shortening

The two-sided orogenic wedges described here (Koons, 1990) are significantly different from the unidirectional accretionary wedge described by others (Davis et al., 1983; Mulugeta and Koyi, 1992; Koyi, 1995). Thrusting in an accreting prism

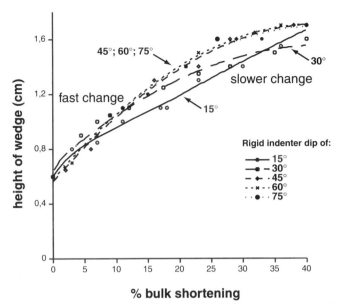

Figure 10. Height of wedge with increasing bulk shortening for models of 0.6 cm initial overburden thickness. Models with rigid indenter dip of 45°, 60°, and 75° show similar slope pattern, whereas models with rigid indenter dips of 15° and 30° have individual slopes.

builds a gravitationally stable foreslope until it reaches a critical taper beyond which the slope slumps. In the two-sided orogenic wedge, the extruded mass spreads over the indenter as well as in the direction of indenting.

As in unidirectional wedges (Mulugeta and Koyi, 1992; Koyi, 1995), the episodic lateral accretion of the widening and rising two-sided wedge was controlled by lateral compaction above a stick-slip basal decollement punctuated by episodes of shear along imbricate foreshears and backshears and gravity collapse of the extruded wedge. Each step predated the formation of a new forekink (Fig. 11), which implies that the orogenic wedge compacts by lateral shortening and extrudes along the imbricate shears until the upward movement locks and overcomes the basal friction so that the basal decollement slips.

The stability of the foreslopes and backslopes flanking the rising plateau (Fig. 12) is controlled by emergence of the hanging walls of kinkbands. The foreslope grows as a basically stable slope that only slumps at its toe, where the hanging walls of successive forekinks are temporarily active (Fig. 12). By contrast, the backslope of the extruded orogenic wedge becomes unstable. This is because the single backshear remains active throughout convergence; consequently the entire backslope

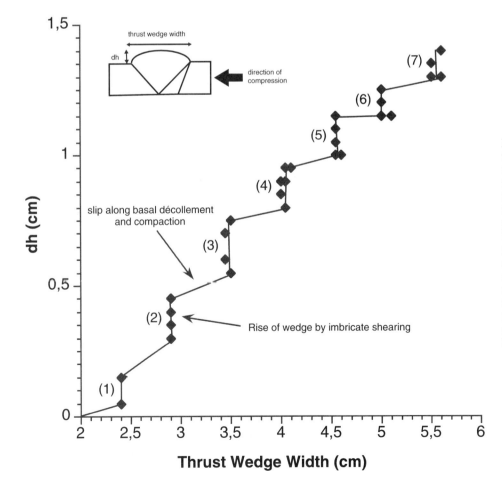

Figure 11. Increase of height (dh) of extruded sand wedge as function of its width, emphasizing episodic development for model with initial sand-pack thickness of 1.0 cm and rigid indenter face dipping 45°. Each imbrication is predated by compaction and slip along basal decollement. Increase in height is mainly due to extrusion along imbricate shears. Increase of wedge height by compaction and decrease by slumping is not clearly visible in diagram. Measurements of height are taken with measurement stick and are therefore not as accurate as measurements with laser beam.

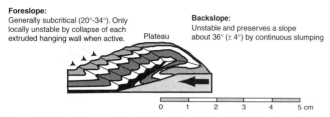

Figure 12. Foreslope and backslope of model wedge. Foreslope is generally subcritical (20°–34°), but never unstable as whole. However, foreslope becomes locally unstable at individual forekinks when these are active. Backslope oversteepened by only backshear preserve slope of ~36° (±4°) by continued gravity collapse.

oversteepens and preserves a slope ~36° (±4°) by continuous slumping (Fig. 12). Gravity collapse of the backslopes plays a significant role in the process by decreasing the load on the backthrust and facilitating further deformation along the existing backshear instead of accommodating strain by building new forethrusts.

In an attempt to check that the dips of forekinks and backkinks depend on the angle of internal friction, some models were built using glass microbeads, which have a lower angle of internal friction. At an initial stage these models also developed effective indenters if the rigid ramp dip did not correspond to the angle of internal friction of the beads. However, gravity collapse was easier and more pronounced down both foreslopes and backslopes of ~24°. Each kink stays active for longer as slumping removes the load on its hanging wall and loads the footwall. With progressive deformation the glass beads favored imbricate shearing on existing kinks instead of slip along the basal decollement, initiating of new kinks, and compaction of the effective indenter. As a result, the effective indenter did not survive. Instead, glass beads are transferred from the initial effective indenter along the long-lived backshear and extrude on the backslope. The physical properties of the material are therefore also crucial for the developing wedge geometry and for determining whether an effective indenter develops.

Models compared to natural orogenic wedges

Both one- and two-sided orogens are recognizable in deep crust seismic profiles. One-sided wedges such as in Japan (Faure et al., 1987, 1995) appear to show only foreshears verging toward the indenter. Such cases appear to be distinct from the indentation of converging continents, perhaps because the mass is significant as well as the stiffness of the potential indenter, so that docking terranes do not indent. In contrast, crustal-scale seismic reflection profiles of some orogens, e.g., the Urals (Juhlin et al., 1998), northwestern Canada (Cook et al., 1998), the British Caledonides (McBride et al., 1996), and the Baltic Svecofennides (BABEL Working Group, 1990, 1993), show geometries suggestive of two-sided wedges.

The major structures deduced from deep seismic data and surface geology studies in the Swiss-Italian Alps (Bernoulli et al., 1990; Pfiffner et al., 1990; Schmid et al., 1996) of Eoalpine-Mesoalpine age (50–100 Ma) were used by Bonini et al. (1999) to illustrate a natural case where an effective indenter developed in front of a rigid indenter dipping 75° (cf. Fig. 5C) after 40% bulk shortening. Seismic profiles of two other natural orogenic wedges are illustrated here as possible examples of the two other cases, first the Urals in Asia, where an effective indenter seems to be lacking (cf. Fig. 5B), and second, the Svecofennian orogen in Baltica, where an effective indenter seems to have developed above a gently dipping initial indenter (cf. Fig. 5A).

The Urals have been described as a late Paleozoic orogenic belt with bivergent thrust geometry where the thrusts are terrane boundaries (Juhlin et al., 1998). Applying the lessons learned from the models reported here to a seismic section in the Middle Urals, the following interpretations are suggested. The Ural orogenic wedge is between an unnamed west-verging forethrust (here called Kvarkush frontal thrust) and the east verging Deevo fault (Friberg, 2000) (called the Trans-Uralian Thrust Zone in Juhlin et al., 1998), interpreted as the backshear (Fig. 13). The Deevo fault has a dip of about 27° and the Kvarkush frontal thrust has a dip of about 35°. A volcanic arc complex is thrust westward onto Precambrian basement (Kvarkush anticline, Fig. 13A) along the Main Uralian thrust fault. Farther west, the Kvarkush anticline is thrust over the East European craton along the Kvarkush frontal thrust (Fig. 13A). The Tagil oceanic and volcanic arc complex (Fig. 13A) and the Salda metamorphic complex (Fig. 13A) accreted to the East European craton during Late Carboniferous to Early Permian time. Comparison with the model in Figure 13B suggests that the final stage of the Ural orogenic contraction was when the Kvarkush frontal thrust deactivated as it was carried up the Deevo fault backshear. It is not clear whether a later forethrust developed from the P point (Fig. 13A). If it did, it could have resulted in the accreted terranes in the west being thrust over the exotic terranes of the Alapaevsk terrane along the Deevo fault (Fig. 13). Such comparisons also suggest that the Main Uralian thrust fault continues all the way down to the Deevo fault, i.e., that the Main Uralian thrust fault could be an earlier foreshear that deactivated as it was carried upward along the Deevo fault when the late Kvarkush frontal forethrust developed. On this model, both the Main Uralian thrust fault and the Kvarkush frontal thrust foreshears would be older structures that were reactivated when indented by the stiffer Alapaevsk terrane (Fig. 13). The presence of high-grade rocks above the Deevo fault to the east and lower metamorphic grade rocks toward the west (Friberg and Petrov, 1998) supports the interpretation based on the sand-box models, which suggests a continuously reactivated backshear and several short-lived foreshears.

Figure 14 suggests an application of the modeling results to part of a deep (southwest-northeast) seismic reflection profile in the Baltic Sea. This profile (BABEL Working Group, 1990, 1993) shows a ca. 1.9 Ga orogenic wedge in the Svecofennian crust of the Baltic shield beneath the Bothnian Bay, where no surface geology is available. The asymmetry of the

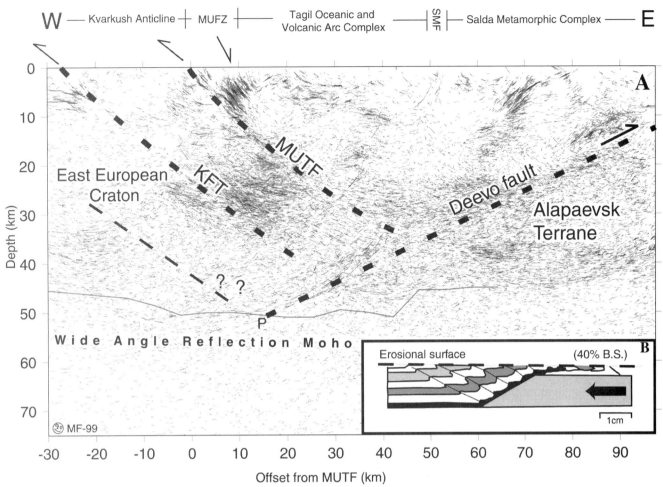

Figure 13. A: Migrated deep seismic reflection profile of Middle Urals (after Juhlin et al., 1998) with depth to Moho only locally defined (as thin line) by wide-angle reflection data (Thouvenot et al., 1995) with interpretations based on models. KFT, Kvarkush frontal thrust; MUFZ, Main Uralian fault zone; MUTF, Main Uralian thrust fault; SMF, Serov-Mauk fault. B: Experimental model with rigid indenter dipping 30° (after 40% bulk shortening, B.S.). Comparing structures in model with seismic profile, model is probably more laterally shortened than Urals. Erosional surface is induced to ease comparison.

seismic structure suggests a series of forethrusts and backthrusts in the natural orogenic wedge. The south-dipping reflector (A-A′) can be interpreted as the backshear (dipping 20°–30°) up which five successive deactivated forethrusts traveled with dips of 28°–32°. The reflectic zone (B-B′) is taken as the face of the underlying indenting continent dipping ~15° to the south. When compared with the models, the area between reflectors A-A′ and B-B′—interpreted as an accreted terrane of metavolcanics, the Skelefte arc (BABEL Working Group, 1990, 1993)—could be reinterpreted as an effective indenter compacted from the indented continent. The coincident P and S points (cf. Figs. 5A and 14B) mark the toe of the indenter from which successive new forethrusts initiated. The profile clearly indicates the former subduction zone in the Moho dipping northeast (Fig. 14), implying a rigid indenter of Svecokarelian crust toward the northeast.

If, as in the models, the dips of natural foreshears and backshears depend on the angle of internal friction of the indented material, then this was 25°–30° in all the natural cases whatever the initial dip of the rigid indenter. Whether an effective indenter develops, the orogenic wedge is so symmetrical that its geometry alone cannot be used to distinguish indenter from indented continents in seismic profiles. However, the asymmetry of foreshears and backshears within the wedge are not symmetrical and can be identified.

There is a striking difference between the natural orogenic wedges illustrated here and the Central Alps (Bonini et al., 1999). The basal decollement from which the wedge rose in the Alps was along a boundary within the crust assumed to be the transition between the brittle upper and ductile lower crust (Bonini et al., 1999, and here). However, seismic profiles across the Proterozoic to Palaeozoic orogens imply basal detachment and orogenic wedges rising from a P = S point on the Moho (as modeled by Beaumont et al., 1994; Fig. 1A). It is not clear whether this difference is real or due to different interpretations of reflections from the deep crust. If it is real, this difference

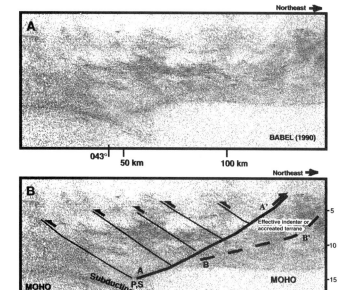

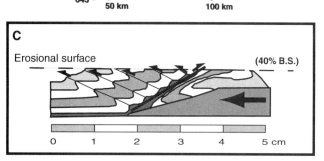

Figure 14. A: Part of BABEL reflection profile (migrated, after BABEL Working Group, 1990). B: Interpretation of BABEL data based on present models. Line A-A' is here backshear, line B-B' is shallow-dipping indenter, P point is where kinks are initiated and S is toe of "rigid" indenter. C: Experimental model resembling structures of BABEL profile. Erosional surface is introduced to ease comparison. B.S. is bulk shortening.

may indicate the level at which tectonic wedging diverge has changed as the lithosphere increased in thickness with time, so that the entire contracted crust escaped vertically from the Proterozoic Svecofennian (ca. 2.0 Ga) to the Paleozoic Urals (250 Ma), whereas only the upper crust escaped in the Phanerozoic Alps (100–50 Ma).

SUMMARY

The geometry of the leading edge of the rigid indenting continent controls not the geometry of the orogenic wedge as a whole, but the number and spacing of internal shears and whether an effective indenter develops. The wedges everywhere are bound by foreshears and backshears with dips corresponding to the internal friction (25°–30°) in both models and nature. Seismic profiles of the Urals and the Baltic shield are interpreted to image two-sided orogenic wedges, one with no effective indenter, the second with the new type of indenter introduced herein.

The dip of the first forekink does not change with the thickness of the initial indented material. The dips of subsequent kinks increase slightly in response to lateral and vertical compaction, which increases the angle of internal friction. Where effective indenters are built, the curvature of long-lived backshears reflects the increase in tectonic compaction (and backrotation).

Rather than affect the deformation mechanism, the initial thickness of the brittle indented crust controls the amount of bulk lateral shortening needed to observe the same structural geometry. The distribution of shear bands, particularly the number of forekinks and whether an effective indenter develops, is an indication of strain partitioning into the most energy-efficient structural geometry. Effective indenters develop and mature when rigid ramp dips depart from the angle of internal friction of the shortened material. The geometry of the complete orogenic wedge is independent of the geometry of the indenter, which appears to control the spacings and rates of imbricate shears and/or kink formation. The development of an effective indenter inactivates the initial suture, which is replaced by one or more temporarily active backshear(s). Imbricate slices of indented continent are easy to mistake for slices of the indenting continent or accreted terranes.

The number of forekinks developed in the indented continent relates to the thickness of the indented continent. Thin indented continents develop a greater number of more closely spaced kinks than a thicker continent after equivalent degrees of lateral bulk shortening. The amount of tectonic exhumation (dependent on gravity collapse of the backslope that controls the weight of the rising wedge) also controls the rate of thrust formation.

Spacings of kinks within the orogenic wedge relate to the initial dip of the indenter. Spacing increases with initial hanging-wall thickness and increasing thickening due to tectonic shortening, which increases the vertical load on where kinks initiate.

Indenters with ramp dips corresponding to the angle of internal friction of the indented material do not build effective indenters and are most efficient in building forekinks and/or forethrusts and raise the orogenic wedge the least. Situations that develop an effective indenter develop fewer kinks in total. Wedges that have risen from a P point in front of a rigid indenter (e.g., the Central Alps) develop fewer forekinks. Where the P point coincides with the S point at the toe of the indenter (e.g., the Svecofennides), more imbricate shears develop than those with rigid ramp dips >45° (e.g., the Central Alps). This is attributed to the need for larger volumes of indented material to laterally compact before slip can occur along the basal decollement for wedges with P points in front of the rigid toe (P ≠ S). The lower the volume of indented material that laterally compacts, the greater the number of kinks that develop. Once the effective indenter has compacted to a rigid mass, new kinks might develop at the same rate independent of the initial rigid indenter geometry. This is not shown in the experiments reported here, where most underwent only as much as 40% bulk shortening.

Two-sided orogenic wedges are never unstable as a whole. Instead, gravity collapse of the extruding hanging walls of foreslopes and backslopes controls how much strain partitions along the relevant shear. Slumping of the foreslope is only local because successive forethrusts are only active for relatively short times. By contrast, the stability of the backslope is controlled by a single continuous backshear and oversteepens regularly to preserve a slope of ~36° (±4°) by essentially continuous gravity collapse. The backshear remains active longer because of continuous unloading by slumping of its extruded hanging wall.

Lowering the angle of internal friction (glass beads) decreases slope stability so that gravity collapse of both foreslopes and backslopes unloads the hanging walls and loads the footwall of active shears so that each remains active longer. Slumping of surficial slopes favors activity on existing imbricate shears over compaction, slip along the basal decollement, and the initiation of new kinks.

Like the dips of foreshears and backshears bounding two-sided orogenic wedges, the dips of the foreslope and backslope reflect the physical properties of the indented material. All the natural and model foreshears and backshears illustrated here are within the range 25°–35°. The models reported here suggest that continuous slumping of the unstable back slope of an orogen results in the exhumation of high-grade metamorphic rocks to exposure along long-lived backshears. Exhumation of deep rocks is more limited along shorter lived foreshears.

ACKNOWLEDGMENTS

I thank Marco Bonini and his coworkers for keeping me up to date with their work, also performed at Hans Ramberg Tectonic Laboratory at Uppsala University. I also thank Dimitrios Sokoutis for advice and constructive criticism; Christopher Talbot for reading and commenting on the many versions of the manuscript; and Olivier Merle, Genene Mulugeta, Geoffrey Milnes, and the two reviewers for their comments on both science and English. Special thanks to the reviewer who suggested additional modeling with other materials, which helped to clarify the importance of slope stability. This work was financed by doktorandtjänst from Uppsala University.

REFERENCES CITED

BABEL Working Group, 1990, Evidence for early Proterozoic plate tectonics from seismic reflection profiles in the Baltic shield: Nature, v. 348, p. 34–38.

BABEL Working Group, 1993, Integrated seismic studies of the Baltic shield using data in the Gulf of Bothnia region: Geophysical Journal International, v. 112, p. 305–324.

Beaumont, C., Fullsack, P., and Hamilton, J., 1994, Styles of crustal deformation in compressional orogens caused by subduction of the underlying lithosphere: Tectonophysics, v. 232, p. 119–132.

Beaumont, C., Ellis, S., Hamilton, J., and Fullsack, P., 1996, Mechanical model for subduction-collision tectonics of Alpine-type compressional orogens: Geology, v. 24, p. 675–678.

Bernoulli, B., Heitzmann, P., and Zingg, A., 1990, Central and southern Alps in southern Switzerland: Tectonic evolution and first results of reflection seismics: Mémoires de la Société Géologique de France, v. 156, p. 289–302.

Bonini, M., Sokoutis, D., Talbot, C.J., Boccaletti, M., and Milnes, A.G., 1999, Indenter growth in analogue models of Alpine-type deformation: Tectonics, v. 18, p. 119–128.

Burg, J.-P. and Ford, M., 1997, Orogeny through time: An overview, in Burg, J.-P., and Ford, M., eds., Orogeny through time: Geological Society [London] Special Publication 121, p. 1–17.

Byerlee, J., 1978, Friction of rocks: Pure and Applied Geophysics, v. 116, p. 615–626.

Cobbold, P.R., and Davy, C., 1988, Indentation tectonics in nature and experiment. 2. Central Asia: Uppsala University Geological Institutions Bulletin, v. 14, p.143–162.

Coletta, B., Letouzey, J., Pinedo, R., Ballard, J.F., and Balé, P., 1991, Computerized X-ray tomography analysis of sandbox models: Examples of thin-skinned thrust systems: Geology, v. 19, p. 1063–1067.

Cook, F.A., Hall, K.W., and Roberts, B.J., 1998, Tectonic delamination and subcrustal imbrication of the Precambrian lithosphere in northwestern Canada mapped by LITHOPROBE: Geology, v. 26, p. 839–842.

Davis, D., Suppe, J., and Dahlen, F.A., 1983, Mechanics of fold-and-thrust belts and accretionary wedges: Journal of Geophysical Research, v. 88, p. 1153–1172.

Davy, C., and Cobbold, P.R., 1988, Indentation tectonics in nature and experiment. 1. Experiments scaled for gravity: Uppsala University Geological Institutions Bulletin v. 14, p. 129–141.

Dennis, J.G., 1967, International tectonic dictionary. American Association of Petroleum Geologists Memoir 7, 196 p.

England, P., Houseman, G., and Sonder, L., 1985, Length scales for continental deformation in convergent, divergent, and strike-slip environments: Analytical and approximate solutions for a thin viscous sheet model: Journal of Geophysical Research, v. 90, p. 3551–3557.

Faure, M., Caridroit, M., Guidi, A., and Charvet, J., 1987, The Late Jurassic orogen of south west Japan: Nappe tectonics and longitudinal displacement: Bulletin de la Société Géologique de France, v. 8, p. 477–485.

Faure, M., Natal'in, B.A., Monié, P., Vrublevsky, A.A., Borukaiev, C., and Prikhodko, V., 1995, Tectonic evolution of the Anuy metamorphic rocks (Sikhote Alin, Russia) and their place in the Mesozoic geodynamic framework of East Asia: Tectonophysics, v. 241, p. 279–301.

Friberg, M., 2000, Tectonics of the Middle Urals [Ph.D. thesis]: University of Uppsala, 32 p.

Friberg, M., and Petrov, G.A., 1998, Structure of the Middle Urals, east of the Main Uralian Fault: Geological Journal v. 33, p. 37–48.

Goodman, R.E., 1988, Introduction to rock mechanics: Chichester, Wiley, 576 p.

Huiqi, L., McClay, K.R. and Powell, D., 1992, Physical models of thrust wedges, in McClay, K.R., ed., Thrust tectonics: London, Chapman and Hall, p. 71–81.

Juhlin, C., Friberg, M., Echtler, H.P., Hismatulin, T., Rybalka, A., Green, A.G., and Ansorge, J., 1998, Crustal structure of the Middle Urals: Results from the (ESRU) Europrobe seismic reflection profiling in the Urals experiments: Tectonics, v. 17, p. 710–725.

Koons, P.O., 1990, Two-sided orogen: Collision and erosion from the sandbox to the Southern Alps, New Zealand: Geology, v. 18, p. 679–682.

Koyi, H., 1995, Mode of internal deformation in sand wedges: Journal of Structural Geology, v. 17, p. 293–300.

Malavieille, J., 1984, Modélisation expérimentale des chevauchements imbriqués: application aux chaînes de montagnes: Bulletin de la Société Géologique de France, v. 7, p. 129–138.

Marshak, S., and Wilkerson, M.S., 1992, Effect of overburden thickness on thrust belt geometry and development: Tectonics, v. 11, p. 560–566.

McBride, J.H., Snyder, D.B., England, R.W., and Hobbs, R.W., 1996, Dipping reflectors beneath old orogens: A perspective from the British Caledonides: GSA Today, v. 6, p. 1–6.

McClay, K.R., and Ellis, P.G., 1987, Analogue models of extensional fault geometries, in Coward, M.P., et al., eds., Continental extensional tectonics: Geological Society [London] Special Publication 28, p. 109–125.

Merle, O., and Abidi, N., 1995, Approche expérimentale du fonctionnement des rampes émergentes: Bulletin de la Société Géologique de France, v. 166, p. 439–450.

Mulugeta, G., 1988, Modelling the geometry of Coulomb thrust wedges: Journal of Structural Geology, v. 10, p. 847–859.

Mulugeta, G., and Koyi, H., 1987, Three-dimensional geometry and kinematics of experimental piggyback thrusting: Geology, v. 15, p. 1052–1056.

Mulugeta, G., and Koyi, H., 1992, Episodic accretion and strain partitioning in a model sand wedge: Tectonophysics, v. 202, p. 319–333.

Pfiffner, O.A., Frey, M., Valasek, M., Stäuble, L., Levato, L., Dubois, S.M., Schmid, S., and Smithson, S.B., 1990, Crustal shortening in the alpine orogen: Results from deep seismic reflection profiling in the eastern Swiss Alps, Line NFP 20-East: Tectonics, v. 9, p. 1327–1355.

Ratschbacher, L., Merle, O., Davy, P., and Cobbold, P., 1991, Lateral extrusion in the eastern Alps, part 1: Boundary conditions and experiments scaled for gravity: Tectonics, v. 10, p. 245–256.

Schmid, S.M., Pfiffner, O.A., Froitzheim, N., Schönborn, G., and Kissling, E., 1996, Geophysical-geological transect and tectonic evolution of the Swiss-Italian Alps: Tectonics, v. 15, p. 1036–1064.

Tapponnier, P., and Molnar, P., 1976, Slip-line field theory and large-scale continental tectonics: Nature, v. 264, p. 319–324.

Tapponnier, P., Pelzer, G., Le Dain, A.Y., Armijo, R., and Cobbold, P., 1982, Propagating extrusion tectonics in Asia: New insights from simple experiments with plasticine: Geology, v. 10, p. 611–616.

Thompson, A.B., Schulmann, K., and Jezek, J., 1997, Extrusion tectonics and elevation of lower crustal metamorphic rocks in convergent orogens: Geology, v. 6, p. 491–494.

Thouvenot, F., Kashubin, S.N., Poupinet, G., Makovsky, V.V., Kashubina, T.V., Matte, P., and Jenatton, L., 1995, The root of the Urals: Evidence from wide-angle reflection seismics: Tectonophysics, v. 250, p. 1–13.

Willett, S., Beaumont, C., and Fullsack, P., 1993, Mechanical model for the tectonics of doubly vergent compressional orogens: Geology, v. 21, p. 371–374.

MANUSCRIPT ACCEPTED BY THE SOCIETY APRIL 12, 2000

Geological Society of America
Memoir 193
2001

Horses and duplexes in extensional regimes: A scale-modeling contribution

Roy H. Gabrielsen*, Jill A. Clausen*
Geological Institute, University of Bergen, Allégaten 41, N-5007 Bergen, Norway

ABSTRACT

Horses and duplexes (rock lenses and overlapping arrays of lenses) are common intrinsic geometrical building stones of larger extensional faults. Such structures have been studied by using plaster analogue models. Study indicates that several mechanisms are active in the initiation and development of extensional horses and duplexes. The following mechanisms are defined and described: tip-line coalescence, segment linkage, tip-line bifurcation, asperity bifurcation, segment splaying, and segment amalgamation. Of these, segment linkage, asperity bifurcation, and footwall and hanging-wall segment splaying and amalgamation are the most commonly observed in the experiments, and are also believed to be most common in natural faults.

The process of generation of extensional duplexes is divided into the following stages: (1) initiation of a through-going fault plane, sometimes characterized by irregularities due to top-line coalescence, segment linkage, and tip-line bifurcation; (2) smoothing of fault-plane irregularities by asperity bifurcation, which is characterized by fault-related cutoff of irregularities in the footwall as well as in the hangingwall; (3) generation of a second (or third) generation of asperity bifurcation horses, or horses created by splaying or coalescence of secondary faults in the immediate hangingwall and footwall of the master fault. At this stage, the potential for producing complete duplexes becomes greater. The next stage is (4) internal shearing of the individual horses and splitting of larger horses; and (5) collapse of the fault zone, and smearing of the fault-zone units to constitute a homogeneous mass along the fault plane.

INTRODUCTION

The need for detailed understanding of fault geometry in hydrocarbon exploration and exploitation has recently motivated intensive studies of extensional faults, many emphasizing the geometric complexity of fault zones. The internal architecture of faults is of particular interest for the study of fluid communication along and across fault zones, because contacts between units of high or low permeability control the fluid flow (Childs et al., 1997; Knipe, 1997; Knipe et al., 1997; Losh et al., 1999).

Primary and secondary shear stresses influence the structural style of the fault zone (Knott, 1994; Lyakhovsky et al., 1997; Gabrielsen et al., 1998). Furthermore, strain caused by isostatic and elastic response of the hanging wall and footwall, as well as stress release associated with the faulting (Kusznir et al., 1987; King et al., 1988; Yielding and Roberts, 1992; Roberts and Yielding, 1991), add to the development of damage zones. Deformation within the fault zone, including splaying, linkage, and ramping (e.g., Bruhn et al., 1994; Gibbs, 1990; Walsh and Watterson, 1991) are the most prominent elements in the generation of its geometric complexities.

The basic internal geometry of extensional faults is commonly that of lensoid rock bodies (horses), sometimes stacked or overlapping to constitute duplexes (Gibbs, 1983,

*E-mails: Gabrielsen, roy.gabrielsen@geol.uib.no;
Clausen, jill.clausen@geol.uib.no

Gabrielsen, R.H., and Clausen, J.A., 2001, Horses and duplexes in extensional regimes: A scale-modeling contribution, *in* Koyi, H.A., and Mancktelow, N.S., eds., Tectonic Modeling: A Volume in Honor of Hans Ramberg: Boulder, Colorado, Geological Society of America Memoir 193, p. 207–220.

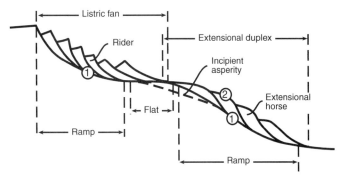

Figure 1. Schematic representation of ramp-flat-ramp fault with associated (internal) structures (hanging-wall fault block is not shown). 1 = floor fault, 2 = roof fault. Sketch and nomenclature are based on Gibbs (1984) and Childs et al. (1997).

1984) (Fig. 1). Such structures are delineated by surfaces or zones of intense shear that constitute the floor fault and roof fault of the duplex (Woodcock and Fisher, 1986; Gabrielsen and Koestler, 1987; Cox and Scholz, 1988; Swanson, 1988; Cruikshank et al., 1991; Childs et al., 1997). The horses may consist of undeformed to heavily fractured rock (Koestler and Ehrmann, 1991; Childs et al., 1996a), or may be entirely dominated by fault rocks (Sibson, 1977).

Listric fans are frequently developed in the upper part of extensional faults. Such structures may develop, into, or be reactivated to constitute, an upper duplex, particularly where sedimentation continues during stretching (Heybroek, 1975; Gibbs, 1983). Field studies suggest that there is a positive correlation between fault displacement and the thickness of the fault zone (Robertson, 1983; Hull, 1988), and it seems clear that the development of extensional horses and duplexes may contribute to fault-zone widening (Childs et al., 1997).

It is generally accepted that single fault surfaces expand by the incremental advancement of the tip line, so that the displacement shifts systematically from a maximum at the fault center to zero at the fault tip (Watterson, 1986; Barnett et al., 1987; Walsh and Watterson, 1989). Further development of faults commonly involves asperity bifurcation and tip-line bifurcation (Fig. 2), followed by segment linkage, where intersection and capture of overlapping en echelon faults occur (Anders and Schlische, 1994; Cartwright et al., 1996; Schlische and Anders, 1996; Childs et al., 1997). These processes are accompanied by the development of horses, which become trapped within the fault zone and contribute to widening of the fault zone. All these processes have been observed in the present experiments, in which further widening of fault zone is supported by segment splaying and segment amalgamation.

EXPERIMENTAL APPROACH

Because of its very fine grained nature, and hence potential to develop and preserve very delicate details, and because it solidifies after deformation, plaster of Paris is well suited to the experimental study of fault-zone geometry and development. In our study a mechanically homogeneous plaster column with passive color markings was extended until faulting occurred, using the method described in Sales (1987) and Fossen and Gabrielsen (1996) (Fig. 3).

A glass box with two fixed glass walls, two moveable short wooden walls, and a hard (wooden) or soft (barite) base of different geometries was used. The fault box can be set up with two different widths (21.5 cm and 11.5 cm), but the wide setup was applied in most of the experiments. The upper surface was unconfined. The walls were sealed by putty, tightened by rubber bands, and lubricated with oil, and passive markers consisting of carbon powder were painted on the sides of the plaster model with a brush. Retracting one of the short walls by hand extended the models. Hence, the strain rate was not constant during the experiment, but this nonsteady strain rate is necessary to keep pace with the rapid solidification of the plaster. During the analysis, continuous strain-rate measurements are obtained from the video recordings. Shifting shear angles can easily be tracked from the videotape.

The plaster is generally homogeneous, but stratification due to slow drainage of expulsion-water may enhance local ductile deformation and favor the development of detachment zones. The disadvantages of this type of experiment are the rapid change in physical character of the plaster during the experiments and the lack of internal marker horizons. Hence, nothing can be gained by study cuts of the plaster models.

The resulting structures are remarkably similar to structures observed in nature. The structures can be studied in the four dimensions, because both sides are videotaped and photographed with regular intervals through the glass wall, and the top of the experiment is easily accessible (Fig. 4). In this study we focus on the vertical section, because our method prevents observations of the horizontal section, except at the free surface of the model.

EXPERIMENTAL RESULTS

Our analysis comprised 14 experiments. Each experiment was documented by the use of still photographs and video recordings. These recordings were used in determining time for fault initiation and disruption of layers, throw, heave, fault dip angle, and time of initiation of splay faults and other accommodation faults. The continued development of single splay faults and the coalescence of splays to form extensional horses (lenses) and duplexes were also recorded.

In all experiments a stiff (wooden) or partly soft (wooden covered by barite or plasticene) basement was used. The profile of the basement varied from highly profiled ramp-flat-ramp to flat (Fig. 5). Key parameters of the experiments are given in Table 1. In the experiments where the profiled basement was used, a primary fault, which was detached from the basement-plaster contact, invariably developed in the plaster sequence, usually 5–7 cm above the contact between the plaster and the profiled, dipping basement front. This fault commonly had an overall geometry similar to, but smoother than, that of the top of the basement.

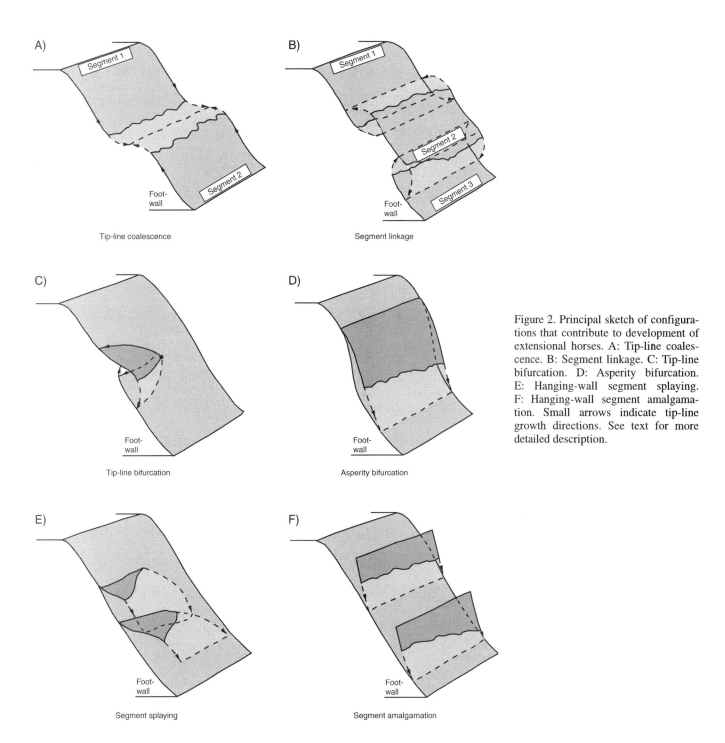

Figure 2. Principal sketch of configurations that contribute to development of extensional horses. A: Tip-line coalescence. B: Segment linkage. C: Tip-line bifurcation. D: Asperity bifurcation. E: Hanging-wall segment splaying. F: Hanging-wall segment amalgamation. Small arrows indicate tip-line growth directions. See text for more detailed description.

The sequence of events in most of the experiments was as follows: Initiation of the primary master fault occurred after ~5%–12% of extension (Fig. 6; model I). This stage was characterized either by one fault strand propagating upward or downward, or two strands growing toward each other and finally coalescing in the central part of the plaster sequence. After 15%–25% of extension, the first asperity bifurcation fault and splay faults associated with the master fault initiated, commonly nucleating at or near the master fault surface. As many as six splay faults were produced in some experiments. Most of the experiments used in the geometric-kinematic analysis were terminated after 55%–80% of extension. In some cases, particularly where extension was continued beyond 70%, a second master fault would be seen farther outward in the hanging-wall fault block. Advanced stages of extension would be associated with internal shearing of the extensional horses of the master fault, and finally collapse of the fault zone.

Horse growth by tip-line coalescence, segment linkage, asperity bifurcation, and segment splaying

An example of a horse generated by segment linkage is shown in Figure 7 (model II). A semisoft basement with a ramp and flat dipping 25° and 7°, respectively, was used. Two separate strands of the master fault nucleated after ~7% of

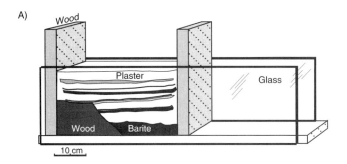

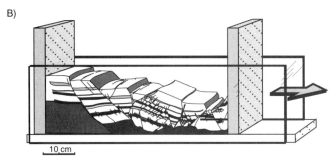

Figure 3. Experimental setup for plaster of Paris extensional fault box. This figure illustrates setup with stiff wooden basement with (soft) barite frontal part. Variety of basement configurations used in our experiments is displayed in Figure 5 and Table 1. Modified from Fossen and Gabrielsen (1996).

extension, growing from the top surface and from the flat contact between the basement and the plaster. At stage 1, the two fault strands (faults 1a and 1b) had not linked completely. After 19% of extension, the tip lines of the two strands overlapped with a right-stepping extensional geometry, and a reorientation of the propagation direction of the tip line became evident in that reorientation of the fault tips occurred (stage 2). By continued extension, the two fault strands became linked, and an incipient horse-structure was generated (stage 3, 40% extension). In many experiments where horses were developed by segment linkage or tip-line bifurcation, the structures were commonly destroyed by asperity bifurcation at a later stage.

Figure 8 (model III) displays a model where both segment linkage and asperity bifurcation occurred. In this experiment, a wooden basement with a large horizontal flat modified by plasticine was used (Fig. 5; Table 1). Here, the primary master fault, which was initiated after only 5% of extension, developed from two separate faults, one originating from the top and one from the base of the plaster sequence (faults 1a and 1b in stages 1 and 2). By the linkage of these strands in the center of the plaster column (stage 3; 14% extension), a minor flat developed. It is noteworthy that the two fault tips did not link up completely, so that the fault tip, which was moving upsection, spilt into two strands, one of which continued to grow.

By continued deformation a horse delineated by a steep floor fault was initiated in the footwall (stage 4, 20% extension; horse I). This can be classified as an asperity bifurcation footwall horse. The fault strand, which had been growing from below, became less steep and linked with the strand growing from above (stage 5, 29% extension; horse II). Hence, two horses were now established, one generated by segment linkage, and one by asperity bifurcation. Together they constitute an overlapping horse configuration, where the floor fault of the foot-

Figure 4. Three-dimensional view of plaster model obtained by experimental setup displayed in Figure 3.

wall horse became the roof fault of the hanging-wall horse. Hence, the structure does not meet the definition of a duplex, which requires that the horses have common floor and roof faults (e.g., Boyer and Elliot, 1982; Nystuen, 1989). By continued extension horse I became passive, and slip was focused on the roof fault of this structure. Thus, the original flat was neutralized at this stage, and horse II was sliding passively above an almost planar fault (41% extension; cf. stages 5 and 6).

In model IV (Fig. 9A) a steep wooden basement draped by a thin layer of barite was used (Fig. 5; Table 1). The development at the earliest stages was comparable to those previously described for model III. Two separate fault segments were simultaneously growing upward and downward (stage 1; fault 0), becoming joined by tip-line coalescence (stage 2; 17.5% extension). By 25% of extension (stage 3), two different horses had developed (Fig. 9B). The major horse, which was situated above the lowermost and steepest part of the master fault (stage 3; horse I), was generated by the initiation of a low-angle fault in the hanging wall of fault 0, which accordingly can be classified as an asperity bifurcation hanging-wall horse. A second extensional horse (stages 3 and 4; horse II) was cut into the hanging wall at the base of the master fault, apparently to smooth the steep fault plane at the base of fault 0. By sliding along this (low angle) fault plane, horse I became displaced on top of horse II, defining a duplex.

At the same stage, the footwall beneath the upper and steepest part of fault 0 became destabilized, and a series of synthetic footwall segment splays were activated (faults 1, 2, and 3). Of these, the one positioned farthest into the footwall, and which displayed the smaller dip angle, became the focus of strain in the latest stage of the experiment, and a third horse (stage 4; horse III) became activated.

Horse growth by footwall and hanging-wall segment splaying and coalescence

Although horses generated by asperity bifurcation are common, the majority of the extensional horses produced in our experiments nucleated from synthetic or antithetic faults in the footwall or hanging wall after the active master fault became established. In principle, such faults can nucleate either at the master fault plane or very close to it. In such cases, continued displacement causes either splaying and/or coalescence of the master fault and the secondary fracture.

Model V (Fig. 10) demonstrates a geometrically simple process for the development of an incipient system of extensional horses by segment splaying. The experiment was performed by the use of a complex, stiff (wooden) basement with two minor flats (Fig. 5; Table 1). In this case, a slightly listric master fault, which was initiated after ~8% of extension (stage 1; fault 0), was accompanied by a synthetic, subparallel extensional fault in the hanging-wall. This fault became visible after 15% of strain (stage 2; fault 1). By progressing deformation, fault 1 became amalgamated to fault 0 to produce a thin rider. After between 25% and

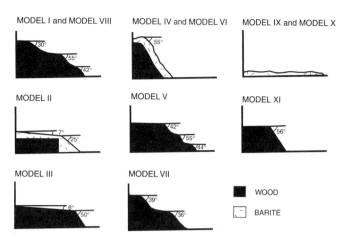

Figure 5. Basement configurations applied in our models. Parameters for each experiment are displayed in Table 1.

TABLE 1. DATA ON EXPERIMENTAL SETUP, FIGURE AND LABORATORY REFERENCES, AND FINAL EXTENSION FOR MODELS I-IX

Model number	Laboratory reference	Figure	Basement type	Extension
I	7-99A	5	W	85%
II	13-99	6	W+B	64%
III	10-98-	7	W	41%
IV	12-98A	8	W+B	56%
V	3-99-	9	W	35%
VI	12-98B	10	W+B	56%
VII	21-98	11	W	69%
VIII	7-99B	11	W	85%
IX	22-98	12	B	72%
X	1-89	12	B	
XI	2-98	12	W	77%

See Figure 5 for basement configuration. Abbreviations: W—wood; B—barite.

35% extension, another synthetic hanging-wall fault (fault 2) was initiated, defining a second (not yet complete) horse in the upper part of the hanging-wall (stage 3; horse I). Thus, the extensional duplex generated in this experiment consists of one rider and one horse, initiated by the coalescence of a synthetic hanging-wall fault and the master fault in two separate stages (Fig. 10B).

Model VI (Fig. 11A) initially displayed a slightly irregular master fault above a hard wooden basement with a pronounced ramp-flat geometry and a frontal dip angle of 62° (Fig. 5; Table 1). By 27% of extension a synthetic extensional fault had been nucleated in the hanging-wall. This fault, which may have been influenced by an irregularity in the basement topography caused by a piece of barite, merged with the master fault, thus defining a footwall horse (stage 3; horse I). At this stage, another footwall splay fault became visible (stage 3; fault 5). It contrast to fault 4, however, fault 5 was nucleated at its primary contact with fault 0, its tip line moving downward. After ~53% of extension (stage 4), the tip of fault 5 came into contact with fault 4, which defined the floor fault

MODEL I

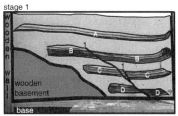

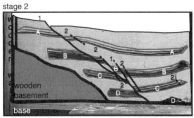

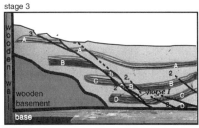

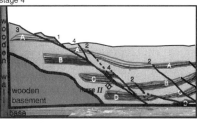

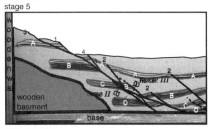

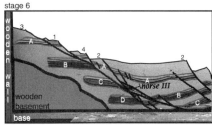

Figure 6. Model I illustrates typical development of experiment run to 70% extension. Plaster sequence was extended above solid wooden basement with ramp-flat-ramp geometry. Three extensional horses and four riders were generated in this example. Numbers indicate sequence of initiation of faults and horses. Layers are identified by uppercase letters.

MODEL II

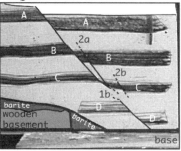

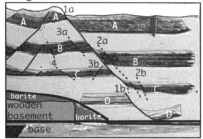

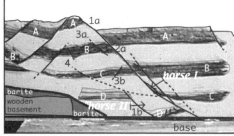

Figure 7. Model II shows development of an extensional horse by segment linkage followed by asperity bifurcation. Note short life of segment link structure before it is transformed into horse I. Numbers indicate sequence of initiation of faults and horses. Layers are identified by uppercase letters. See text for more detailed description.

of horse I, and a duplex consisting of two separate horses (one derived from the footwall and one from the hanging-wall) came into existence.

The amalgamation of footwall and hanging-wall splay faults is obviously an efficient way of producing arrays of horses and, hence, complete extensional duplexes. In the two examples described here, the progress in faulting was such that

MODEL III

Stage 1
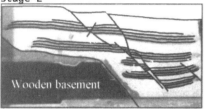
5% extension

Stage 2
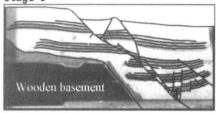
9% extension

Stage 3
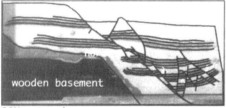
14% extension

Stage 4
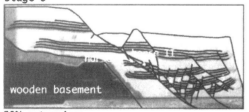
20% extension

Stage 5
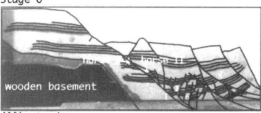
29% extension

Stage 6
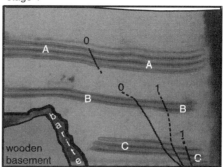
41% extension

Figure 8. Example of asperity bifurcation in footwall (horse I) and hanging wall (horse II), following segment linkage (model III). Numbers indicate sequence of initiation of faults and horses. Layers are identified by uppercase letters. See text for more detailed description.

MODEL IV

stage 1
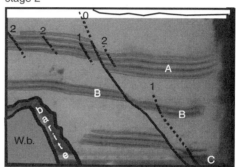
6% extension

stage 2
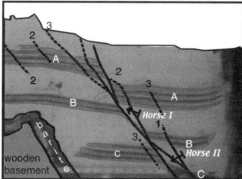
17.5% extension

stage 3

27% extension

stage 4

33% extension

Figure 9. A: Tip-line coalescence followed by development of incipient extensional duplex by combination of segment amalgamation and asperity bifurcation (model IV). Numbers indicate sequence of initiation of faults and horses. Layers are identified by capital letters. See text for more detailed description.

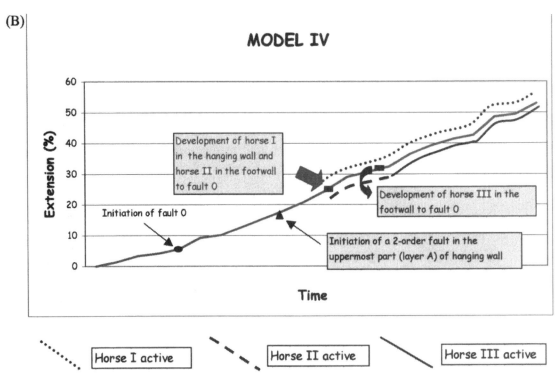

Figure 9. B: Time-strain-event diagram for model IV.

development continued both in the hanging-wall and in the footwall of the master fault. Consequently, these experiments demonstrate that it is fully possible that splay faults may carve lenses from both the footwall and the hanging wall during a sequential evolution, resulting in duplexes, which vary in terms of their geometrical complexity (Fig. 12).

Horse collapse by continued faulting

Although transfer of strain to the smoother master fault surface (roof fault or floor fault of the duplex) is the most common style of deformation to continue beyond the establishment of duplexes, internal shearing of lenses is seen in some cases (Fig. 13A). The process of internal shearing of lenses generally seems to follow well-established principles of shear, so that the dominant zones of intense strain are focused along the (synthetic) Riedel and Y-shear directions (Tchalenko, 1970; Christie-Blick and Biddle, 1985). With continued deformation, complete collapse of the entire duplex may occur (Fig. 13B), followed by smearing out of the duplex along the entire fault plane (Fig. 13C). It seems reasonable to attribute the two latter stages in the development to massive strain softening along the entire plane of the master fault.

INITIATION AND GROWTH OF EXTENSIONAL HORSES AND DUPLEXES

By analysis of still photographs and video recordings, the initiation and growth of splay faults, horses, and duplexes for 14 experiments were recorded and plotted as a function of strain (Fig. 14). In all our experiments, the first fault was recorded after 5%–12%, most commonly after 8%–12% of strain (see also Odinsen, 1992; Fossen and Gabrielsen, 1996).

In some experiments where the asperity bifurcation mechanism prevailed, the first horse became active after 10%–30% of extension (Fig. 14A). Where horses developed from riders or splay faults, the first horse (delineated by complete floor and roof faults) was established after 18% of extension in some rare cases, whereas more common values would be 20%–40% (Fig. 14B). The second horse commonly was seen after 30%–45% of strain, although examples of both earlier and later developments were seen (Fig. 14C). This implies that horses were most commonly activated at one time and in isolation, and that duplexes are typical of a more advanced stage of deformation (i.e., after more than 40% of extension). This stage may be characterized by two or more duplexes being brought into contact by transport along the master fault plane, or by splay faulting taking place both in the footwall and the hanging-wall margins of the master fault. The potential for activation of asperity bifurcation horses is greater at an early stage of development. In contrast, hanging-wall and footwall splaying and coalescence were common over an extended range of stages of deformation.

In the next stages, an array of horses would develop to different degrees of complexity before size reduction of horses took place by intrinsic shearing, and complete collapse of the fault zone occurred, commonly after more than 40% of extension (Fig. 14D).

MODEL V

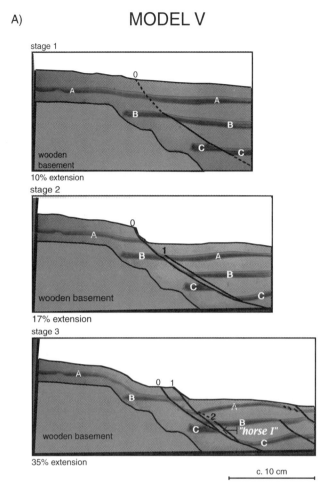
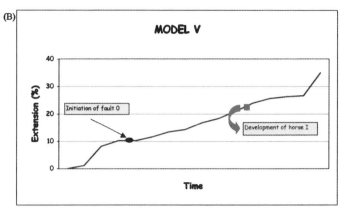

Figure 10. A: Three stages of development of horse due to collapse of hanging-wall rider by antithetic segment splaying in model V. Numbers indicate sequence of initiation of faults and horses. Layers are identified by capital letters. See text for more detailed description. B: Time-strain-event diagram for model V.

MECHANISM FOR DEVELOPMENT OF DUPLEXES

Experiments of extensional faulting in plaster of Paris above a basement with different configurations and mechanical properties produced a variety of fault patterns, which are strikingly similar to those seen in nature. Several of these patterns involve geometries that contribute to the initiation of extensional horses and, finally, extensional duplexes (Fig. 2). We distinguish between two principal groups of patterns: the primary, which are affiliated with the establishment of the fault, and the secondary, which are associated with the development of the fault zone after a continuous fault plane is established.

The general model of fault growth (Barnett et al., 1987; Walsh and Watterson, 1989) implies that fault surfaces may be regarded as elliptical surfaces, indicating that the tip line progresses at different rates, as seen in a horizontal section and the vertical sections. In this study, we focused on the vertical section, because the method does not allow observations to be made in the horizontal section, except for the free surface of the model. The direction of fault-tip growth considered is indicated in each case in Figure 2.

Tip-line coalescence (Fig. 2A) is a primary feature, i.e., related to the establishment of the fault. It is generated from separate, parallel, or subparallel fault segments that link directly, without the development of a relay ramp in the vertical section. The result is a continuous, wavy fault plane, which includes small flats in the master fault plane. In our experiments, these flats commonly became nuclei for horses at the more advanced stages of fault development.

Segment linkage (Fig. 2B) involves intersection and capture of overlapping en echelon faults (Anders and Schlische, 1994; Cartwright et al., 1996). Segment linkage is related to the (primary) establishment of the fault plane, and is perhaps the most common mechanism for the development of larger fault zones. The individual segments may be true en echelon separate or en echelon branching (Ferrill et al., 1999). Horses are generated by segment linkage of separate fault branches, which are bent toward each other and coalesce due to mutually interfering tip-line stresses (e.g., Olson and Pollard, 1989; Engelder, 1993, p. 49–50). Such structures are common in the plaster models, where segment linkage results from interconnection of fault traces growing upsection from the basement and downsection from the surface. The fault traces are first seen to curve slightly as the tip lines start to overlap, before linkage causes a pronounced ramp-flat-ramp geometry. Segment links are commonly short lived features, because the irregularity caused by the linkage is quickly obliterated by asperity bifurcation.

Tip-line bifurcation (Fig. 2C) is characterized by the interference of the primary tip line of two subparallel fault strands originating from one fault (Childs et al., 1996b). The tip-line embayment may originate from a mechanical heterogeneity (Huggins et al., 1995; Childs et al., 1997). In this model, lozenge-shaped rock units are generated in areas where the propagating fracture tips come sufficiently near each other for the stress fields associated with the tip lines to interfere. By continued tip-line propagation, the two fault strands will reunite to define a relay structure (Olson and Pollard, 1989; Engelder, 1993, p. 49–50;

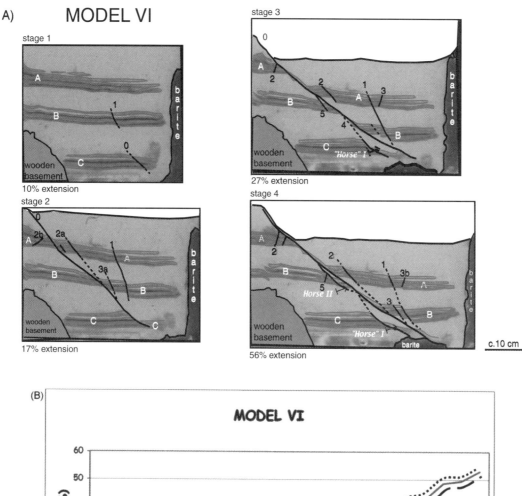
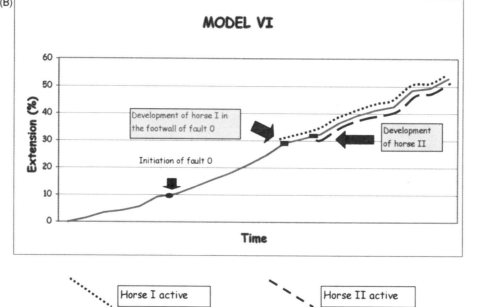

Figure 11. A: Horses generated by footwall segment splaying (model VI). Numbers indicate sequence of initiation of faults and horses. Layers are identified by uppercase letters. See text for more detailed description. B: Time-strain-event diagram for model VI.

Cruikshank et al., 1991; Peacock and Sanderson, 1991, 1994; Cruikshank and Aydin, 1995; Childs et al., 1997; Fossen and Hesthammer, 1997; Walsh et al., 1999). The tip-line bifurcation structures observed in our models most probably result in connection with horizontal fault growth (Fig. 2C). This implies that the fault split occurs inside the model, and that the two branches propagate horizontally, hitting the glass surface at different stages of interference, hence contributing to the first generation of horses. In our experiments, such structures are tiny and short-lived, and have not been easily documented in videos or still photographs.

Asperity bifurcation (Fig. 2D) is characterized by shearing off of preexisting fault-plane irregularities to produce a smooth fault surface. Accordingly, the asperity bifurcation is a secondary process. In several experiments where flats were produced during the early stage by tip-line coalescence or segment linkage, the fault plane soon became smoothened by asperity bifurcation. The cutoff may occur both in the footwall and the hanging wall, so that the cutoff constitutes either the floor fault or the roof fault of the horse. Several generations of asperity bifurcation horses may be generated at later stages of the fault process where continued ex-

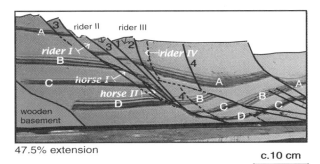

Figure 12. Complex horse configurations. A: Stacked horses derived from footwall and hanging wall (model VII). B: Duplex generated by stacking of horses generated by asperity bifurcation and collapse of riders of upper listric fan (model VIII). Numbers indicate sequence of initiation of faults and horses. Layers are identified by uppercase letters.

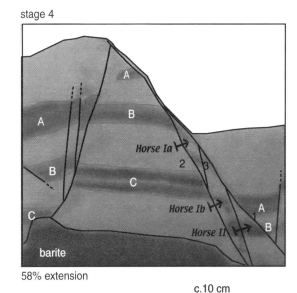

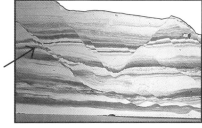

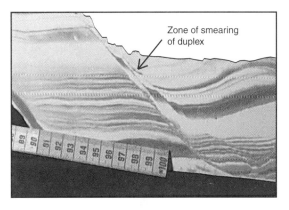

Figure 13. Shearing and collapse of horses. A: Internal shear in extensional horse (model IX). B: Collapsed duplex (model X). C: Completely smeared-out duplex (model XI). Numbers indicate sequence of initiation of faults and horses. Layers are identified by uppercase letters.

tension creates an uneven geometry of the master fault, or when the first generation of horses is insufficient to smooth the master fault plane. In some cases the first-generation asperity bifurcation horses have even contributed to enhanced fault-plane relief, promoting a second or even third generation of asperity bifurcation structures. In natural faults, bedding-parallel slip may contribute to asperity bifurcation (Barnett et al., 1987; Childs et al., 1997).

Segment splaying (Fig. 2E) is regarded as a secondary process, because it commonly becomes important after establishment of the master fault. This mechanism is characterized by development of fault strands that originate at the master fault surface. These structures will commonly grow synthetically to the master fault, and may eventually link up to form an array of horses. In nature, such structures are commonly nucleated by inhomogeneities situated at the fault plane (Vermilye and Scholz, 1999).

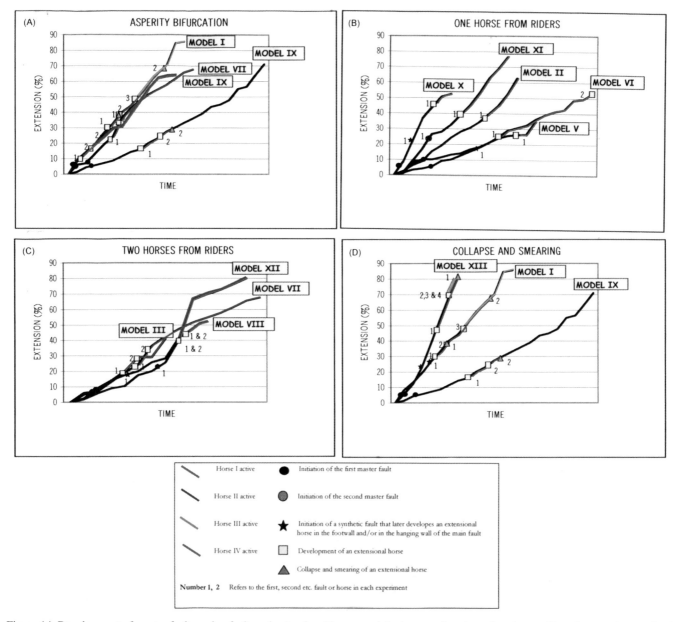

Figure 14. Development of master faults, splay fault, and extensional horses and duplexes as function of strain rate. Experiments characterized by A, asperity bifurcation; B, development of one horse; C, multiple horses; and D, collapse of horses followed by smearing along master fault zone. Experiment identification is given in boxes in diagram.

It is possible to distinguish between two types of horses associated with segment splays. Hanging-wall splay horses (Fig. 2E) are generated by the synthetic splaying of secondary faults. These nucleate at the master fault surfaces, and are distinguished from the asperity bifurcation horses in that they are not associated with irregularities in the master fault surface. In the earliest stage of development, the splay faults are less steep than the master fault plane. During continued strain the splay faults either become steeper so that their tip lines again merge with the master fault, or their tip lines coalesce with another secondary fault, which may grow in an antithetic or synthetic manner to the master fault, to produce a horse. Duplexes may be generated directly by simultaneous initiation and growth of several fault branches.

Segment amalgamation (Fig. 2F) is also typical for the secondary stage of fault development. This mechanism becomes active when faults, which nucleate in the hanging wall or the footwall, grow downward to approach and finally to coalesce with the master faults. Such structures seem to be related either to high strain zones or damage zones associated with the master fault or to synthetic accommodation structures. Hanging-wall amalgamation horses (Fig. 2F) arise from hanging-wall riders, and are distinguished from the asperity bifurcation horses in that they are unconstrained by irregularities in the master fault surface. In cases where the riders are situated so deep that they have no contact with the surface, or when material is added at the surface to bury the surface trace of the rider, a complete horse may be generated by activation of an upper fault strand (Heybroek, 1975; Gibbs, 1983). This strand may grow either in a synthetic fashion from the master fault plane, or antithetically from the original (blind) tip line of the roof fault. In some cases, several faults develop simultaneously, so that duplexes are generated directly.

Structures developed in the secondary stage by asperity bifurcation, segment splaying and segment coalescence may affect both the footwall and the hanging wall. Furthermore, secondary faults may propagate in synthetic or antithetic directions as related to the master fault. Hence, both footwall splay horses and footwall amalgamation horses may exist. These are principally equal to their hanging-wall counterparts, but cut into the footwall at a steeper angle than that of the master fault. Footwall horses may form duplexes by mechanisms similar to those previously described herein for the hanging-wall duplexes.

CONCLUSIONS

The development of the fault zones can be divided into several steps. These are distinguished by characteristic geometry and fault-zone structure.

The first stage of fault development produced a through-going fault plane, which, in some cases, was characterized by irregularities due to tip-line coalescence, segment linkage, and, in some cases, tip-line bifurcation. These processes produce irregularities and minor horses that are destroyed after a short time of continued extension.

In the second stage, larger fault-plane irregularities are smoothed by asperity bifurcation, which includes fault-related cutoff of irregularities in the footwall as well as in the hanging wall. The horses of the first two stages most commonly form in isolation, so that complete duplexes seem to be rare at this stage.

The third stage is characterized by the initiation of a new generation of horses. These are partly generated by a second (or third) generation of asperity bifurcation horses, or by horses created by splaying or coalescence of secondary faults in the hanging wall and footwall of the master fault. At this stage, the potential for producing complete duplexes becomes larger.

In the fourth stage, internal shearing of the individual horses becomes evident, the larger horses split into several minor ones, and complex duplexes become more common.

From stage 1 through to stage 4 a bulk widening of the fault zone occurs, mainly due to the development and transposition of horses, and finally, stacking of horses into duplexes. Locally, however, temporary thinning takes place, mainly by the same processes.

Very few experiments were run until the fifth and final stage was reached. By extreme extension, however, the collapse of the fault zone is complete, and the horses are smeared out to constitute a homogeneous mass along the fault plane. In general, this stage is characterized by a bulk thinning of the fault zone along its full length.

Thus, tip-line bifurcation processes may be active immediately after establishment of the master fault (typically 5%–12% of extension in our experiments), whereas the asperity bifurcation mechanism becomes active during continued fault-plane shearing (typically 10%–25% in our experiments). Several first-generation asperity bifurcation horses may be seen in cases where the primary fault plane has more than one irregularity.

ACKNOWLEDGMENTS

Our work was done at the Structural Geology Laboratory at the Geological Institute, University of Bergen, with financial support from Elf Aquitaine Inc. via the project The Effect of Clay Smear on Sealing Capacity of Faults. We thank Elf Aquitaine for permission to publish the results. Haavard Enge, Tone Gjelsvik, Ane E. Lothe, Trond Olav Sygnabere, Oddvar Bøe, Irene Moldal Bøe, and Terje Bremnes assisted in the experimental work, and Michael R. Talbot gave advice concerning the text. Terje Bremnes and Jane Ellingsen helped in the preparation of the figures. The manuscript has benefited from the very constructive referee comments from Wayne Baily and Thomas P.-O. Mauduit.

REFERENCES

Anders, M.H., and Schlische, R.W., 1994, Overlapping faults, intrabasin highs, and the growth of normal faults: Journal of Geology, v. 102, p. 165–180.

Barnett, J.A.M., Mortimer, J., Rippon, J.H., Walsh, J.J., and Watterson, J., 1987, Displacement geometry in the volume containing a single normal fault: American Association of Petroleum Geologists Bulletin, v. 71, p. 925–937.

Boyer, S.E., and Elliot, D., 1982, Thrust systems: American Association of Petroleum Geologists Bulletin, v. 66, p. 1196–1230.

Bruhn, R.L., Parry, W.T., Yonkee, W.A., and Thompson, T., 1994, Fracturing and hydrothermal alteration in normal fault zones: Pure and Applied Geophysics, v. 142, p. 609–644.

Cartwright, J.A., Manfield, C., and Trudgill, B., 1996, The growth of normal faults by segment linkage, in Buchanan, P.G., and Nieuwland, D.A., eds., Modern developments in structural interpretation, validation and modelling: Geological Society [London] Special Publication 99, p. 163–177.

Childs, C., Nicol, A., Walsh, J.J., and Watterson, J., 1996a, Growth of vertically segmented normal faults: Journal of Structural Geology, v. 18, p. 1389–1397.

Childs, C., Watterson, J., and Walsh, J.J., 1996b, A model for the structure and development of fault zones: Geological Society of London Journal, v. 153, p. 337–340.

Childs, C., Walsh, J.J., and Watterson, J., 1997, Complexity in fault zone structure and implications for fault seal prediction, in Møller-Pedersen, P., and Koestler, A.G., eds., Hydrocarbon seals: Importance for exploration and production: Norwegian Petroleum Society Special Publication 7, p. 61–72.

Christie-Blick, N., and Biddle, K.T., 1985, Deformation and basin formation along strike-slip faults, in Biddle, K.T., and Christie-Blick, N., eds., Strike-slip deformation, basin formation and sedimentation: Society of Economic Mineralogists and Palaeontologists Special Publication, v. 37, p. 79–103.

Cox, S.D.J., and Scholz, C.H., 1988, On the deformation and growth of faults: An experimental study: Journal of Structural Geology, v. 10, p. 413–430.

Cruikshank, K.M., and Aydin, A., 1995, Unweaving the joints in Entrada Sandstone, Arches National Park, Utah, U.S.A.: Journal of Structural Geology, v. 17, p. 409–421.

Cruikshank, K.M., Zhao, G., and Johnson, A.M., 1991, Duplex structures connecting fault segments in Entrada Sandstone: Journal of Structural Geology, v. 13, p. 1185–1196.

Engelder, T., 1993, Stress regimes in the lithosphere: Princeton, New Jersey, Princeton University Press, 467 p.

Ferrill, D.A., Stramatakos, J.A., and Sims, D., 1999, Normal fault corrugation: Implications for growth and seismicity of active normal faults: Journal of Structural Geology, v. 21, p. 1027–1038.

Fossen, H., and Gabrielsen, R.H., 1996, Experimental modeling of extensional fault systems by use of plaster: Journal of Structural Geology, v. 18, p. 673–687.

Fossen, H., and Hesthammer, J., 1997, Geometric analysis and scaling relations of deformation bands in porous sandstone: Journal of Structural Geology, v. 19, p. 1479–1493.

Gabrielsen, R.H., and Koestler, A.G.,1987, Description and structural implications of fractures in the Late Jurassic sandstones of the Troll Field, northern North Sea: Norsk Geologisk Tidsskrift, v. 67, p. 371–381.

Gabrielsen, R.H., Aarland, R.-K., and Alsaker, E., 1998, Distribution of tectonic and non-tectonic fractures in siliciclastic porous rocks, in Coward, M.P., et al., eds., Structural geology in reservoir characterization: Geological Society [London] Special Publication 127, p. 49–64.

Gibbs, A.D., 1983, Balanced cross-section construction from seismic sections in areas of extensional tectonics: Journal of Structural Geology, v. 5, p. 153–160.

Gibbs, A.D., 1984, Structural evolution of extensional basin margins: Geological Society of London Journal, v. 141, p. 609–620.

Gibbs, A.D., 1990, Linked fault families and basin formation: Journal of Structural Geology, v. 12, p. 795–803.

Heybroek, P., 1975, On the structure of the Dutch part of the central North Sea Graben, in Woodland, A.W., ed., Petroleum and the continental shelf of north-west Europe: London, Institute of Petroleum Geology, p. 338–352.

Huggins, P., Watterson, J., Walsh, J.J., and Childs, C., 1995, Relay zone geometry and displacement transfer between normal faults recorded in coalmine plans: Journal of Structural Geology, v. 17, p. 1741–1756.

Hull, J., 1988, Thickness-displacement relationships for deformation zones: Journal of Structural Geology, v. 10, p. 431–435.

King, G.C.P., Stein, R.S., and Rundle, J.B., 1988, The growth of geological structures by repeated earthquakes. 1. Conceptual framework: Journal of Geophysical Research, v. 93, p. 761–815.

Knipe, R.J., 1997, Juxtaposition and seal diagrams to help analyze fault seals in hydrocarbon reservoirs: American Association of Petroleum Geologists Bulletin, v. 81, p. 187–195.

Knipe, R.J., Fisher, Q.J., Jones, G., Clennell, M.R., Farmer, A.B., Harrison, A., Kidd, B., McAllister, E., Porter, J.R, and White, E.A., 1997, Fault seal analysis: Successful methodologies, application and future directions, in Møller-Pedersen, P., and Koestler, A.G., eds., Hydrocarbon seals: Importance for exploration and production: Norwegian Petroleum Society Special Publication 7, p. 15–40.

Knott, S.D., 1994, Fault zone thickness versus displacement in the Permo-Triassic sandstones of NW England: Geological Society of London Journal, v. 151, p. 17–25.

Koestler, A.G., and Ehrmann, W.U., 1991, Description of brittle extensional features in chalk on the crest of a salt ridge, in Roberts, A.M., et al., eds., The geometry of normal faults: Geological Society [London] Special Publication 56, p. 113–124.

Kusznir, N.J., Karner, G.D., and Egan, S., 1987, Geometric, thermal and isostatic consequences of detachments in continental lithosphere extension and basin formation, in Beaumont, C., and Tankard, A.J., eds., Sedimentary basins and basin formation: Canadian Society of Petroleum Geologists Memoir 12, p. 185–203.

Losh, S., Eglinton, L., Schoell, M., and Woods, J., 1999, Vertical and lateral flow related to a large growth fault, South Eugene Island Block 330 Field, offshore Louisiana: American Association of Petroleum Geologists Bulletin, v. 83, p. 244–276.

Lyakhovsky, V., Ben-Zion, Y., and Agnon, A., 1997, Distributed damage, faulting and friction: Journal of Geophysical Research, v. 102, p. 27635–27649.

Nystuen, J.P., ed., 1989, Rules and recommendations for naming of geological units in Norway: Norsk Geologisk Tidsskrift, v. 69, supplement 2, 111 p.

Odinsen, T., 1992, Modellering av normalforkastninger: analoge ekstensjonsgipsmodeller og balansering [thesis]: Bergen, University of Bergen, 155 p.

Olson, J., and Pollard, D.D., 1989, Inferring paleostress from natural fracture patterns: A new method: Geology, v. 17, p. 345–348.

Peacock, D.C.P., and Sanderson, D., 1991, Displacements, segment linkage and relay ramps in normal fault zones: Journal of Structural Geology, v. 13, p. 721–733.

Peacock, D.C.P., and Sanderson, D.J., 1994, Geometry and development of relay ramps in normal fault systems: American Association of Petroleum Geologists Bulletin, v. 78, p. 147–165.

Roberts, A.M., and Yielding, G., 1991, Deformation around basin-margin faults in the North Sea/mid-Norway rift, in Roberts, A.M., et al., eds., The geometry of normal faults: Geological Society [London] Special Publication 56, p. 61–78.

Robertson, E.C., 1983, Relationship of fault displacement to gouge and breccia thickness: American Institute of Mining Engineers, Society of Mining Engineers Transactions, v. 35, p. 1426–1432.

Sales, J.K., 1987, Tectonic models, in Seyfert, C.K., ed., Encyclopedia of structural geology and tectonics: New York, Van Nostrand Reinhold, p. 785–794.

Schlische, R.W., and Anders, M.H., 1996, Stratigraphic effects and tectonic implications of the growth of normal faults and extensional basins, in Beratan, K.K., ed., Reconstructing the history of Basin and Range extension using sedimentology and stratigraphy: Geological Society of America Special Paper 303, p. 183–203.

Sibson, R.H., 1977, Fault rocks and fault mechanisms: Geological Society of London Journal, v. 133, p. 191–213.

Swanson, M.T., 1988, Pseudotachylyte-bearing strike-slip duplex structures in the Fort Foster brittle zone, S. Maine: Journal of Structural Geology, v. 10, p. 813–828.

Tchalenko, J.S., 1970, Similarities between shear zones of different magnitudes: Geological Society of America Bulletin, v. 81, p. 1625–1640.

Vermilye, J.M., and Scholz, C.H., 1999, Fault propagation and segmentation: Insight from the microstructural examination of a small fault: Journal of Structural Geology, v. 21, p. 1623–1636.

Walsh, J.J., and Watterson, J., 1989, Displacement gradients on fault surfaces: Journal of Structural Geology, v. 11, p. 307–316.

Walsh, J.J., and Watterson, J., 1991, Geometric and kinematic coherence and scale effects on estimates of fault-related regional extension, in Roberts, A.M., et al., eds., The geometry of normal faults: Geological Society [London] Special Publication 56, p. 193–203.

Walsh, J.J., Watterson, J., Bailey, W.R., and Childs, C., 1999, Fault relays, bends and branch-lines: Journal of Structural Geology, v. 21, p. 1019–1026.

Watterson, J., 1986, Fault dimensions, displacements and growth: Pure and Applied Geophysics, v. 124, p. 365–373.

Woodcock, N.H., and Fisher, M., 1986, Strike-slip duplexes: Journal of Structural Geology, v. 8, p. 725–735.

Yielding, G., and Roberts, A.,1992, Footwall uplift during normal faulting—Implications for structural geometries in the North Sea, in Larsen, R.M., et al., eds., Structural modelling and its application to petroleum geology: Norwegian Petroleum Society Special Publication 1, p. 289–304.

MANUSCRIPT ACCEPTED BY THE SOCIETY APRIL 12, 2000

Printed in the U.S.A.

Crustal rheology and its effect on rift basin styles

Anthony P. Gartrell
Tectonics Special Research Centre, Department of Geology and Geophysics, The University of Western Australia, Nedlands WA 6907, Australia

ABSTRACT

Rheological stratification within the crust has important implications for the structural development of extensional basins. A variety of rift basin architectures was developed in analog models simulating intracrustal extension by varying proportions of rheological layers, including or excluding a topographic high, using different density synrift sands and by changing the symmetry of slab movement. Localized rift systems developed when an initial perturbation (topographic high) in the stress field encouraged necking instabilities in the upper and middle crust; otherwise distributed rift systems developed.

Where localized rift systems developed, a range of master fault styles was generated, including low-angle planar, listric, and high-angle planar cross sections. The style of master fault was determined by the effectiveness of a horizontal shear couple, developed between the strong middle and weak lower crustal layers, in rotating the principal stress axes. High-angle master faults developed in experiments with thick upper and middle crustal layers. The extra thickness of these layers results in an increased vertical load, which decreases the effectiveness of the basal shear zone. Listric master faults formed in experiments with intermediate upper and middle crustal layer thickness, where the effects of a basal shear zone decrease toward the surface. Decreasing the density of synrift sediments resulted in listric master faults with decreased curvature. Low-angle planar master faults developed where thin upper and middle crustal layers were extended and low-density sediments were added. The direction of slab movement affects the symmetry of the localized rift basins developed in the experiments. Asymmetric rift basins developed where traction was applied from one end of the model (while the other end was pinned). Symmetric rift basins formed where traction was applied to both ends of the model.

Distributed rifting was characterized by an array of small faults over a broad area of the model. No discrete master faults developed during distributed rifting. However, at high strain, deformation became localized at the margins of the distributed rift basin and localized rifts developed.

INTRODUCTION

Most rheological profiles for the lithosphere indicate that weak ductile layers will occur at lower crustal levels (e.g., Kusznir and Park, 1987; Ranalli and Murphy, 1987; Ord and Hobbs, 1989). When the lower crust is weak, necking of the more competent upper and middle crustal layers may be compensated for by flow in the lower crust (e.g., Royden and Keen, 1980; Block and Royden, 1990; Fig. 1). Evidence for this behavior comes from the relative uniformity in crustal thickness, gravity, Moho topography, and surface elevations across highly extended and less extended regions of the Basin and Range Province of the United States (Gans, 1987; Wernicke, 1990; Kruse et al., 1991; Wernicke, 1992). Deformation in the crust may be mechanically decoupled from the upper mantle during extension to some degree under these conditions (e.g., Hopper and Buck, 1998).

Gartrell, A.P., 2001, Crustal rheology and its effect on rift basin styles, *in* Koyi, H.A., and Mancktelow, N.S., eds., Tectonic Modeling: A Volume in Honor of Hans Ramberg: Boulder, Colorado, Geological Society of America Memoir 193, p. 221–233.

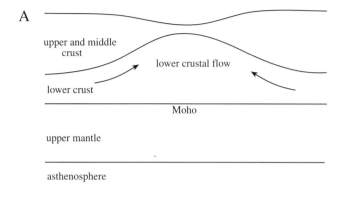

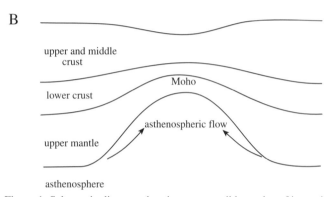

Figure 1. Schematic diagram showing two possible modes of isostatic compensation for extension. A: Intracrustal extension, where flow in lower crust compensates for extension and thinning of more competent middle and upper crust. Section of Moho underlying thinned crust remains relatively flat. Extension in upper mantle may be laterally offset from extension in crust. B: Whole lithosphere extension, where asthenospheric flow compensates for extension and thinning of whole lithosphere.

Numerical and analog modeling studies have provided a basis for the development of primary low-angle (<30°) detachment faults where decoupled extension occurs (Yin, 1989; Melosh, 1990; Westaway, 1998; Brune and Ellis, 1997; Gartrell, 1997). Decoupling between rheological layers promotes horizontal shear gradients (basal shear zones), which can rotate the stress field and allow low-angle normal faults to form without contravening fault mechanics laws.

Gartrell (1997) discussed the initial results of an analog modeling technique that physically demonstrated the development of low-angle detachment faults in a rheologically stratified crustal system. Horizontal shear at the interface between a strong middle crustal stress guide and a weak ductile lower crust was interpreted to be responsible for allowing low-angle faults (or sections of faults) to form in association with necking instabilities.

Further modeling using similar techniques was performed to test the effects of varying experimental parameters and boundary conditions on extensional styles, and to put greater constraints on the interpretations made in Gartrell (1997). The modeling results presented in this paper show that rift basins characterized by low-angle detachments (as in Gartrell, 1997) are only part of a spectrum of structural styles possible during extension of a three-layer crust. Factors that determine whether a rift will be localized or distributed, symmetric or asymmetric, or bounded by high-angle or low-angle faults are investigated. The analog modeling also addresses discrepancies in numerical modeling results regarding the sense of principal stress axes rotation with respect to basal shear.

RHEOLOGY OF THE CONTINENTAL LITHOSPHERE AND STRENGTH PROFILES

Geological, geophysical, and laboratory evidence all suggest that the lithosphere is rheologically stratified (Goetze and Evans, 1979; Brace and Kohlstedt, 1980; Chen and Molnar, 1983; Kirby, 1983; Ranalli and Murphy, 1987; Carter and Tsenn, 1987; Ord and Hobbs, 1989; Ranalli, 1997; Fernandez and Ranalli, 1997). This layering reflects changes in mechanical behavior and flow process of continental lithospheric rocks as determined by the depth-dependant physical (temperature, pressure) and chemical (mineralogy, presence of water) environment (Carter and Tsenn, 1987). The variation of strength with depth in the lithosphere can be modeled, as a first approximation, by means of rheological strength profiles (Ranalli, 1997). These profiles allow the identification of deformation mechanism and give an estimation of strength, providing a guide for modeling the mechanical behavior of the lithosphere (Fig. 2).

To construct strength profiles, the first-order rheological behavior (brittle or ductile) at any given depth is determined by the relative magnitude of frictional and creep strength (Fig. 2A). Where brittle deformation occurs, the fracture and sliding strength of most rocks is assumed to be proportional to depth, according to the Byerlee law (Byerlee, 1968). Where rocks begin to deform by ductile creep, a power law relationship is assumed (e.g., Kirby, 1983). Strength is temperature dependant (exponential decrease in strength with temperature) rather than pressure dependant, and lower in quartzo-feldspathic crustal rocks than in olivine-rich mantle rocks. In some profiles, mechanical layering within the crust is included, quartz being the strength-controlling mineral in the upper crust, plagioclase in the middle crust, and mafic minerals in the lower crust. For the sake of simplicity, only strength profiles in which quartz is the strength-controlling mineral in the crust and olivine is the strength-controlling mineral in the mantle are considered here.

Strength profiles of this kind have been used successfully in analog modeling of lithospheric processes (e.g., Davy and Cobbold, 1991; Benes and Davy, 1996). There are many assumptions and uncertainties associated with the construction of strength profiles. Fernandez and Ranalli (1997) divide the uncertainties into two groups. Operational uncertainties arise from an imperfect knowledge of the composition and structure of the lithosphere, errors in estimated temperature distribution, scatter in experimentally determined rheological parameters, and lack

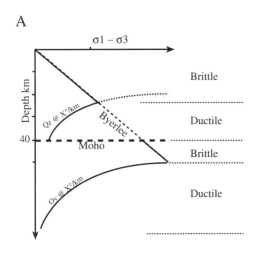
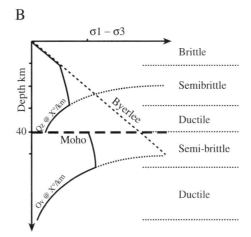

Figure 2. Construction of strength profiles for continental lithosphere. A: Construction of strength profile in which Byerlee relationship for brittle deformation is assumed to hold at all depths. B: Construction of strength profile in which semibrittle rheologies are recognized, as Byerlee law breaks down at depths below 10–15 km. Qz is quartz, Ov is olivine.

of constraints on pore-fluid pressure. Methodological uncertainties relate to the basic assumptions used in the construction of the strength profiles.

A major methodological assumption is that the Byerlee relationship holds for brittle parts of the lithosphere at all depths, although the linear Byerlee relationship probably breaks down at moderate depths (10–15 km) due to an increase in confining pressure (Carter and Tsenn, 1987; Ord and Hobbs, 1989; Fernandez and Ranalli, 1997). At these higher pressures, transitional zones that show a nonlinear increase in strength with depth (only weakly dependant on pressure) and deformation by both brittle and ductile processes (semibrittle behavior) is predicted (Kirby, 1983; Carter and Tsenn, 1987). The effect of taking this transitional behavior into account is a reduction in the maximum strength of the lithosphere and the inclusion of semibrittle layers (Fig. 2B). Transitional rheologies are often overlooked in analog modeling studies but, being the strongest layers, they should play a crucial role in the way the lithosphere deforms.

EXPERIMENTAL PRINCIPLES AND METHODS

The results of an analog model will be dynamically similar to a natural prototype if scaling of the geometry, density ratios, and intrinsic rheology is sufficient (Hubbert, 1937; Ramberg, 1981; Weijermars et al., 1993). The materials used in the experiments to represent the crustal layers are commonly used in analog modeling programs, and their value as scaled materials has been discussed in detail therein (Weijermars, 1986; Davy and Cobbold, 1991; Weijermars et al., 1993). However, the lithospheric strength profiles being scaled for modeling are at best order of magnitude estimations of lithosphere rheology (Ranalli, 1997). With this in mind, and considering the limitations of the analog technique, it seems unrealistic to expect that the models will be precisely scaled. However, analog models continue to generate structures consistent with data from natural systems, suggesting that many of the processes involved are somewhat independent of scale. The models are presented as qualitative demonstrations of elementary deformation behaviors, applicable to some degree at lithospheric scale. An approximate working scale of $1:10^6$ is used (Fig. 3).

For the purposes of the experiments, crustal deformation is considered in isolation. A basic crustal profile with a brittle upper crust, strong transitional middle crust, and weak ductile lower crust was used in all experiments (Fig. 3A).

Dry quartz sand (1.2 gm/cm^3) is used for the upper crust (Fig. 3A) because it obeys Coulomb failure criteria, has low cohesion, and has an internal angle of friction of about 30°. A product known as Pozosphere microballoons (an additive for concrete and fibre-glass moldings) was added to sand in order to generate low-density (0.7 gm/cm^3) synrift sediments where required.

Two types of silicone putty were used for the middle and lower crustal layers (Fig. 3A). Silicone putties are power law fluids available in a wide range of viscosities and power law exponents. Rhodorsil Gomme's silicone gum 70009 was used to model a low-strength ductile lower crust. Gum 70009 has a viscosity of ~2×10^4 Pa.s at room temperature and a power law coefficient of 1 (i.e., Newtonian fluid; Wejermars, 1986) at the experimental strain rates. Dow Corning's Silastic silicone putty was used as a strong middle crustal layer with transitional rheology. This silicone putty is a power-law pseudoplastic with a viscosity of ~1×10^5 Pa.s and a power law coefficient of about 9 (Weijermars, 1986).

The apparatus and procedures used are similar to those in Gartrell (1997). However, refinements in the technique were made in order to obtain more consistent results and to address the objectives of this study. A simple base plate has a fixed horizontal central section and flexible sides that can be inclined (Fig. 4). The multilayer models were placed on this base plate with the flexible sides initially set in the horizontal position. Extension is achieved by inclining one (Fig. 4A) or both (Fig. 4B) of the flexible sides, depending on whether unidirectional or bidirectional traction is required. Gravity acts on the inclined side(s) of the slab and generates extension across the model. Unlike previous gravity-driven extension experiments (e.g., Vendeville and Cobbold, 1988), the deforming area of interest remains on a horizontal

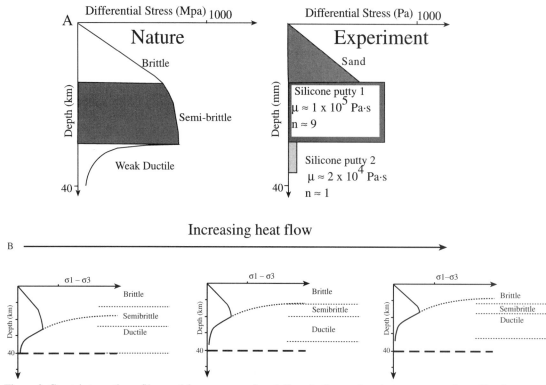

Figure 3. Crustal strength profiles used for purposes of modeling. A: Comparison between strength profiles for natural system and as modeled for mean strain rate of 10^{-3} s^{-1}. B: Effect of increasing heat flow on distribution of rheological layers in simple crustal system. Middle crustal layer (semibrittle) changes thickness, whereas thickness of upper and lower crustal layer remains relatively unchanged. Thickness of middle crustal layer was varied in experiments to test effect on extensional style.

surface. Deformation on inclined surfaces is suitable for special cases, such as salt gliding, but leads to unrealistic stress field relationships for the more general case of rift development. Rift basins developed during extension are filled with colored layers of sand, or a sand and Pozoshpere mixture, at regular intervals in order to simulate synrift sedimentation and to maintain fault-scarp stability. Still photographs and time-lapse video are used to record the development of surface structure. Extension values are determined by measurements taken between two fixed surface markers, compared to measurements of the change in whole slab

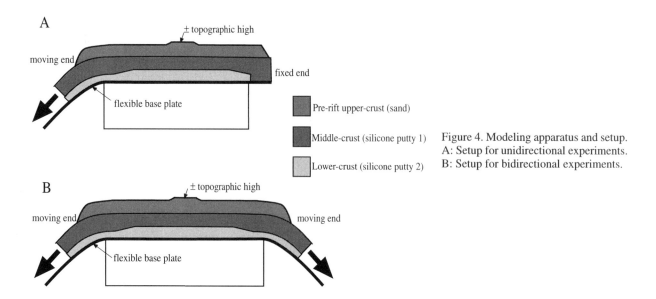

Figure 4. Modeling apparatus and setup. A: Setup for unidirectional experiments. B: Setup for bidirectional experiments.

length, as in Gartrell (1997). Vertical slices are photographed following the completion of an experiment.

The parameters tested were the thickness of the middle crustal layer, the presence or absence of a small topographic high in the prerift section, the symmetry of traction applied to the model, and the density of synrift sediments added. A description of the experiments presented is summarized in Table 1.

Heat flow strongly affects the distribution of the rheological layers in the lithosphere. In particular, the thickness of the semibrittle layer will decrease with increasing heat flow, whereas the thickness of the upper and lower crustal layers should remain relatively constant (Fig. 3B). For the experiments, an upper crustal thickness of 1.5 cm and a lower crustal thickness of 1 cm were used for most experiments (except experiment F). The initial thickness of the middle crustal layer varied between 0.6 and 2 cm.

A topographic high (~2 mm high and 100 mm wide) was built into the prerift sediments of some experiments in order to help localize extension into rift basins in a more consistent manner (Fig. 4). A similar technique is used in numerical models to initiate necking instabilities by perturbing the stress conditions across the model (e.g., Bassi et al., 1993).

In order to test the effect of transport direction, some of the experiments were repeated with applied traction from either one end (unidirection) or from both ends (bidirectional) of the model (Fig. 4). Reduced synrift loading may occur where extension outpaces sedimentation or where relatively low density sediments are deposited. The experiments do not give good results in the absence of synrift sedimentation, because fault scarps collapse in the unconsolidated sand if they are allowed to grow too large. Therefore, the addition of low-density synrift sediment is used to simulate decreased synrift loading conditions in specified experiments.

EXPERIMENTAL RESULTS

Two fundamental modes of extension were developed in the experiments. Localized rift basins (Fig. 5) formed when necking instabilities localized extension into discrete zones. If necking instabilities did not develop at an early stage, then distributed rift basins formed.

Localized rifts

Discrete rift basins were consistently produced in experiments that included a topographic high (Fig. 6, experiments A–H). A general evolution similar to that described in Gartrell (1997) occurred for experiments in which localized rift basins formed (Fig. 5). An initial stage of distributed extension is followed by the development of discrete rift basins as necking instabilities developed. As necking proceeds, the middle crustal layer thins and eventually tears. Sections of the upper crustal layer are subsequently placed directly on the lower crustal layer (Fig. 5). The manner in which this occurs depends on the style of master fault(s) generated during extension.

Rift basins dominated by high-angle planar master faults tended to form when the middle crustal layer was relatively thick (≥2 cm; Fig. 6, experiments A, B). High-angle master fault–bounded basins tend to be associated with sets of planar antithetic minor faults, which help accommodate horizontal displacement. Synrift horizons are generally not rotated to any great extent.

Listric master faults dominated in experiments with upper and middle crustal layers of thin to intermediate thickness (0.5–1.5 cm; Fig. 6, experiments C–H). The listric master faults tend to develop long horizontal sections that are able to accommodate large amounts of horizontal displacement. Synrift beds are highly rotated above the listric detachment faults and sets of synthetic listric faults are characteristic of this style of rift basin.

The use of low-density synrift sediments resulted in listric master faults that have a lower degree of curvature (Fig. 6, experiments E, F, G). A consequence of this is that bedding is less rotated. In the extreme case, where both the upper crust and middle crustal layers are thin and low-density synrift was used, low-angle planar master faults formed (Fig. 6, experiment F). These master faults are associated with closely spaced sets of planar antithetic faults.

TABLE 1. DESCRIPTION OF EXPERIMENTS A-J

Experiment no	Layer thickness (cm)			Topographic high	Traction symmetry	Synrift
	Upper Crust	Middle Crust	Lower Crust			
A	1.5	2	1	yes	bidirectional	normal
B	1.5	2	1	yes	unidirectional	normal
C	1.5	1.5	1	yes	unidirectional	normal
D	1.5	1	1	yes	bidirectional	normal
E	1.5	1.5	1	yes	unidirectional	low-density
F	1.2	0.6	1	yes	unidirectional	low-density
G	1.5	1.2	1	yes	unidirectional	low-density
H	1.5	1	1	yes	bidirectional	normal
I	1.5	2	1	no	bidirectional	normal
J	1.5	1	1	no	bidirectional	normal

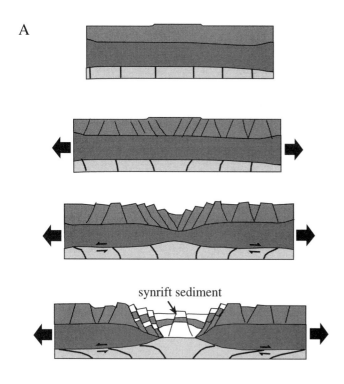

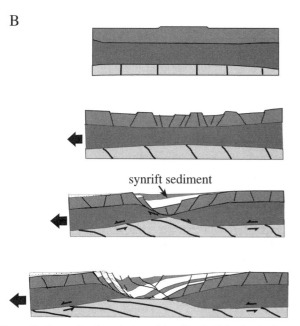

Figure 5. Generalized evolution of localized rift basin. A: Symmetric rift basin development. B: Modification to Figure 5 from Gartrell (1997), showing correct orientation of basal shear zone with respect to listric master fault systems for asymmetric rift development. Legend for layers as in Figure 4. Lines in lower crustal layer show shear strain in this layer.

The symmetry of the localized rift basins formed during the experiments was determined by whether the extension was unidirectional or bidirectional. Bidirectional extension resulted in relatively symmetric rift basins (Fig. 6, experiments B, D, H). Similar amounts of displacement occur on master faults on either side of the rift basin. Master faults dip away from the slab transport direction. Strongly asymmetric basins developed in unidirectional experiments, the majority of strain being accommodated on the master fault closest to the moving slab (Fig. 6, experiments B, C, E, F). Faults dipping toward the transport direction tend to be relatively steep.

Distributed rifts

Distributed rifts were generated in experiments that did not have an initial topographic high (Fig. 6, experiments H, I). This style of extension is characterized by more uniform thinning over a relatively broad area. Deformation tended to initiate in the central section of the model, subsequently migrating outward as extension progressed. The extensional strain is distributed over an array of many small faults in the brittle layer. However, necking instabilities eventually develop at one or both of the margins of the distributed rift basin at high strain. Localized rift basins may then grow (Fig. 7; Fig. 6, experiment I).

DISCUSSION

Primary differences in deformation styles are related to the potential of the lithosphere to localize or distribute deformation (Benes and Davy, 1996), as determined by whether the growth of lithospheric necking instabilities is enhanced or suppressed. Necking instability is related to the contrasting rheological behavior of layers of different viscosity and results in pinch-and-swell instabilities growing on the more competent layer(s) during layer-parallel extension (Ramberg, 1955; Smith, 1977). The magnitude of instabilities is a function of the strength contrast between layers and the degree to which the rocks involved are non-Newtonian (power law coefficient >1; e.g., sand and strong putty layers). Localized zones of deformation, and in general, nonhomogeneous deformation develop more easily than homogeneous stretching or shortening in such a layer (Smith, 1977). With a small perturbation in the topography of a layer interface, a small increase in extension is accompanied by a large growth in the disturbance amplitude. In contrast, perturbations within Newtonian layers (e.g., weak basal putty) grow very slowly and tend to decay over time. At given values of temperature, pressure, and strain rate, the more competent brittle and semibrittle layers tend to deform in a plastic manner and will control the development of instabilities. The weaker ductile layers tend to behave like a viscous fluid and will passively follow the induced perturbation, and even tend to inhibit their development (Martinod and Davy, 1992).

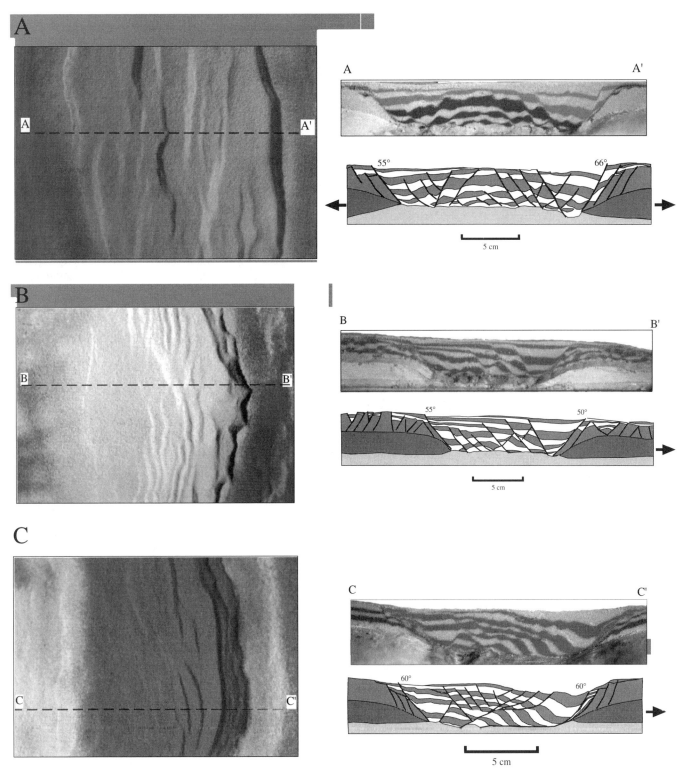

Figure 6 (on this and next two pages). Results from key experiments A-I (see Table 1 for details of experiments). Photograph of final surface structure shown on left. Photograph of cross section and interpretation shown on right. Arrows indicate applied tractional forces. All faults display normal sense of movement. A: Symmetric localized rift with high-angle master faults (220% extension). B: Asymmetric localized rift with high-angle master faults (210% extension). C: Asymmetric localized rift with listric master fault (190% extension). D: Symmetric localized rift basin with listric master faults (220% extension). E: Asymmetric localized rift basin with listric master fault at lower strain (137% extension). F: Asymmetric localized rift with low-angle planar master fault (220% extension). G: Asymmetric localized rift with listric master fault showing markers in middle and lower putty layers indicating basal shear direction (210% extension). H: Symmetric localized rift basin with listric master faults with markers in lower putty layer (210% extension). I: Distributed rift basin (30% extension). J: Distributed rift basin with localized rift on one margin (50% extension). Legend for layers as in Figures 4 and 5.

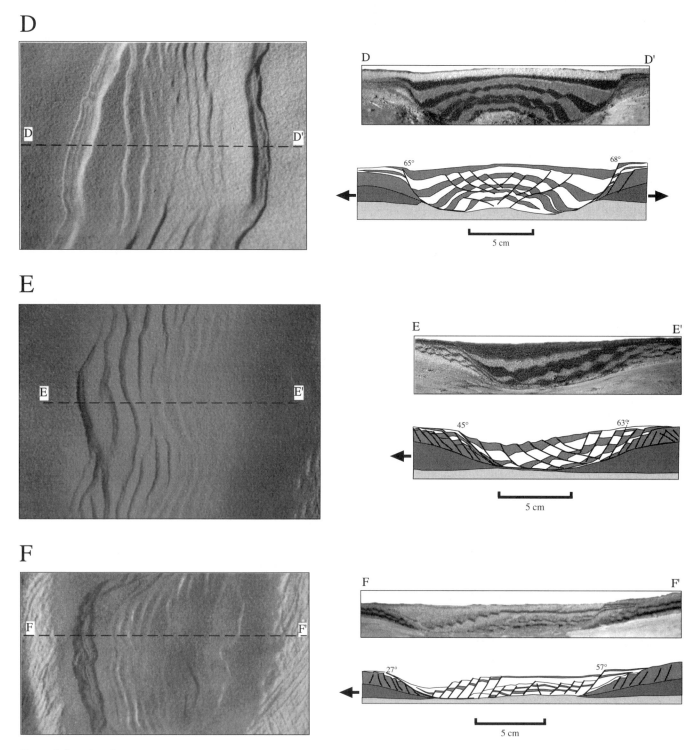

Figure 6. (*continued*)

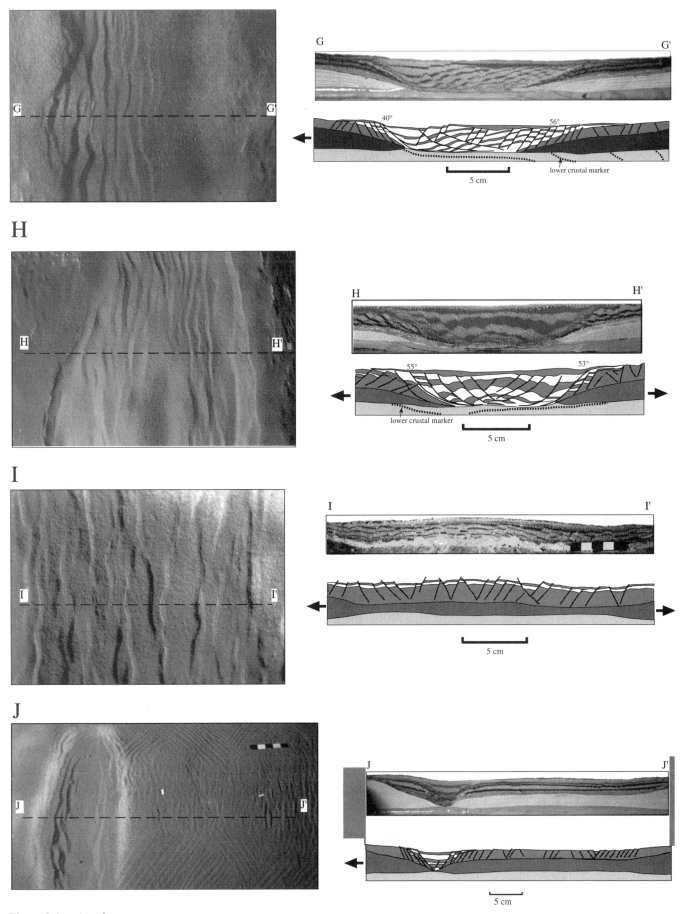

Figure 6. (*continued*)

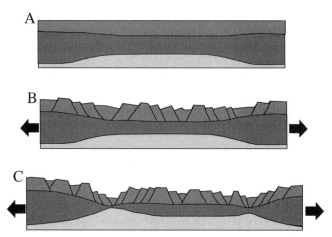

Figure 7. Generalized evolution of distributed rift basin. A: Predeformation state. B: Deformation distributed over broad area of model. C: Localization of necking instabilities on margins of distributed rift basin at higher strain. Legend for layers as in Figure 4.

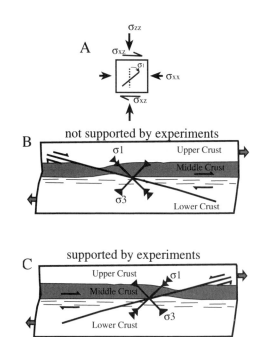

Figure 8. Rotation of principal stress axes and orientation of low-angle normal faults where basal shear occurs at interface between middle and upper crust during extension. A: Components of stress and rotation of maximum principal stress axis. B: Relationship as predicted in Yin (1989), Lister and Davis (1989), Melosh (1990), and Wernicke (1992). C: Relationship as predicted by Westaway (1998) and demonstrated in analog modeling results.

In the experiments described here, the main control on whether necking instabilities were strongly developed was the presence of an initial topographic high. The topographic high increases the local stress difference and focuses deformation. A tendency to distribute deformation occurred in the absence of a topographic high. In some cases, the strong development of a dominant fault in the early stages of extension caused sufficient perturbation to localize deformation. The thickness of the layers played a secondary role, in that localized rifts formed more rapidly with greater amplitude and shorter wavelengths of necking in experiments where the brittle upper crustal layer was thick relative to the more ductile middle and lower crustal layers. It is reasonable to predict that localized rifts would also be more likely if the middle crustal layer had greater non-Newtonian tendencies; however, this was not tested.

Several authors have recognized that basal shear will alter the attitude of normal faults developed during extension (Yin, 1989; Lister and Davis, 1989; Melosh, 1990; Wernicke, 1992; Gartrell, 1997; Brune and Ellis, 1997; Westaway, 1998). More specifically, rotation of principal stress axes (to 45°) occurs in the vicinity of the basal shear zone (Fig. 8A). Normal faults may then form at low dip angles and meet the requirement that they develop at about 30° to the inclined axis of maximum principal stress. Buck (1990) and Wills and Buck (1997) criticized this concept by suggesting that the shear tractions modeled, particularly in Yin (1989), are too small to influence fault orientations. Westaway (1998) addressed this problem, citing errors in the manner in which rock mechanics data were used in Yin (1989), and considering other factors (e.g., decreased depth to basal shear zone) that may contribute to an increased effectiveness of the basal shear zone. There also appears to be some confusion over the direction of rotation of stress axes with respect to basal shear direction. For example, Yin (1989), Melosh (1990), Lister and Davis (1989), and Wernicke (1992) all suggested that the stress axes should rotate in the opposite direction to the sense of basal shear (Fig. 8B), whereas Westaway (1998) suggested that the stress axes will rotate in the same direction as the sense of basal shear (Fig. 8C).

The results of the analog modeling shown here appear to support the conclusions of Westaway (1998). Low-angle normal faults and detachments display the opposite sense of shear to the basal shear zone developed in the lower crustal layer, implying that the stress axis has rotated in the same direction as the basal shear zone (Figs. 6 [G, H], 8, and 9). Steep normal faults tend to form in the opposite dip orientation. The interpretation of the relationship between low-angle detachments and basal shear shown in Figure 5 of Gartrell (1997) was based on the earlier papers. The results of the modeling shown here and the results of Westaway (1998) require that the sense of basal shear with respect to low-angle detachments and necking instabilities should be as shown in Figure 5.

Westaway (1998) also showed that a relationship exists between the amount of vertical loading and basal shear, which determines the degree to which the principal stress axes within the middle and upper crustal layers will rotate. Where the ratio between basal shear stress and vertical loading tends to zero, the maximum principal stress axis will be vertical. As the ratio tends to infinity, the maximum principal stress axis tends to plunge at 45°. This relationship is physically demonstrated in the experiments (Fig. 9). High-angle planar master faults form when the upper and middle crustal layers are thick in the experiments (greater vertical loading), whereas low-angle and listric master faults form when these layers are thinner or low-density synrift sediments are used (decreased vertical loading). In contrast to Westaway (1998), however, the experiments suggest that listric faults will be an important part of the spectrum of master geometries. Listric master fault profiles are generated as the effects of the basal shear zone diminish toward the surface, where the maximum principal stress approach vertical (Fig. 9). A reduction in synrift sediment loading (using low-density synrift sand) allows the effects of the basal shear zone to be felt closer to the surface. This resulted in lower exit angles on the master faults and a reduction in fault curvature and hanging-wall rotation.

Numerical modeling (e.g., Buck, 1991; Bassi et al., 1993) and previous multilayer analog modeling (Benes and Davy, 1996) shows that initial rheological configuration at the lithospheric scale will determine the primary mode of deformation during extension (Fig. 10). Three modes are predicted: metamorphic core complex mode (thick crust and high heat flow), wide rift mode (intermediate crustal thickness and heat flow), and narrow rift mode (thin crust and low heat flow). The analog modeling presented here shows that, at a finer scale, interactions between rheological layers within the crust may exert further import controls, particularly on predominant fault styles. The modeling results are consistent with the numerical models and natural observations that suggest that low-angle detachments are most probable (but not guaranteed) within the thermo-mechanical regime of metamorphic core complex mode extension. Marked decoupling between the crust and the upper mantle, thin upper and middle crustal layers, and low sedimentation rates all contribute to detachment development in this mode. High-angle planar master faults may be more characteristic of narrow rift extension, as minimal decoupling between layers (stronger lower crust) should occur and the upper and middle crustal layers will be relatively thick. Sedimentation rates may also be somewhat higher in a narrow rift setting due to greater subsidence on the more steeply dipping faults.

The development of distributed rifts in the experiments is geometrically similar to wide rift mode extension. However, the process is different in that cooling is responsible for increased strength and density of the lithosphere in the extended region in the numerical models. In the analog experiments we are concerned with the crust, and outward migration of deformation is probably the result of a decrease in vertical loading in the extended region in the absence of any strong tendency for strain weakening (necking instabilities). In nature both processes may work in conjunction. The numerical models of Bassi et al. (1993) also predict a similar evolution to that shown in the distributed rift analog models, where localization occurs on the margins at high strain (Fig. 7).

The analog experiments suggest that the symmetry of a rift basin will be strongly influenced by the symmetry of the applied tractional forces and the associated sense of basal shear. However, the situation is likely to be much more complicated in nature, as interactions between adjacent zones of extension, preexisting perturbations (e.g., preexisting faults, changes in rock type) and syndeformational perturbations (e.g., magmatism, lower crustal flow) may cause interference patterns.

Variation in the symmetry and dip characteristics of master fault systems developed in natural examples of localized rift basins may in part be the result of the processes discussed herein. For example, master faults in highly extended parts of the Basin and Range Province may have relatively steep dips through much of the crust (e.g., the Wasatch fault zone; Zoback, 1983; Wernicke and Axen, 1988), consistently low dip angles through the crust (e.g., the Sevier and Eagle Pass detachment faults; Wernicke and Axen, 1988; Kruger and Johnson, 1994), or have listric profiles (e.g., the Shell Creek fault; Gans et al., 1985). Interpreted sections from the Nile Rifts (e.g., Bosworth, 1992) and the East African rifts (e.g., Rosendahl, 1987) show similar combinations of symmetry and master fault style as developed in the analog experiments. Wide rift margins, such as the Orphan Basin (eastern Canadian margin; Bassi et al., 1993), the Campos Basin (southern Brazilian margin; Davison, 1997), and the North West Shelf (northwest Australian margin), may have developed in a manner similar to that shown in the high strain distributed rifts.

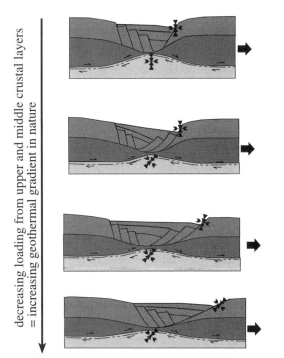

Figure 9. Effect of upper and middle crustal loading on localized rift basin style (asymmetric). Basal shear zones developed at interface between strong middle crust and weak lower crust cause rotation of principal stress axes. Effect of rotated stress axes is diminished by higher overburden pressures. Inset shows rotation of stress axis within sheared system. Legend for layers as in Figures 4 and 5. Double-headed arrows indicate $\sigma 1$ and single-headed arrows indicate $\sigma 3$.

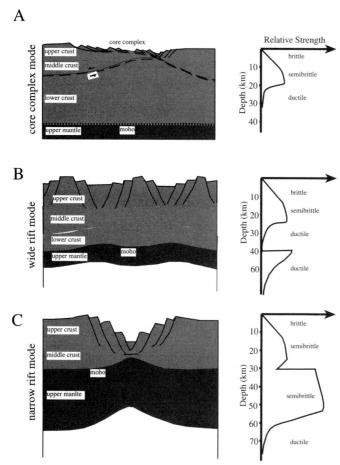

Figure 10. Three modes of extension as predicted by numerical modeling of Buck (1991). A: Metamorphic core complex mode predicted for thickened crust and high heat-flow conditions. B: Wide rift mode predicted for intermediate crustal thickness and heat flow where ductile deformation processes dominate. C: Narrow rift mode predicted for thin crust and low heat-flow conditions where brittle deformation processes dominate. Plots (schematic only) on right side of diagram show yield strengths with depth for hypothetical geothermal gradients and crustal thickness (modified after Buck, 1991).

CONCLUSION

The results of the analog modeling show that a range of rift basin styles can be developed during extension of a multilayer crust. Factors such as the thickness ratio of rheological layers, initial perturbations (e.g., topography), sedimentary loading, basal shear, and the symmetry of tractional forces all contribute to variation in rift architectures.

Contrasting rheological behavior of the brittle upper crust, semibrittle middle crust, and weak ductile lower crust may result in necking instabilities within the crust. Low-angle detachments are mechanically feasible in such a system, but only for a limited set of conditions. Necking instabilities localize deformation into discrete rift basins, while the development of a basal shear zone at the interface between the middle and lower crustal layers causes rotation of the stress field so that low-angle normal faults can develop. However, this process is critically dependent on the depth to the basal shear zone, which is largely determined by heat flow in nature. Low overburden pressures, associated with shallow depths to the basal shear zone (thin middle crustal layer), combined with strong shear gradients between rheological layers favor low-angle detachment development. These conditions are likely for higher heat flow environments in nature. Low synrift sediment loading may also contribute to low-angle detachment development. High overburden pressures, where the depth to the basal shear zone is greater (thick middle crustal layer), will limit the ability of low-angle detachments to develop and favor high-angle master faults. These conditions are likely for lower heat flow conditions. Lower heat flow conditions will also tend to decrease decoupling between rheological layers due to a strength increase in the lower crust. A further alternative is that distributed rifting may develop if necking instabilities are not established.

Explanations for the structures developed in the analog experiments are found in the results of various numerical modeling studies. However, the analog modeling offers a higher degree of geometric detail, which is easily recognized and assimilated into conceptual geological models. The analog experiments physically demonstrate a link between rheological layering, crustal necking, basal shear, sedimentation, and rift-basin fault styles. Hence, the models represent a unification of concepts derived from numerical modeling studies that consider necking (e.g., Buck, 1991; Bassi et al., 1993) and faulting (e.g., Yin, 1989; Melosh, 1990; Westaway, 1998) separately.

ACKNOWLEDGMENTS

I was supported by an Australian Postgraduate Award scholarship in conjunction with a Minerals and Energy Research Institute of Western Australia top-up scholarship. This chapter represents part of my Ph.D. thesis under the supervision of Lyal Harris and Mike Dentith of The University of Western Australia, whose advice, support, and editing of the original manuscript is gratefully acknowledged. This is Tectonics Special Research Center Publication 81.

REFERENCES CITED

Bassi, G., Keen, C.E., and Potter, P., 1993, Contrasting styles of rifting: Models and examples from the eastern Canadian margin: Tectonics, v. 12, p. 639–655.

Benes, V., and Davy, P., 1996, Modes of continental lithospheric extension: Experimental verification of strain localisation processes: Tectonophysics, v. 254, p. 69–87.

Block, L., and Royden, L.H., 1990, Core complex geometries and regional scale flow in the lower crust: Tectonics, v. 9, p. 557–567.

Bosworth, W., 1992, Mesozoic and early Tertiary rift tectonics in East Africa: Tectonophysics, v. 209, p. 115–137.

Brace, W.F., and Kohlstedt, D.L., 1980, Limits on lithospheric stress imposed by laboratory experiments: Journal of Geophysical Research, v. 85, p. 6248–6252.

Brune, J.N., and Ellis, M.A., 1997, Structural features in a brittle-ductile model of continental extension: Nature, v. 387, p. 67–70.

Buck, W.R., 1990, Origin of regional, rooted low-angle normal faults: A mechanical model and its tectonic implications: Comment: Tectonophysics, v. 9, p. 545–546.

Buck, W.R., 1991, Modes of continental lithospheric extension: Journal of Geophysical Research, v. 96, p. 20161–20178.

Byerlee, J.D., 1968, Brittle-ductile transition in rocks: Journal of Geophysical Research, v. 73, p. 4741–4750.

Carter, N.L., and Tsenn, M.C., 1987, Flow properties of continental lithosphere: Tectonophysics, v. 136, p. 27–63.

Chen, W.P., and Molnar, P., 1983, Focal depths of intracontinental and intraplate earthquakes and their implications for the thermal and mechanical properties of the lithosphere: Journal of Geophysical Research, v. 88, p. 4183–4214.

Davison, I., 1997, Wide and narrow margins of the Brazilian South Atlantic: Geological Society of London Journal, v. 154, p. 471–476.

Davy, P., and Cobbold, P.R., 1991, Experiments on shortening of a 4-layer model of the continental lithosphere: Tectonophysics, v. 188, p. 1–25.

Fernandez, M., and Ranalli, G., 1997, The role of rheology in extensional basin formation modeling: Tectonophysics, v. 282, p. 129–145.

Gans, P.B., 1987, An open-system, two-layer crustal stretching model for the eastern Great Basin: Tectonics, v. 6, p. 1–12.

Gans, P.B., Miller, E.L., McCarthy, J., and Ouldcott, M.L., 1985, Tertiary extensional faulting and evolving ductile-brittle transition zones in the northern Snake Range and vicinity: New insights from seismic data: Geology, v. 13, p. 189–193.

Gartrell, A.P., 1997, Evolution of rift basins and low-angle detachments in multilayer analog models: Geology, v. 25, p. 615–618.

Goetze, C., and Evans, B., 1979, Stress and temperature in the bending lithosphere as constrained by experimental rock mechanics: Royal Astronomical Society Geophysical Journal, v. 59, p. 463–478.

Hopper, J.R., and Buck, W.R., 1998, Styles of extensional decoupling: Geology, v. 26, p. 699–702.

Hubbert, K.M., 1937, Theory of scale models as applied to the study of geologic structures: Geological Society of America Bulletin, v. 48, p. 1459–1520.

Kirby, S.H., 1983, Rheology of the lithosphere: Reviews of Geophysics and Space Physics, v. 21, p. 1458–1487.

Kruger, J.M., and Johnson, R.A., 1994, Raft model of crustal extension: Evidence from seismic reflection data in southeast Arizona: Geology, v. 22, p. 351–354.

Kruse, S., McNutt, M., Phipps-Morgan, J., Royden, L., and Wernicke, B., 1991, Lithospheric extension near Lake Mead, Nevada: A model for ductile flow in the lower crust: Journal of Geophysical Research, v. 96, p. 4435–4456.

Kusznir, N.J., and Park, R.G., 1987, The extensional strength of the continental lithosphere: Its dependance on geothermal gradient, and crustal composition and thickness, in Coward, M.P., et al., eds., Continental extension tectonics: Geological Society [London] Special Publication 28, p. 35–52.

Lister, G.S., and Davis, G.A., 1989, The origin of metamorphic core complexes and detachment faults formed during Tertiary continental extension in the northern Colorado River region, U.S.A.: Journal of Structural Geology, v. 11, p. 65–94.

Martinod, J., and Davy, P., 1992, Periodic instabilities during compression or extension of the lithosphere 1. Deformation modes from an analytical perturbation method: Journal of Geophysical Research, v. 97, p. 1999–2014.

Melosh, H.J., 1990, Mechanical basis for low-angle normal faulting in the Basin and Range Province: Nature, v. 343, p. 331–335.

Ord, A., and Hobbs, B.E., 1989, The strength of the continental crust, detachment zones and the development of plastic instabilities: Tectonophysics, v. 158, p. 189–269.

Ramberg, H., 1955, Natural and experimental boudinage and pinch-and-swell structures: Journal of Geology, v. 63, p. 512–526.

Ramberg, H., 1981, Gravity, deformation and the Earth's crust: London, Academic Press, 214 p.

Ranalli, G., 1997, Rheology of the lithosphere in space and time, in Burg, J.-P., and Ford, M., eds., Orogeny through time: Geological Society [London] Special Publication 121, p. 19–37.

Ranalli, G., and Murphy, D.C., 1987, Rheological stratification of the lithosphere: Tectonophysics, v. 132, p. 281–296.

Rosendahl, B.R., 1987, Architecture of continental rifts with special reference to East Africa: Annual Review of Earth and Planetary Sciences, v. 15, p. 445–503.

Royden, L., and Keen, C.E., 1980, Rifting process and thermal evolution of the continental margin of eastern Canada determined from subsidence curves: Earth and Planetary Science Letters, v. 51, p. 343–361.

Smith, R.B., 1977, Formation of folds, boudinage and mullions in non-Newtonian materials: Geological Society of America Bulletin, v. 88, p. 312–320.

Vendeville, B., and Cobbold, P., 1988, How normal faulting and sedimentation interact to produce listric faults profiles and stratigraphic wedges: Journal of Structural Geology, v. 10, p. 649–659.

Weijermars, R., 1986, Flow behavior and physical chemistry of bouncing putties and related polymers in view of tectonic laboratory applications: Tectonophysics, v. 124, p. 325–358.

Weijermars, R., Jackson, M.P.A., and Vendeville, B., 1993, Rheological and tectonic modeling of salt provinces: Tectonophysics, v. 217, p. 143–174.

Wernicke, B., 1990, The fluid crustal layer and its implications for continental dynamics, in Salisbury, M., and Fountain, D., eds., Exposed cross-sections of the continental crust: Boston, Massachusetts, Kluwer Academic, 317 p.

Wernicke, B., 1992, Cenozoic extensional tectonics of the U.S. Cordillera, in Burchfiel, B.C., et al., eds., The Cordilleran orogen: Conterminous U.S.: Boulder, Colorado, Geological Society of America, Geology of North America, v. G3, p. 553–581.

Wernicke, B.P., and Axen, G.J., 1988, On the role of isostacy in the evolution of normal fault systems: Geology, v. 16, p. 848–851.

Westaway, R., 1998, Dependence of active normal fault dips on lower crustal flow regimes: Geological Society of London Journal, v. 155, p. 233–253.

Wills, W., and Buck, W.R., 1997, Stress-field rotation and rooted detachment faults: A Coulomb failure analysis: Journal of Geophysical Research, v. 102, p. 20503–20514.

Yin, A., 1989, Origin of regional, rooted low-angle normal faults: A mechanical model and its tectonic implications: Tectonics, v. 8, p. 469–482.

Zoback, M.L., 1983, Structure and Cenozoic tectonism along the Wasatch fault zone, Utah, in Miller, D., et al., eds., Tectonic and stratigraphic studies in the eastern Great Basin: Geological Society of America Memoir 157, p. 3–27.

MANUSCRIPT ACCEPTED BY THE SOCIETY APRIL 12, 2000

New apparatus for controlled general flow modeling of analog materials

Sandra Piazolo, Saskia M. ten Grotenhuis, and Cees W. Passchier
Tectonophysics, Department of Geosciences, University of Mainz, 55099 Mainz, Germany

ABSTRACT

We present a new deformation apparatus to model homogeneous deformation in general flow regimes, in which all combinations of simple shear parallel to the xy-plane and pure shear parallel to the x-, y-, and z-axes can be realized. With this apparatus it is possible to control the kinematic vorticity number of monoclinic flow during progressive deformation. The user defines the type of deformation by a set of parameters such as kinematic vorticity number, strain rate, and duration of the experiment. The apparatus consists of a set of mobile pistons on a low friction sole and is open at the top. All pistons are flexible to ensure homogeneity of deformation in a major part of the sample. The corners of the box are connected to four sliding carriages, which themselves are sliding on another set of four carriages positioned at right angles to the first set. This setup and the controlled movement of the sliding carriages allow the user to model any type of monoclinic transtension and transpression. A computer program controls six stepping motors used to move the different carriages simultaneously and accurately. In the apparatus materials with a viscosity range of 10^3 to 10^6 Pa s can be used. A set of pilot experiments investigating the rotation of mica fish in different general regimes is presented as an example for the use of this apparatus.

INTRODUCTION

Research in recent years has shown that many shear zones cannot be explained with a simple shear model but that combinations of simple and pure shear are likely to represent the true character of flow in natural shear zones. Therefore, noncoaxial monoclinic and/or triclinic shear zones have been the focus of recent work, including both analytical and numerical studies (e.g., Ghosh and Ramberg, 1976; Sanderson and Marchini, 1984; Weijermars, 1991, 1993, 1997; Jezek et al., 1994, 1996; Robin and Cruden, 1994; Dewey et al., 1998, and reference therein; Fossen and Tikoff, 1998; Passchier, 1998), field studies (e.g., Druguet et al., 1997; Tikoff and Greene, 1997; Krabbendam and Dewey, 1998), and analog modeling (e.g., Giesekus, 1962; Weijermars, 1998; Cruden and Robin, 1999; Griera and Carreras, 1999).

However, the effect of different flow geometries on the development of structural elements within and at the boundaries of shear zones is still unclear. Data from the field and field-derived samples are the most important source of information on shear zones. Nevertheless, small-scale structures in shear zones can only be correctly interpreted if their development is modeled under controlled conditions. An important complementary tool to field data and analytical work is experimental rock deformation (e.g., Price and Torok, 1989; Tullis and Yund, 1991) and analog modeling using paraffin wax (e.g., Abbassi and Mancktelow, 1990; Grujic, 1993), modeling clay and bouncing putties (e.g., Ramberg, 1955; Ghosh and Ramberg, 1976), and crystalline materials (e.g., Bons and Urai, 1995). Analog modeling using viscous materials such as polymers and bouncing putties offers the opportunity to investigate the development of structural elements such as folds, shear sense indicators, and stretching lineations in three-dimensions during progressive deformation up to relatively high finite strain. Several deformation machines have been designed to model flow

*E-mail: Piazolo, piazolo@mail.uni-mainz.de; ten Grotenhuis, saskiatg@geo.uu.nl; Passchier, cpasschi@mail.uni-mainz.de.

in shear zones using analog materials; these are used to model simple shear flow (e.g., Robertson and Acrivos, 1970; Price and Torok, 1989), circular simple shear or Couette flow (e.g., Passchier and Sokoutis, 1993), and general flow, i.e., combinations of pure and simple shear (e.g., Giesekus, 1962; Weijermars, 1998; Cruden and Robin, 1999; Griera and Carreras, 1999). We designed an apparatus in which flow is homogeneous over a large part of the sample. In this apparatus it is possible to control strain rates along three perpendicular directions, and therefore all types of shear zones in monoclinic flow such as transpression and transtension (Fig. 1) can be modeled. Extending and contracting flexible walls of the deformation box are used to obtain homogeneous deformation. The described apparatus, therefore, offers the possibility to perform progressive homogeneous deformation at a chosen constant or changing strain rate and kinematic vorticity number W_n (Means et al., 1980) during progressive deformation in monoclinic flow. A large number of controlled inhomogeneous flow types and histories can also be modeled. A general description of the new apparatus, its possibilities and limitations, suitable materials, and experimental procedure is presented here. A set of pilot experiments is described to illustrate the scope of the apparatus.

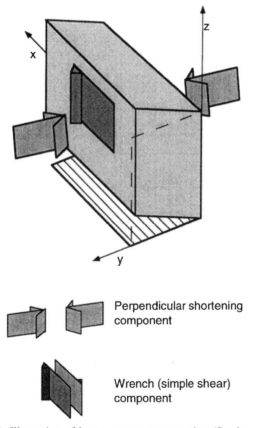

Figure 1. Illustration of homogeneous transpression (Sanderson and Marchini, 1984) as combination of simple shear (wrench) component and simultaneous coaxial shortening component perpendicular to vertical shear plane. Example here is in plane strain (modified from Sanderson and Marchini, 1984).

DESCRIPTION OF THE APPARATUS

The apparatus assembly consists of four main parts: the deformation apparatus, a computer, a power driver subsystem, and a camera placed 50 cm above the deformation apparatus (Figs. 2 and 3).

The deformation apparatus consists of a four-sided deformation box that contains the sample and that we describe in an x-y-z reference frame (Fig. 2A). Two opposing sides are always parallel to the x-direction and the two other sides can rotate about the z-axis. The walls of the box are constructed from 1 cm × 12 cm Plexiglas segments (30 on the sides parallel to the x-direction, 20 on the other two sides), which are connected with flexible plastic to corrugated deformable pistons. Every second Plexiglas piece is connected at its back (side facing away from inner part of the box) to two metal springs. This construction ensures homogeneous contraction and extension of the walls. The range of the length of the sides parallel to the x-direction is between 15 and 30 cm and of the other two sides is between 5 and 20 cm. The springs are connected to four aluminum plates at the corners of the deformation box. A sheet made of 0.35 mm thick elastic latex forms the bottom of the deformation box. Its corners are fixed to the four aluminum plates. This construction results in a deformation box that slides with low friction on the basal plate and that is open at the top. Each one of the four aluminum plates (P) is attached to a sliding carriage (C1). These four sliding carriages are arranged two by two at each long side of the box, parallel to the x-direction. The sliding carriages sit on two PVC boards (B1) that are parallel to the long side of the box and attached to another set of four sliding carriages (C2) oriented parallel to the y-direction. These sliding carriages (C2) are attached to a 1 m × 1 m basal plate (B2). The contraction or extension of the flexible sides of the deformation box is controlled by six stepping motors (M). Four motors drive the movement of the corners of the box (P) in the x-direction via shafts and are fixed to the two PVC boards (B1). Two other motors are attached with shafts to the basal plate (B2) and control movement in the y-direction. This configuration of motors and sliding carriages is chosen to meet two requirements: (1) the center of the deformation box has to remain in one place so that objects of interest in the deforming sample do not move with respect to the camera (Fig. 3); (2) all possible monoclinic flow types should be attainable. For simple shear, motors M1, M2, M3, and M4 move the corners of the deformation box at the same constant velocity but in different directions; M1 and M4 in the one direction and M2 and M3 in the opposite (Fig. 4). For pure shear M1 and M3 drive their corners of the deformation box in the x-direction and M2 and M4 move their corners in the y-direction. M5 and M6 control movement in the y-direction, and their velocity is determined by the velocity of the other 4 motors (Fig. 4) if flow is to be plane strain. Deformation in monoclinic flow at a constant kinematic vorticity number requires movement to be transmitted by all motors and recalculation of the velocity generated by the six motors for each time step, because the ratio of pure and simple

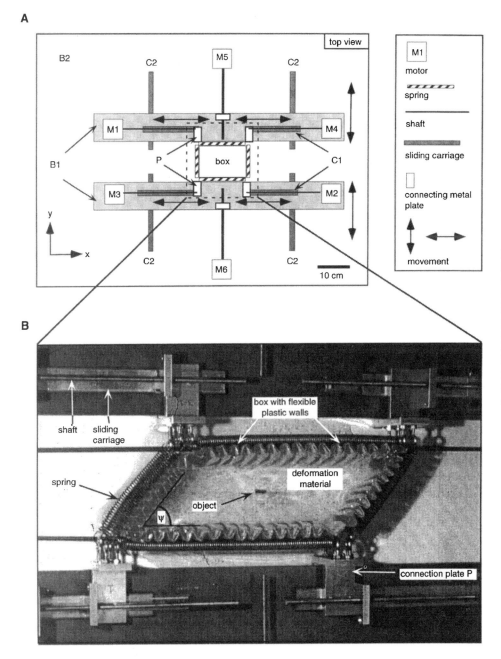

Figure 2. A. Schematic drawing of deformation apparatus (view from top) where x and y are along symmetry axes of apparatus. B1 are PVC boards, B2 is base plate, C1 is set of 4 sliding carriages parallel to x-direction, C2 is a set of 4 sliding carriages parallel to y-direction, P is connecting aluminum plates, and M1–M6 are motors. B. Close-up of deformation box with flexible walls (view from top). Angle ψ is angle between sides of deformation box.

shear is constantly changing during progressive deformation if the kinematic vorticity number W_n is to remain constant. Corresponding particle paths (Ramberg, 1975) or flow lines (Fig. 4) are shown for the coordinate system used here, because the geometry of the flow lines is characteristic for each flow type (Passchier, 1998).

A control panel allows the user to set specific parameters, such as the kinematic vorticity number and strain rate along a specific axis and duration of the experiment. Each second signals are passed on from the PC to the stepping motors via the power driver subsystem. The rotation angle per step is 1.8° with an accuracy of 0.05°. The velocity range is 1–10000 steps/s, which result in 0.005–50 rotations/s and a displacement of 0.005–50 mm/s. The range of possible displacement rates guarantees an accurate control of the movement of the four sides of the box, bulk strain rate, and progressive deformation type. The maximum shear strain that can be reached with our apparatus is restricted by the geometry of the shear box. For one experimental run, the maximum attainable R_{xy} value, defined as the strain axis ratio in the xy-plane, is 10 in simple shear and 4 in pure shear. Corresponding maximum finite shear strain γ for simple shear is 3, and for pure shear the maximum k-value is 2. In other flow types an R_{xy} of at least 4 can always be obtained. In practice, this means that the maximum duration of one experimental run achieving maximum finite shear strain is 180 min.

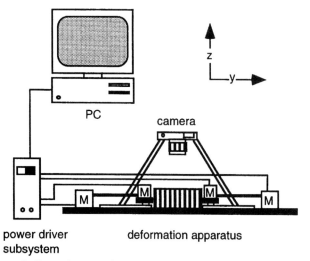

Figure 3. General set-up of apparatus; y and z are along symmetry axes of apparatus.

Higher finite strains than the values mentioned here can be achieved with our machine, but only by running series of experiments. After each experimental run the precise orientation of the object is measured and photographed. After returning the box to the starting position the object is placed in the corresponding measured orientation and another experimental run under the same conditions as the previous run can be carried out. This technique is particularly useful when studying one or a few rigid objects (cf. Passchier and Simpson, 1986).

EXPERIMENTAL MATERIALS

The range of suitable analog materials for our apparatus is restricted by two factors: leakage at the bottom of the deformation box and strength of the stepping motors. Leakage problems of the deformation box limits suitable materials to those with viscosities of at least 10^3 Pa s. The strength of the motors limits suitable materials to those with viscosities $\leq 10^6$ Pa s, leaving a viscosity range of 10^3–10^6 Pa s. This means that most crystalline analog materials, as well as paraffin wax, cannot be deformed in our apparatus. Polymers such as Polydimethylsiloxane (PDMS; trade name SGM 36; produced by Dow Corning, UK, a transparent polymer with a density of 0.965 gcm^{-3} consisting

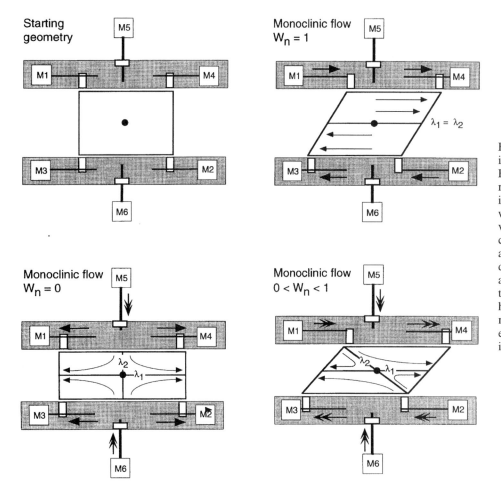

Figure 4. Schematic diagrams illustrating different flow types in shear box (see Fig. 2). Arrows indicate direction of movement of pistons and specific parts in reference frame of machine; arrows with bold arrowheads signify constant velocity; double arrow heads indicate continuously changing velocity. Longer arrow indicates higher velocity. Within deformation box, bold lines signify flow apophyses (λ_1 and λ_2), and thin lines trace particle paths where small arrowheads point in direction of particle movement. Black dot is center of reference frame. M1–M6 are motors. Figure is not to scale.

of repetitive chains; the longer the chain the higher the viscosity [Weijermars, 1986]), Rhodorsil Gomme (a pinkish opaque bouncing putty produced by the Société des Chemiques Rhône-Poulenc, France), and mixtures of Rhodorsil Gomme and Plastilina (Swedish version of Harbutt's Plasticine; McClay, 1976) are ideally suitable for use in our shear box. For the first tests, which were performed to establish the properties of this shear box and for pilot experiments, we used PDMS and Rhodorsil Gomme. The flow behavior of both materials is Newtonian for strain rates below 10^{-1} s^{-1} (Fig. 5). At room temperature viscosities of PDMS and Rhodorsil Gomme are 5.0×10^4 Pa s and 2.9×10^4 Pa s (Weijermars, 1986), respectively. A major advantage of PDMS over other materials commonly used in geological modeling is its transparency. Furthermore, it is nontoxic and relatively cheap. The flow behavior of Plastilina is highly non-Newtonian (Fig. 5) with n = 7.5 (Weijermars, 1986), where n is the stress exponent of power law flow.

TYPES OF FLOW MODELED BY NEW APPARATUS

In order to discuss the flow types that can be modeled with the new apparatus we use the terminology of Passchier (1998) to describe monoclinic flow. Passchier (1998) defined monoclinic flow using the instantaneous stretching axes, defined as vectors $\mathbf{a_I}$, $\mathbf{a_{II}}$, and $\mathbf{a_{III}}$, where a_I, a_{II}, and a_{III} are the magnitudes of stretching rates of material lines instantaneously parallel to the instantaneous stretching axes. The vorticity vector $\mathbf{w}$ is parallel to $\mathbf{a_I}$; w, the vorticity, is the sum of the angular velocity of material lines instantaneously parallel to $\mathbf{a_{II}}$ and $\mathbf{a_{III}}$ (Passchier, 1997). Any type of monoclinic flow can now be described by the four numbers w, a_I, a_{II}, and a_{III}. Alternatively, three normalized numbers W_n, A_n, and T_n can be defined that describe the geometry of monoclinic flow completely. These numbers are defined as the sectional kinematic vorticity number

$$W_n = w/(a_{II} - a_{III}), \quad (1)$$

the sectional kinematic dilatancy number (representing the instantaneous area change in the x-y plane during progressive deformation):

$$A_n = (a_{II} + a_{III})/(a_{II} - a_{III}), \quad (2)$$

and the sectional kinematic extrusion number (representing the shortening or extension in the z-direction):

$$T_n = a_I/(a_{II} - a_{III}). \quad (3)$$

In addition, a kinematic volume change number can be defined as

$$V_n = T_n + A_n. \quad (4)$$

In our case V_n is approximately zero because of the incompressibility of the sample materials under our experimental conditions, and therefore $T_n = -A_n$. This means that all possible monoclinic flow types at $V_n = 0$ can be depicted in a plane plotting W_n against A_n or T_n (Fig. 6).

As many as three nonrotating material lines can be defined in monoclinic flow, the flow eigenvectors or apophyses. For our apparatus eigenvector λ_3 is always parallel to the z-axis of our reference frame and λ_1 is always parallel to the x-axis. The third eigenvector λ_2 coincides with λ_1 in simple shear flow, and otherwise λ_2 is situated somewhere in the xy-plane. Two different main groups of monoclinic model shear regimes can be defined for shear boxes of this type, based on the magnitude of λ_1 and λ_2. If $|\lambda_1| > |\lambda_2|$, the shear zone is of group 1, and if $|\lambda_1| < |\lambda_2|$, it is of group 2 (Fig. 6). Material lines parallel to the flow apophyses can be either instantaneously extending, shortening, or not changing in length. When there is no stretching or shortening along λ_3 (plane strain flow), A_n and T_n are zero. Material lines parallel to either λ_1 or λ_2 are not deforming when $A_n^2 + W_n^2 = 1$ (Passchier, 1991, 1997), and these situations are represented by the circular curves in Figure 6. All types of flow shown in Figure 6 can be modeled with our apparatus.

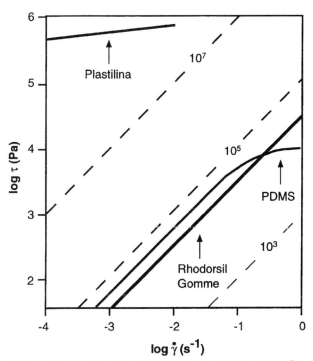

Figure 5. Graph illustrating shear stress and shear strain rate of some materials suitable for use in apparatus (modified from Weijermars 1986). Thin diagonal lines represent viscosity contours in Pa s. See text for discussion.

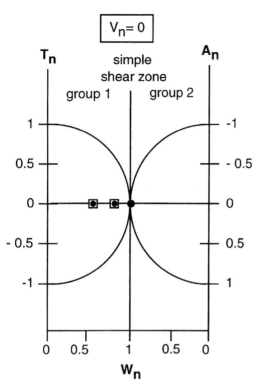

Figure 6. Representation of all types of homogeneous constant-volume monoclinic flow (modified from Passchier, 1998). W_n is sectional kinematic vorticity number, A_n is sectional kinematic dilatancy number, T_n is sectional kinematic extrusion number. V_n, kinematic volume change number, is 0. With apparatus it is possible to model all shear zone types in this graph; open squares represent test experiments and black dots represent pilot experiments.

FLOW TESTS AND BOUNDARY CONDITIONS

In order to test flow conditions that can be obtained in the shear box, we (1) checked the flow patterns in the shear box using particle paths, (2) obtained the distribution of R_{xy} throughout the sample, and (3) investigated the rotation rate of a sphere in simple shear flow.

To check the flow patterns several experimental runs were performed with different kinematic vorticity numbers and strain rates. We then compared flow lines observed in the matrix material with flow lines predicted by theory (Ramberg, 1975) and the movement of particle in the central area of the box. The shear box was filled with PDMS. After a few hours of settling carbon powder was sprinkled on top of the material. An additional 5-mm-thick layer of PDMS was then put on top of the carbon powder. Again a few hours of settling were needed to ensure a flat surface, which is necessary to observe the exact movement of the marker particles. All experiments were performed with a dextral shear sense. For the experiments, the initial dimensions of the shear box were 150×100 mm and the angle ψ between the sides (Fig. 2B) was 135° at the beginning of each experiment. The test experiments presented here were performed at constant kinematic vorticity numbers (W_n) of 0.8 and 0.6, a strain rate in the x-direction of 3.3×10^{-4} s^{-1}, and plane strain flow ($T_n = A_n = 0$). Photographs were stacked on top of each other in order to determine the flow pattern (ten Brink, 1996; Weijermars, 1998). The acute angle α between the two flow apophyses (Fig. 4) during homogeneous deformation can be expressed as $\alpha = \arccos W_n$ for plane strain flow (Bobyarchick, 1986; Passchier, 1986). At $W_n = 0.8$ the angle α is 36.9° and at $W_n = 0.6$ it is 53.1°. The observed α values in the test experiments are 37° ± 0.5° at $W_n = 0.8$ (Fig. 7A) and 53° ± 0.5° at $W_n = 0.6$ (Fig. 7B). This demonstrates that the apparatus is well suited to model homogeneous flow in monoclinic shear zones. The experiments also indicate that the flow pattern is constant in time, as particle paths do not intersect during progressive deformation (Fig. 7). Hence, time-independent plane strain flow is realized in at least part of the apparatus.

To quantify the area of homogeneous deformation in the sample we performed a number of plane strain test experiments with different vorticities ($W_n = 0.8, 0.6$), strain rates (stretching rate of the x-axis: 2.0×10^{-4} s^{-1}, 3.3×10^{-4} s^{-1}, 4.7×10^{-4} s^{-1}) and matrix materials (PDMS, Rhodorsil Gomme). The preparation and procedure of the experimental runs were identical to those described for the other test experiments. We used the computer program PatMatch (Bons and Jessell, 1995), which performs strain analysis by analyzing displacements between two images to obtain the distribution of R_{xy} within a deformed sample. Results reveal that a large area of the samples deform homogeneously and the R_{xy} values closely correspond to theoretical values (Fig. 8). Inhomogeneous flow occurs only in a narrow zone adjacent to the corrugated walls. The width of this zone depends on the viscosity of the matrix material and is independent of the kinematic vorticity number and strain rate (within the accuracy of our measurements). For PDMS the width of the zone with inhomogeneous flow is 0.8–1.3 cm measured from the tips of the corrugated wall; for Rhodorsil Gomme this width is 0.6–1.1 cm.

A third set of experiments investigated the behavior of a rigid sphere in plane strain simple shear flow at a simple shear strain rate of 1×10^{-3} s^{-1} to a finite strain γ of 2. We inserted a Plexiglas sphere with a diameter of 0.8 cm in the middle of the deformation box and took a number of consecutive photos during progressive deformation. Figure 9 shows the change in orientation of the sphere (θ) against simple shear strain (γ) observed in the experiment and the theoretical values of rotation (Ghosh and Ramberg, 1976). Experimental values show good agreement with theoretical values.

The flow tests presented here are all in plane strain ($A_n = T_n = 0$), where the specimen surface remains flat. Under these conditions a large volume of homogeneous flow can be obtained. In experiments with extrusion of material parallel to the z-axis ($A_n \neq 0$ and $T_n \neq 0$), the rigidity of the pistons in the z-direction causes the sample surface to attain a parabolic shape, which has an effect on the flow homogeneity. Accordingly, any user of our apparatus working at $T_n \neq 0$ should test the effect of boundary conditions for the specific experimental setup used.

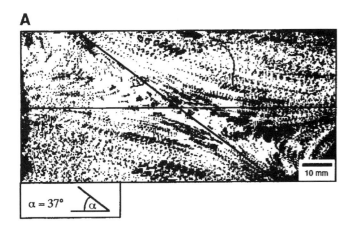
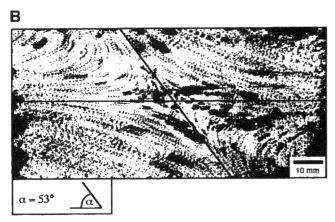
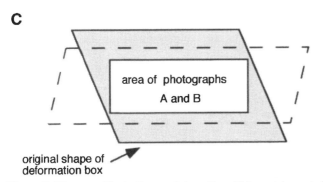

Figure 7. Diagrams illustrating particle paths within matrix material PDMS during progressive deformation at $W_n = 0.8$ (A) and at $W_n = 0.6$ (B). Series of photographs taken during test experiments were stacked onto each other in order to determine flow pattern during progressive deformation. Insets in A and B show theoretical acute angle α between two flow apophyses (see Fig. 4) during progressive deformation. C: Position of area photographed during deformation (white box) inside deformation box. Original shape of deformation box is shown in gray and final shape is shown by dashed line.

Leakage of material with a viscosity of 5.0×10^4 Pa s out of the deformation box is <0.5 vol% during one experimental run if the experimental duration is <4 hr. Leakage does not exceed 1 vol% in 24 hr if the sample material (viscosity: 5.0×10^4 Pa s) is left in the deformation box without performing experiments.

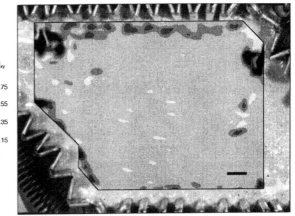

Figure 8. Photograph of deformation box. Inset represents distribution of R_{xy} values over analyzed area of sample. Inset was generated using pattern-matching program PatMatch (Bons and Jessell, 1995). Different gray shades represent different ranges of R_{xy} values. Dashed line in legend represents theoretical value of R_{xy} ($R_{xy} = 1.25$) and at same time measured average of analyzed area (excluding narrow zone close to pistons). Most of analyzed area exhibits R_{xy} values that correspond to theoretical value of R_{xy}. Note that on upper left corner and on upper right side reflections cause significant errors in PatMatch routine, because program cannot identify patterns in reflected areas. Experimental conditions are: $W_n = 0.6$, stretching rate in x-direction is 3.3×10^{-4} s^{-1}, and sample material is Rhodorsil Gomme. Scale bar = 1 cm.

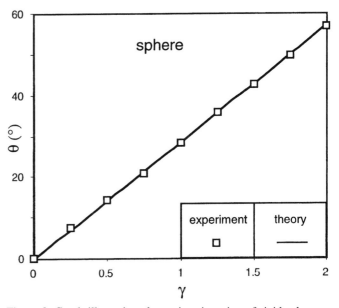

Figure 9. Graph illustrating change in orientation of rigid sphere as function of shear strain in simple shear flow ($W_n = 1$). Solid line represents theoretical values (after Ghosh and Ramberg, 1976) and squares represent experimental values.

PILOT EXPERIMENTS: ROTATION OF MICA-FISH IN PLANE STRAIN

As an example of the practical application of our apparatus, we present the modeling of mica fish. Mica fish are commonly observed in mylonites and their asymmetrical form and the stair-stepping of their tails are often used to determine shear sense (Lister and Snoke, 1984; Passchier and Trouw, 1996). Study of natural mica fish shows that in the plane parallel to the stretching lineation and perpendicular to the foliation, their shape is predominantly monoclinic. All studied examples lie with their long axis tilted 2°–40° with respect to the inferred flow plane (ten Grotenhuis and Passchier, 1999). This could mean that mica fish are in a stable irrotational position with respect to the apophyses of bulk flow for at least part of their evolution. In addition, it is not clear to what extent object shape, flow partitioning, W_n, or other factors influence the development of mica fish. Our deformation apparatus is suitable to study kinematic aspects of the development of such mica fish.

Analytical work of Ghosh and Ramberg (1976) showed that in two-dimensional homogeneous flow with a particular kinematic vorticity number, the rate of rotation of a rigid elliptical particle in a Newtonian viscous fluid varies in a systematic manner depending on the orientation and the axial ratio (R_{ob}) of the inclusion. This means that the orientation of the particle is a function of the initial angle, the aspect ratio of the particle, and strain. Until now, experiments investigating the rotation of rigid objects in homogeneous flow were mainly restricted to simple shear (Ghosh and Ramberg, 1976; ten Brink, 1996). With our apparatus, we can expand such work to homogeneous monoclinic flow experiments.

To investigate the effect of object shape on the rotation behavior we performed experiments with a monoclinic object in a Newtonian viscous fluid (PDMS). The object is made of rigid India rubber and has a long side A of 10 mm, a short side B of 7 mm, and height h of 16 mm (Fig. 10). The angle between A and B is 135°. The aspect ratio of the object (R_{ob}) in the xy-plane is three. The angle θ is between the long axis of the object and the y-direction of the reference frame. Angles are measured clockwise. A and B are parallel to the z-axis. The object is placed in the middle of the box with its top 5 mm below the surface of the PDMS and parallel to the y-direction, $\theta_0 = 0°$. The experiments model plane strain monoclinic flow ($T_n = 0$) with kinematic vorticity numbers (W_n) of 1 (simple shear flow), 0.8, and 0.6 (Fig. 6). In the simple shear experiment the simple shear strain rate is 1×10^{-3} s^{-1} and the final shear strain γ is 3. For the flow experiments at $W_n = 0.8$ and $W_n = 0.6$ the stretching rate along the x-axis is held constant at 3.3×10^{-4} s^{-1}. After one experimental run the maximum R_{xy} value reached was 3.58 at $W_n = 0.8$ and 2.68 at $W_k = 0.6$. For the experiments with $W_n = 0.8$ and 0.6, seven consecutive runs were performed.

Figure 11 shows the experimental results and the theoretical orientation of the long axis for ellipses with $R_{ob} = 3$ against the simple shear component of strain (γ) for $W_n = 1$, $W_n = 0.8$, and $W_n = 0.6$ (equations 11–13 in Ghosh and Ramberg, 1976). We chose to plot the simple shear component of strain (γ) and not R_{xy} in Figure 11 because γ increases linearly with progressive deformation. The highest γ value for one experimental run was 1.08 at $W_n = 0.8$ and 0.61 at $W_n = 0.6$.

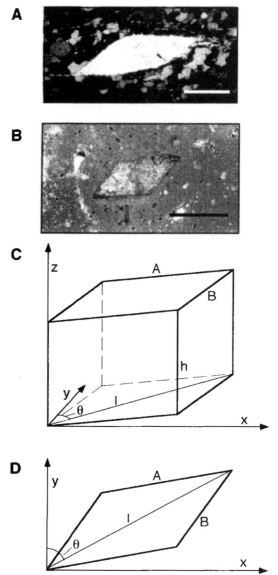

Figure 10. A: Natural example of muscovite mica fish (Ribeira belt, Brazil); crossed polars; scale bar = 1 mm. B: Photograph of object used in experiments, inserted in its matrix (PDMS); scale bar = 10 mm. C: Schematic three-dimensional drawing of rigid monoclinic object used as analog for mica fish; x, y, and z correspond to reference frame of apparatus, h is height, l is longest axis of object, and θ is angle between longest axis (l) and y-axis of reference frame. D: Schematic drawing of AB-plane of object.

The experiments show a deviation in the orientation of the lozenge object from theoretical values for ellipses. In simple shear the rotation curves of the object and the theoretical curve of an ellipse with $R_{ob} = 3$ differ considerably (Fig. 11A). The theoretical curve for an ellipse of $R_{ob} = 2.7$ fits the experimental results better with regard to the amount of shear strain necessary for a 180° rotation of the object, but the shape of the curves (θ versus γ) is systematically different (Fig. 11A). In monoclinic flow with $W_n = 0.8$ and $W_n = 0.6$ experimental results show that the object attains a semistable position

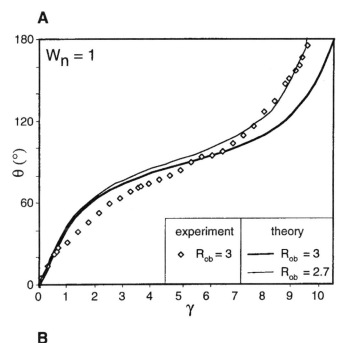

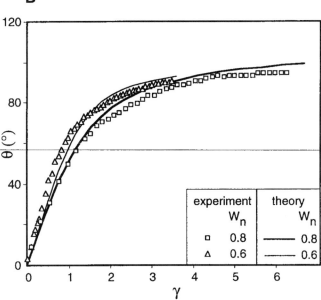

Figure 11. A: Graph illustrating change in orientation of rigid monoclinic object (θ) as function of simple shear component of strain in monoclinic flow with $W_n = 1$. Solid line represents theoretical values (after Ghosh and Ramberg, 1976) and symbols represent experimental values. R_{ob} is aspect ratio of object. B: Graph illustrating change in orientation of rigid monoclinic object (θ) with aspect ratio of 3 as function of simple shear strain in monoclinic flow with $W_n = 0.8$ and 0.6. Solid lines represent theoretical values (after Ghosh and Ramberg, 1976) and symbols represent experimental values.

(Fig. 11A). This position is 95° at $W_n = 0.8$ and 90° at $W_n = 0.6$ (Fig. 11B). Theory predicts 5° higher values for the semistable position of an ellipse of $R_{ob} = 3$ in both experimental flow types. Theoretical values for an ellipsoid with $R_{ob} = 2.7$ show even stronger deviations from experimental results. We suggest that the observed deviation from theory is due to the shape of the rigid object. Both the axial ratio of an object and its shape—i.e., elliptical, rectangular, or lozenge shaped—influence the behavior of such an object in monoclinic flow.

Nevertheless, the preferred orientation (θ) of mica fish in nature is 50°–88° using the reference frame of the experiments. Therefore, monoclinic flow with a kinematic vorticity number smaller than 1 cannot be the only explanation for the orientation of mica fish in shear zones, and further experiments are needed to explain the behavior of these structures.

CONCLUSIONS

The newly developed apparatus described in this paper can combine simple and pure shear independently and simultaneously and allows the modeling of structures in all types of homogeneous monoclinic flow, as illustrated in Passchier (1998). Test and pilot experiments show that flow in the apparatus is homogeneous in plane strain except for a small zone along the pistons, and that particle paths, R_{xy}, and rotation rates of spheres are comparable to theoretical values in monoclinic shear zones. Limitations of the apparatus are the maximum finite strain of one experimental run, availability of suitable analog materials, and the limited range of attainable strain rates. For one experimental run, the maximum strain axis ratio in the xy-plane R_{xy} is at least 4 for all types of monoclinic flow that can be modeled in the machine. The setup of the apparatus and technical limitations restrict suitable materials to those with a viscosity in the range of 10^3 to 10^6 Pa s. The attainable range of strain rate is strongly dependent on the analog material used.

ACKNOWLEDGMENTS

Technical assistance from the workshop of the Department of Earth Sciences (University of Mainz, Germany) is gratefully acknowledged. We thank A. Klügel (Max Planck Institute, Mainz, Germany) for his support during the construction of the apparatus, D. Sokoutis and G. Mulugeta (University of Uppsala, Sweden) for making samples of Plastilina available, A.R. Cruden and R. Weijermars for helpful reviews, and Paul Bons for constructive discussions. Piazolo and ten Grotenhuis acknowledge the financial support of the German Research Foundation (GRK 392/1), and ten Grotenhuis thanks the Deutscher Akademischer Austauschdienst for financial support (415-br-probral).

REFERENCES CITED

Abbassi, M.R., and Mancktelow, N.S., 1990, The effect of initial perturbation shape and symmetry on fold development: Journal of Structural Geology, v. 15, p. 293–307.

Bobyarchick, A.R., 1986, The eigenvalues of steady flow in Mohr space: Tectonophysics, v. 122, p. 35–51.

Bons, P.D., and Jessell, M.W., 1995, Strain analysis in deformation experiments with pattern matching or a stereoscope: Journal of Structural Geology, v. 17, p. 917–921.

Bons, P.D., and Urai, J.L., 1995, An apparatus to experimentally model the dynamics of ductile shear zones: Tectonophysics, v. 256, p. 145–164.

Cruden, A., and Robin, P.-Y., 1999, Analogue models of ductile transpression zones: Journal of Conference Abstracts, v. 4, p. 604.

Dewey, J., Holdesworth, R.E., and Strachan, R.A., 1998, Transpression and transtension zones, in Holdesworth, R.E., et al., eds., Continental transpressional and transtensional tectonics: Geological Society [London] Special Publication 135, p. 1–14.

Druguet, E., Passchier, C.W., Carreras, J., Victor, P., and den Brok, S., 1997, Analysis of a complex high-strain zone at Cap de Creus, Spain: Tectonophysics, v. 280, p. 31–45.

Fossen, H., and Tikoff, B., 1998, Extended models of transpression and transtension, and application to tectonic settings, in Holdesworth, R.E., et al., eds., Continental transpressional and transtensional tectonics: Geological Society [London] Special Publication 135, p. 15–33.

Ghosh, S.K., and Ramberg, H., 1976, Reorientation of inclusions by combination of pure shear and simple shear: Tectonophysics, v. 34, p. 1–70.

Giesekus, H., 1962, Strömungen mit konstantem Geschwindigkeitsgradienten und die Bewegung von darin suspendierten Teilchen, Teil II: Ebene Strömungen und eine experimentelle Anordnung zu ihrer Realisierung: Rheologica Acta, v. 2, p. 112–122.

Griera, A., and Carreras, J., 1999, Mechanical instabilities associated to rod development: Preliminary results from analogue modelling: Journal of Conference Abstracts, v. 4, p. 603.

Grujic, D., 1993, The influence of initial fold geometry on Type 1 and Type 2 interference patterns: An experimental approach: Journal of Structural Geology, v. 15, p. 293–307.

Jezek, J., Melka, R., Schulmann, K., and Venera, Z., 1994, The behaviour of rigid triaxial ellipsoidal particles in viscous flows—Modeling of fabric evolution in a multiparticle system: Tectonophysics, v. 229, p. 165–180.

Jezek, J., Schulmann, K., and Segeth, K., 1996, Fabric evolution of rigid inclusions during mixed coaxial and simple shear flows: Tectonophysics, v. 257, p. 203–221.

Krabbendam, M., and Dewey, J.F., 1998, Exhumation of UHP rocks by transtension in the Western Gneiss Region, Scandinavian Caledonides, in Holdesworth, R.E., et al., Continental transpressional and transtensional tectonics: Geological Society [London] Special Publications 135, p. 159–181.

Lister, G.S., and Snoke, A.W., 1984, S-C mylonites: Journal of Structural Geology, v. 6, p. 617–638.

McClay, K.R., 1976, The rheology of plasticine: Tectonophysics, v. 33, p. T7–T15.

Means, W.D., Hobbs, B.E., Lister, G.S., and Williams, P.F., 1980, Vorticity and non-coaxiality in progressive deformation: Journal of Structural Geology, v. 2, p. 371–378.

Passchier, C.W., 1986, Flow in natural shear zones—The consequences of spinning flow regimes: Earth and Planetary Science Letters, v. 77, p. 70–80.

Passchier, C.W., 1991, The classification of dilatant flow types: Journal of Structural Geology, v. 13, p. 101–104.

Passchier, C.W., 1997, The fabric attractor: Journal of Structural Geology, v. 19, p. 113–127.

Passchier, C.W., 1998, Monoclinic model shear zones: Journal of Structural Geology, v. 20, p. 1121–1137.

Passchier, C.W., and Simpson, C., 1986, Porphyroclast systems as kinematic indicators: Journal of Structural Geology, v. 8, p. 831–844.

Passchier, C.W., and Sokoutis, D., 1993, Experimental modeling of mantled porphyroclasts: Journal of Structural Geology, v. 15, p. 895–909.

Passchier, C.W., and Trouw, R.A.J., 1996, Microtectonics: Berlin, Springer, 289 p.

Price, G.P., and Torok, P.A., 1989, A new simple shear deformation apparatus for rocks and soils: Tectonophysics, v. 158, p. 291–309.

Ramberg, H., 1955, Natural and experimental boudinage and pinch-and-swell structures: Journal of Geology, v. 63, p. 512–526.

Ramberg, H., 1975, Particle paths, displacement and progressive strain applicable to rocks: Tectonophysics, v. 28, p. 1–37.

Robertson, C.R., and Acrivos, A., 1970, Low Reynolds number shear flow past a rotating circular cylinder. Part 1. Momentum transfer: Journal of Fluid Mechanics, v. 40, p. 685–704.

Robin, P-Y.F., and Cruden, A.R., 1994, Strain and vorticity patterns in ideally ductile transpression zones: Journal of Structural Geology, v. 16, p. 447–466.

Sanderson, D.J., and Marchini, W.R.D., 1984, Transpression: Journal of Structural Geology, v. 6, p. 449–458.

ten Brink, C.E., 1996, Development of porphyroclast geometries during non-coaxial flow [Ph.D. thesis]: Utrecht, Utrecht University, 163 p.

ten Grotenhuis, S.M., and Passchier, C.W., 1999, Mica fish and other fish-shaped shear sense indicators: Journal of Conference Abstracts, v. 4, p. 828.

Tikoff, B., and Greene, D., 1997, Stretching lineations in transpressional shear zones: An example from the Sierra Nevada Batholith, California: Journal of Structural Geology, v. 19, p. 29–39.

Tullis, J., and Yund, R.A., 1991, Diffusion creep in feldspar aggregates: Experimental evidence: Journal of Structural Geology, v. 13, p. 987–1000.

Weijermars, R., 1986, Flow behaviour and physical chemistry of bouncing putties and related polymers in view of tectonic laboratory applications: Tectonophysics, v. 124, p. 325–358.

Weijermars, R., 1991, Progressive deformation in anisotropic rocks: Journal of Structural Geology, v. 14, p. 723–742.

Weijermars, R., 1993, Pulsating strains: Tectonophysics, v. 220, p. 51–67.

Weijermars, R., 1997, Pulsating oblate and prolate three-dimensional strains: Mathematical Geology, v. 29, p. 17–41.

Weijermars, R., 1998, Taylor-mill analogs for patterns of flow and deformation in rocks: Journal of Structural Geology, v. 20, p. 77–92.

MANUSCRIPT ACCEPTED BY THE SOCIETY APRIL 12, 2000

New apparatus for thermomechanical analogue modeling

Elmar M. Wosnitza, Djordje Grujic
Geologisches Institut der Universität Freiburg, Albertstraße 23 B, D-79104 Freiburg, Germany
Robert Hofmann
Gundrebenstrasse 3, CH-8932 Mettmenstetten, Switzerland
Jan H. Behrmann
Geologisches Institut der Universität Freiburg, Albertstraße 23 B, D-79104 Freiburg, Germany

ABSTRACT

A new apparatus designed for thermomechanical analogue modeling to investigate tectonic crustal-scale processes, in particular the role of rheology in the distribution and propagation of deformation, is presented. By inducing a vertical thermal gradient through suitable analogue materials and controlling heat flow into the bottom and out of the top of the model, a precise viscosity gradient can be obtained. For the temperature range employed, paraffin wax shows a viscosity range of more than 11 orders of magnitude. A maximum of ~60% shortening or 200% extension is possible in the horizontal direction. The machine can determine the applied force, the horizontal shortening, and the model height at two locations. The bulk rheology is estimated from the calculated stress and the imposed constant strain rate, which can be varied between 10^{-6} and 4×10^{-5} s^{-1}. The two-dimensional temperature distribution of the glass confining side walls is measured with a thermal infrared camera and is proportional to the temperature distribution within the model. Deformation of the isotherms can thus be monitored during the experiment. By scaling down linear dimensions by a factor of 10^6, time by 10^{10}, viscosity by 10^{16}, and temperature by 10, we obtain a consistent set of scaling factors. This allows geometrical, kinematic, dynamic, and rheological similarity with lithospheric processes to be attained.

INTRODUCTION

To simulate the rheological stratification of the Earth's crust in physical models, it is necessary to take into account the variations in mechanical properties induced by variations in temperature. Until now this has been done experimentally by using different materials, such as sand and silicone putty, to model brittle and ductile behavior, respectively (e.g., Davy and Cobbold, 1991). Major progress has been made in this way in understanding crustal and lithospheric processes (e.g., Chemenda et al., 1995). However, the major drawback with such models is that any material point within the model crust retains its physical properties throughout the experiment, regardless of its changing position within the model. Thermal readjustment is not taken into account, and proper scaling with respect to gravity is therefore not achieved.

Analogue experiments to model crustal processes using temperature dependent material properties have been made before. However, none of them has fully exploited the possibilities. Oldenburg and Brune (1972, 1975) focused on freezing the material, but did not use the temperature dependency of viscosity. Shemenda and Grocholsky (1994) and Brune and Ellis (1997) used a temperature-sensitive viscosity, but did not map the temperature distribution in the material. Nataf and Richter (1982) examined convection in fluids with temperature-dependent viscosity and mapped isolines of the thermal gradient, applying their results to the evolution of planets.

The advantage of building models with a temperature-sensitive viscosity is that the mechanical consequences of thermal readjustment during the experiment can be reproduced. Our approach was to apply this concept to crustal rheological stratification and to map the temperature distribution. This represents a major improvement in analogue models of plate tectonic processes. It is particularly significant for experiments investigating lithospheric stretching and the stability of mountain belts. For example, in subduction zones the material balance is strongly controlled by the rheology of the subducted sediments (e.g., Mancktelow, 1995). This concept has also been applied to exhumation in collision zones (e.g., Grujic et al., 1996).

To investigate convergence processes, in particular the role of rheology in the distribution and propagation of deformation, a deformation rig was designed especially for thermomechanical modeling. Analogue materials with a temperature-sensitive viscosity are used to simulate the change of mechanical properties with depth. Correlating the thermal field with the displacement field and the material distribution allows the rheological structure of the model to be deduced.

APPARATUS

Overview

The usable model dimensions of the rig are 35 cm width, 25 cm height, and 50 cm length. Models can be either shortened by 60% or extended by 200% under plane strain (Fig. 1). The upper surface of the model is unconfined. The force is applied to the model through two Teflon-coated wooden pistons. This choice of material offers the most economical combination of strength and thermal insulation. Due to its high thermal conductivity, a metal piston would lead to significant boundary effects. A slit below the pistons allows material to flow behind the pistons, maintaining isostatic equilibrium. The pistons move symmetrically, but they can be independently activated. The side walls are made of glass with a thermal diffusivity similar to that of the analogue materials used in our experiments.

To apply a thermal gradient, the temperature has to be controlled at two levels within the model. The model is therefore both heated from below and cooled from above by two independent thermostats. The apparatus and thermostats are controlled, and data are quasicontinuously acquired with a microcomputer and a 12-bit analogue to digital converter card, using software developed under LabVIEW® Version 5 (Jamal and Pichlik, 1997).

Controlled and measured parameters

The graphical user software interface displays all measured properties, both numerically and graphically. Three parameters, i.e., the desired strain rate and the upper and lower temperatures, are controlled during the experiments. They can be set at the start of an experiment, and maintained constant during the run.

Strain rate and velocity. Most motors suitable for deformation experiments generate constant velocities only for a given time interval. Preliminary calibration experiments showed that the motor adequately maintained the desired velocities, and that correction for this error is unnecessary.

During the test phase a constant strain rate $\dot{\varepsilon}$ was maintained to remove one of the variables controlling viscosity. In these experiments, methods of motor control based on the variation of the model length l with time t can be obtained by integration of

$$\dot{\varepsilon} = \frac{\mathrm{d}l(t)/l(t)}{\mathrm{d}t}, \qquad (1)$$

which leads to

$$l(t) = l(0)e^{\dot{\varepsilon}t}. \qquad (2)$$

Because the velocity v was constant over the time interval Δt, this results in a deviation of the actual strain rate $\dot{\varepsilon}_a$ from the required strain rate $\dot{\varepsilon}$. The velocity v was recalculated at the beginning of each Δt using l_0, the length of the model at that time. The actual strain rate is calculated as

$$\dot{\varepsilon}_a(t) = \frac{v}{l_0 + vt}. \qquad (3)$$

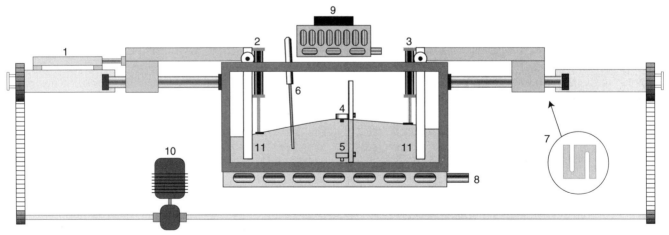

Figure 1. Machine design. 1–3—linear potentiometers, 4–6—temperature probes, 7—strain gauge, 8—heat exchanger, 9—cooling fan, 10—motor, 11—pistons.

To minimize the error, we use the secant of $l(t)$ to determine the desired velocity. From the required lengths at the beginning and the end of the time step, we obtain

$$v(t) = \frac{l(0)e^{\dot{\varepsilon}(t+\Delta t)} - l(0)e^{\dot{\varepsilon}t}}{\Delta t} = \frac{l(t)}{\Delta t}(e^{\dot{\varepsilon}\Delta t} - 1). \quad (4)$$

The mean actual strain rate $\bar{\dot{\varepsilon}}_a$ during Δt, obtained from integration of equation 3 over Δt, is

$$\bar{\dot{\varepsilon}}_a = \frac{e^{\dot{\varepsilon}\Delta t} - 1}{(\Delta t)^2} \int_0^{\Delta t} \frac{1}{1 + (e^{\dot{\varepsilon}\Delta t} - 1)\frac{t}{\Delta t}} \, dt$$

$$= \frac{1}{\Delta t}\ln(e^{\dot{\varepsilon}\Delta t}) = \dot{\varepsilon}. \quad (5)$$

Thus, the mean actual strain rate corresponds to the required strain rate, independent of the duration of the time step.

The motor and the gears allow the strain rate to be set between 2×10^{-6} s^{-1} and 2×10^{-4} s^{-1}. Strain rates can be maintained constant to within 5%. With a linear potentiometer the position of one of the pistons is monitored with an accuracy of <1 μm, thus allowing actual strain rates to be determined using the internal clock of the controlling computer.

Temperatures. The temperature field is controlled at two levels within the model. Two temperature probes at the base and at the top of the model are connected to a heating and a cooling thermostat, respectively. The model can be heated to temperatures up to 150 °C from below using a thermostat and silicone oil for the heat transfer. A cooling fan above the model connected to a second thermostat allows the model surface to be maintained between 20 and 80 °C. The thermostats are able to maintain the temperatures to within 0.1 °C in the steady state. To achieve a stable temperature gradient the model must be preheated for at least 10 hr.

Once they are set by the computer, the thermostats automatically maintain the temperature at their respective probes by using fuzzy-logic software. The computer receives, displays, and stores the temperature time series at the desired sampling rate.

The entire machine is insulated by a box of styrofoam to keep overall temperature conditions constant during experiments. A horizontal layer of styrofoam around the model separates the upper part from the lower part of the box, insulating the higher temperatures at the bottom of the model from the lower temperatures at the top. In addition to the temperature probes mentioned above, a third probe is embedded in the model. All the temperature probes are calibrated to ~0.05 °C.

Stress. To determine the bulk rheology of the model, the force applied to the model during the experiment is monitored by a strain gauge with a tolerance of 0.1%. In addition, the material height, at two positions close to the pistons, is measured with linear potentiometers with an accuracy of <1 μm. This information is required to determine the area of the side of the model, which is used to calculate the flow stress.

Thermal imaging

In addition to the bulk rheology, we plan to map the detailed rheology of the models. The material distribution and the temperature field are essential for this purpose. The former is determined by monitoring passive material markers with an optical camera; the latter is recorded using a thermal infrared camera.

Because the confining glass side walls have thermal diffusivities similar to the analogue materials, the temperature distribution can be mapped through the glass. The thermal energy (between wavelengths of 8 and 12 μm) radiated from the glass is recorded by the camera with a resolution of ~36 temperature pixels/cm^2. Assuming black body radiation, these energy values can be converted to apparent surface temperature.

In our model, the glass wall is far from being a black body; glass is rather reflective in the infrared part of the spectrum. To calibrate the temperature scale, we used a probe to measure the temperature at various positions in the middle of the model, in a section parallel to the glass walls (Fig. 2A). The measured temperatures were then correlated with the values measured by the camera to produce the colorscale for the infrared images given in Figure 2B.

SCALING

Analogue experiments must be designed in a way that they are similar to nature in the relevant aspects. To model nature, the equations that represent these aspects must be accurately scaled, a concept described by Hubbert (1937). Here we present a brief analysis of the significant equations. A more detailed analysis was given by Cobbold and Jackson (1992).

Time

In our experiments we take advantage of the differences in rheology induced by the thermal gradient within the material. The equation governing heat transfer in thermally isotropic materials, without internal heat production, is

$$\frac{dT}{dt} = \frac{\partial\left(a\frac{\partial T}{\partial x_i}\right)}{\partial x_i}, \quad (6)$$

where T is the temperature field, x_i are the three dimensions of space, and a denotes the thermal diffusivity.

During deformation, thermal and mechanical velocities must be treated equally. Thus, we have to ensure that the time needed by material particles or heat to move over a certain distance is scaled by the same factor. For the scaling factors of length l, time t, and thermal diffusivity a, we therefore obtain

$$S_t = \frac{S_l^2}{S_a}, \quad (7)$$

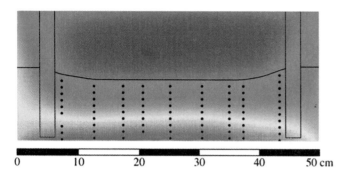

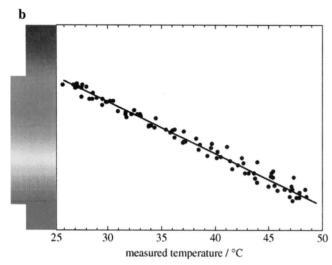

Figure 2. Temperature calibration. A: Scale of infrared camera (obtained from outside surface of side wall) calibrated by measuring actual temperature within model, using temperature probe at positions shown. Material surface and pistons are indicated. B: Temperatures measured at positions given in A correlated with energy values obtained from camera (color). Correlation is linear. Clustering of values is due to homogeneous temperature gradient.

where the scaling factors S_X of a property X represent the ratio of laboratory values to natural values. Equation 7 couples the length and time scales to the material-dependent scale of thermal diffusivity.

Stress

In the model, as in nature, gravity (g) acts on a material unit with a density ρ and a thickness h, resulting in gravitational stresses σ_{grav}. Gravitational stresses need to have the same scaling factor as viscous stresses σ_{visc}, where the latter result from deforming a material with a viscosity η at a strain rate $\dot{\varepsilon}$. Because

$$\sigma_{visc} = \eta \dot{\varepsilon} \qquad (8)$$

and

$$\sigma_{grav} = \rho g h, \qquad (9)$$

the model is only scaled to gravity when

$$S_\eta = S_\rho S_g S_l S_t. \qquad (10)$$

Temperature

To check the scaling factor for the absolute temperature T we have to consider the Arrhenius equation

$$\eta(T) = \eta_0 e^{H/(RT)} \qquad (11)$$

(after Pahl et al., 1991), where R is the universal gas constant and H is the activation energy. This equation does not influence the scale factor for viscosity, but it does require the exponent to be dimensionless. Because $S_R = 1$,

$$S_H = S_T. \qquad (12)$$

FIRST EXPERIMENTS

Setup

The aim of the first experimental runs was to test the possibility of modeling the deformation of isotherms in thrust tectonics. Appropriate analogue materials for which viscosity depends on strain rate and temperature are, e.g., colophony (Cobbold and Jackson, 1992) or paraffin wax (e.g., Mancktelow, 1988; Rossetti et al., 1999). We decided to use paraffin wax because it is easier to work and layers with different rheologies can be constructed. The experiment presented here consisted of two 5-cm-thick layers of paraffin wax with different melting points. The melting points, T_m, for the lower and upper layers were 53 ± 1 and 43 ± 1 °C, respectively. Both paraffin waxes were supplied by MERCK, Darmstadt.

To produce overthrusting, the model was cut in the middle at an angle of 30°. To prevent the cut from annealing during the preheating phase, it was lubricated with Vaseline. Without introducing such an inhomogeneity, the model would deform homogeneously, or the deformation would concentrate at the pistons.

The model was set up in such a way that the underthrust slab would start melting after moving about halfway down the thrust. To keep the experiment as simple as possible, both layers were kept above the phase-transition temperature for the corresponding paraffin wax (e.g., Mancktelow, 1988; Rossetti et al., 1999). This led to a base temperature of 51 °C, a surface temperature of 25 °C, and the wax layer interface temperature of 38 °C. Neglecting strain rate and assuming that the rheology is only dependant on the homologous temperature T/T_m, this corresponds to the rheology profile given in Figure 3.

Scaling

The thermal diffusivity of natural rocks is about 10^{-6} m^2 s^{-1} (e.g., Turcotte and Schubert, 1982; Fowler, 1990), whereas that

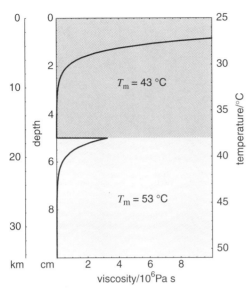

Figure 3. Rheology profile of initial state of two-layer model. Viscosities of lower layer are obtained using data from Rossetti et al. (1999), who described similar wax. For upper layer it was assumed that viscosities only depend on homologous temperature T/T_m. Viscosity at surface is 6.8×10^7 Pa s. Laboratory and natural depth scales are given.

TABLE 1. EXPERIMENTAL PARAMETERS

Quantity		Nature	Laboratory	Scale
Thermal diffusivity	a	10^{-6} m²s⁻¹	8×10^{-8} m²s⁻¹	8×10^{-2}
Strain rate	$\dot{\varepsilon}$	10^{-14} s⁻¹	9.5×10^{-5}/s	9.5×10^{9}
Length	l	170 km	50 cm	2.9×10^{-6}
Time	t	1.16 Ma	3800 s	1.05×10^{-10}
Density	ρ	2700 kg m⁻³	900 kg m⁻³	0.33
Viscosity	η	10^{22} Pa s	3×10^6 Pa s	10^{-16}
Activation energy	H	200 kJ/mol	1000 kJ/mol	5
Temperature	T	740 °C	51 °C	
		980 K	324 K	0.3

of paraffin wax is about 8×10^{-8} m²s⁻¹ (Rossetti et al., 1999). This results in a scaling factor of 8×10^{-2}. Orogenic processes have strain rates on the order of 10^{-14} s⁻¹ (Pfiffner and Ramsay, 1982). For the experiment, we decided to use the fastest possible strain rate of 9.5×10^{-5} s⁻¹. From the resulting scale factor for time of 1.05×10^{-10} and using equation 7, the scale factor for length is found to be 2.9×10^{-6}. Therefore, the initial model length of 50 cm corresponds to a length in nature of about 170 km.

The activation energy for paraffin waxes is ~1000 kJ/mol (Rossetti et al., 1999). For crustal rocks such as quartzites and granites it is around 150–200 kJ/mol (Carter and Tsenn, 1987). This results in a scale factor for the activation energy of 5–6. The temperature at the model base is 51 °C. The natural temperature at a depth of 34 km would be ~700 °C, assuming a geothermal gradient of 20 °C/km (e.g., Decker et al., 1988). These values lead to a scale factor of 0.3 for the temperature. If equation 12 were fulfilled, the model temperatures would have to be higher than the temperatures in nature. Because this is not possible, viscosity in the model depends too strongly on temperature. Therefore, for correct scaling, the rheology profile shown in Figure 3 should be flatter. Following an argument of Cobbold and Jackson (1992, p. 257), only the first two or three orders of magnitude of the overall strength variation are likely to have significant mechanical consequences. Therefore, true thermal scaling is probably unnecessarily rigorous.

Applying equation 10 to the scale factors for time and length and to the densities given in Table 1, we obtain a scaling factor for viscosities of 10^{-16} for the case of $S_g = 1$. According to Figure 3, the viscosity at the wax interface should be around 3×10^6 Pa s. Published flow parameters of rocks yield a broad range of viscosities. On the basis of parameters from Carter and Tsenn (1987) and assuming a temperature of 370 °C, crustal rocks at a depth of 17 km have viscosities ranging from 2×10^{20} Pa s to 4×10^{21} Pa s. Values taken from Kirby and Kronenberg (1987) give viscosities between 2×10^{22} Pa s and 10^{24} Pa s. This leads to actual factors for viscosity of 2×10^{-14} to 3×10^{-18}, which includes the required value of 1×10^{-16}. At greater depths these relations do not hold because of the reasons given herein.

Results

Analysis of the data recorded during the deformation of the model shows that the strain rate was constant. The mean strain rate over the entire experiment was 9.5×10^{-5} s⁻¹, with a standard deviation of 0.8×10^{-5} s⁻¹ (Fig. 4). The deviations are purely statistical and depend only on the sampling rate. No trend can be observed. We can therefore assume constant rheology, at least locally, during the experiment.

The recorded force applied to the model (Fig. 5) increased to a peak value of almost 60 N, followed by a decrease to around 40 N, caused by strain softening.

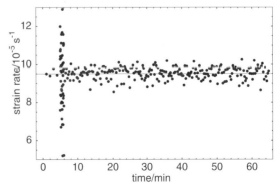

Figure 4. Strain rate obtained from model length and time intervals. Note that short intervals at beginning (every second, as opposed to every minute later on) lead to broader scatter of values. Thin line indicates required value.

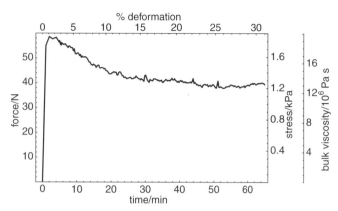

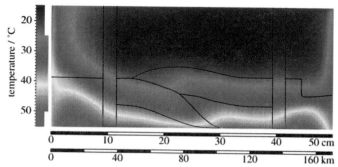

Figure 6. Infrared image of final state of described experiment. Material boundaries and pistons are shown.

Figure 5. Force applied to model during deformation versus time and deformation. After initial rise, strain softening occurs. Stress and bulk viscosity are proportional to force due to constant strain rate and surface. Total area in x-y plane did not change due to isostatic equilibration.

Figure 6 shows the infrared image of the final stage of the experiment. The infrared image is overlain by outlines of the pistons and of the analogue-material boundaries obtained from an optical photograph (Fig. 7). It can be seen that the overthrusting material transports heat upward, while the underthrusting slab causes a downward bend in the isotherms. Due to the strong deflection of the isotherms, the subducting slab in the thickened crust did not cross the melting isotherm. For the given parameters, the advection of heat due to movement of the material is faster than equilibration by conduction.

CONCLUSIONS AND OUTLOOK

We have not attempted to interpret these results in a geological context. The primary aim of these initial simple experiments was to test the feasibility of thermotectonic modeling under controllable conditions.

Experiments such as the one presented here offer an unparalleled possibility to calibrate numerical models involving heat flow (e.g., Jamieson et al., 1996). These models could be calibrated against our results and then readjusted accordingly. Numerical models are limited by the exactness of the equations involved and their thermomechanical coupling. However, they overcome the shortcomings of our apparatus, especially the inability to obtain the exact scaling of all aspects of the experiment with the analogue materials used. Both techniques suffer from poorly defined flow laws of rocks under natural conditions (Handy, 2000).

The apparatus described here can perform thermomechanical experiments because a steady-state thermal gradient is induced by controlling heat flow into the bottom and out of the top of the model. This yields precise and controllable viscosity gradients within the analogue material. The detailed tempera-

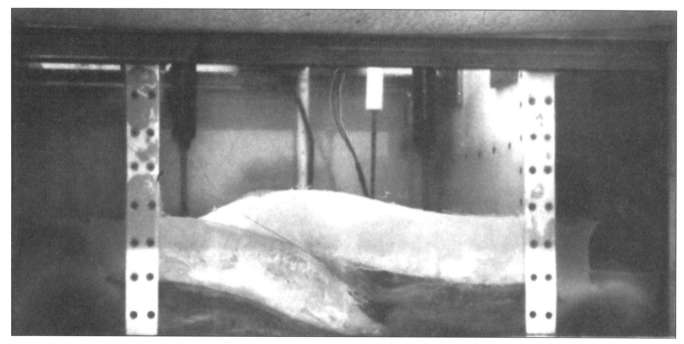

Figure 7. Photograph of final stage of described experiment. Two originally horizontal layers of paraffin wax were shortened by ~30%. Width of photograph is 50 cm.

ture distribution and hence the rheology can be obtained using a thermal infrared camera. From the temperature field and the material distribution, the viscosity can be calculated for each pixel, allowing an estimation of the time-dependent viscosity distribution. In future experiments we will introduce a brittle layer with Mohr-Coulomb behavior to represent the upper crust. Plastic powder (Jet-Plast) has an angle of internal friction of 35° and its bulk density of 0.75 kg/l allows gravity to be scaled correctly for the entire three-layer model. Experiments with constant convergence rates will be used to analyze geologically relevant scenarios. The relationship between the structures produced and the convergence rate will also be examined. The results obtained will further improve our understanding of the interaction between deformation and temperature distribution.

ACKNOWLEDGMENTS

We greatly appreciate the assistance we received from many colleagues and students. In particular, we thank Sigrid Dachnowsky, Alfred Immler, and Neil Mancktelow. We are indebted to Friedemann Burr for lease of the infrared camera. Image processing was done by Dennis Gross. The manuscript benefited from the thoughtful reviews of Phillipe Davy, Benoît Ildefonse, and Neil Mancktelow. The English was greatly improved by David Tanner. Financial support of the project by the Deutsche Forschungsgemeinschaft (Project BE 1041/11) is gratefully acknowledged.

REFERENCES CITED

Brune, J.N., and Ellis, M.A., 1997, Structural features in a brittle-ductile wax model of continental extension: Nature, v. 387, p. 67–70.
Carter, N.L., and Tsenn, M.C., 1987, Flow properties of continental lithosphere: Tectonophysics, v. 136, p. 27–63.
Chemenda, A.I., Mattauer, M., Malavieille, J., and Bokun, A.N., 1995, A mechanism for syn-collisional rock exhumation and associated normal faulting: Results from physical modeling: Earth and Planetary Science Letters, v. 132, p. 225–232.
Cobbold, P.R., and Jackson, M.P.A., 1992, Gum rosin (colophony): A suitable material for thermomechanical modeling of the lithosphere: Tectonophysics, v. 210, p. 255–271.
Davy, P., and Cobbold, P.R., 1991, Experiments on shortening of a 4-layer model of the continental lithosphere: Tectonophysics, v. 188, p. 1–25.
Decker, E.R., Heasler, H.P., Buelow, K.L., Baker, K.H., and Hallin, J.S., 1988, Significance of past and recent heat-flow and radioactivity studies in the southern Rocky Mountains region: Geological Society of America Bulletin, v. 100, p. 1851–1885.
Fowler, C.M.R., 1990, The solid Earth: Cambridge, Cambridge University Press, 472 p.
Grujic, D., Casey, M., Davidson, C., Hollister, L.S., Kündig, R., Pavlis, T., and Schmid, S., 1996, Ductile extrusion of the Higher Himalayan Crystalline in Bhutan: Evidence from quartz microfabrics: Tectonophysics, v. 260, p. 21–43.
Handy, M.R., ed., 2000, Rheology and geodynamic modelling: The next step forward: International Journal of Earth Sciences (Geologische Rundschau), (in press).
Hubbert, M.K., 1937, Theory of scale models as applied to the study of geologic structures: Geological Society of America Bulletin, v. 48, p. 1459–1519.
Jamal, R., and Pichlik, H., 1997, LabVIEW: Programmiersprache der vierten Generation: Munich, Prentice-Hall, 532 p.
Jamieson, R.A., Beaumont, C., Hamilton, J., and Fullsack, P., 1996, Tectonic assembly of inverted metamorphic sequences: Geology, v. 24, p. 839–842.
Kirby, S.H., and Kronenberg, A.K., 1987, Rheology of the lithosphere: Selected topics: Reviews of Geophysics, v. 25, p. 1219–1244.
Mancktelow, N.S., 1988, The rheology of paraffin wax and its usefulness as an analogue for rocks: Uppsala University Geological Institutions Bulletin, v. 14, p. 181–193.
Mancktelow, N.S., 1995, Nonlithostatic pressure during sediment subduction and the development and exhumation of high pressure metamorphic rocks: Journal of Geophysical Research, v. 100, p. 571–583.
Nataf, H.C., and Richter, F.M., 1982, Convection experiments in fluids with highly temperature-dependent viscosity and the thermal evolution of the planets: Physics of the Earth and Planetary Interiors, v. 29, p. 320–329.
Oldenburg, D.W., and Brune, J.N., 1972, Ridge transform fault spreading pattern in freezing wax: Science, v. 178, p. 301–304.
Oldenburg, D.W., and Brune, J., 1975, An explanation for the orthogonality of ocean ridges and transform faults: Journal of Geophysical Research, v. 80, p. 2575–2585.
Pahl, M.H., Geißle, W., and Laun, H.-M., 1991, Praktische Rheologie der Kunststoffe und Elastomere: Düsseldorf, Germany, VDI-Verlag, 428 p.
Pfiffner, O.A., and Ramsay, J.G., 1982, Constraints on geological strain rates: Arguments from finite strain states of naturally deformed rocks: Journal of Geophysical Research, v. 87, p. 311–321.
Rossetti, F., Ranalli, G., and Faccenna, C., 1999, Rheological properties of paraffin as an analogue material for viscous crustal deformation: Journal of Structural Geology, v. 21, p. 413–417.
Shemenda, A.I., and Grocholsky, A.L., 1994, Physical modeling of slow seafloor spreading: Journal of Geophysical Research, v. 99, p. 9137–9153.
Turcotte, D.L., and Schubert, G., 1982, Geodynamics. Application of continuum physics to geological problems: New York, John Wiley & Sons, 450 p.

MANUSCRIPT ACCEPTED BY THE SOCIETY APRIL 12, 2000

Modeling of temperature-dependent strength in orogenic wedges: First results from a new thermomechanical apparatus

Federico Rossetti, Claudio Faccenna
Dipartimento di Scienze Geologiche, Università "Roma Tre," Largo S.L. Murialdo, 1-00146 Rome, Italy
Giorgio Ranalli
Department of Earth Sciences and Ottawa-Carleton Geoscience Centre, Carleton University, Ottawa, Ontario K1S 5B6, Canada
Renato Funiciello, Fabrizio Storti
Dipartimento di Scienze Geologiche, Università "Roma Tre," Largo S.L. Murialdo, 1-00146 Rome, Italy

ABSTRACT

Temperature-induced variations in mechanical properties within orogenic wedges require analogue materials with temperature-sensitive viscosity. A new computer-controlled, thermomechanical experimental apparatus has been set up. The thermomechanical apparatus permits the scaling of the mechanical properties of thick orogenic wedges in terms of the temperature-dependent viscosity of 52/54 EN-type commercial paraffin. A 3 °C vertical temperature difference (from 37 to 40 °C) in the paraffin model results for an experimental strain rate of 10^{-6}–10^{-5} s^{-1} in a stress drop of about two orders of magnitude (from 10^2 to 1 Pa). For length, stress, and time-scale factors of 6×10^5, 2×10^6 and 3×10^{10}, respectively, these stress values are correctly scaled to the natural prototype for the ductile portion of the strength envelope. As an application of this modeling technique, we present results of one experiment in which the paraffin wedge is accreted against a rigid vertical backstop. The main deformation mechanism corresponds to upward flow at the rear of the wedge and frontward migration of the deformation front. For a fixed strain rate, model viscous wedges grow in a self-similar way, in a manner analogous to that predicted by the critical taper theory for Coulomb wedges.

INTRODUCTION

Orogenic wedges have been usually described as bodies resting upon a basal decollement and with a rigid buttress at the back (e.g., see Chapple, 1978; Cowan and Silling, 1978; Davis et al., 1983; Stockmal, 1983; Platt, 1986; Molnar and Lyon-Caen, 1988; Beaumont et al., 1994). The plane-strain critical taper theory, based on a Coulomb failure criterion applied throughout the wedge (Davis et al., 1983; Dahlen, 1984), has been successfully employed to infer the geometry, state of stress, thermal, and mechanical energy balance in thin-skinned thrust belts in the external parts of major mountain belts (cf. Dahlen, 1990, and references therein). On this basis, sandbox models have been widely used to provide further insights into the deformation and kinematics of orogenic wedges and fold and thrust belts (e.g., Davis et al., 1983; Malavieille, 1984; Mulugeta, 1988; Liu et al., 1992; Lallemand et al., 1994; Koyi, 1995; Storti and McClay, 1995; Wang and Davis, 1996). However, the applicability of critically tapered frictional wedge models to thick orogenic wedges is limited by the assumption of uniform Coulomb plasticity, because thermally activated ductile flow is dominant at depth (cf. Pavlis and Bruhn, 1983; Platt, 1986; Jamieson and Beaumont, 1988; Barr and Dahlen, 1989; Willett et al., 1993). Consequently, properly scaled analogue models should account for the variations in the rheological properties induced by the temperature gradient within the wedge. This problem can be approached in the laboratory using either two different materials (e.g., silicone putty and sand, Davy and

Cobbold, 1991) or a single material with temperature-sensitive viscosity (e.g., Cobbold and Jackson, 1992). The latter approach has the advantage of reproducing the mechanical consequences of thermal readjustments within the crust. Here we follow the second methodology by using 52/54 EN-type commercial paraffin as analogue material, the temperature-dependent viscosity of which satisfies the scaling requirements for modeling crustal rheology in the ductile regime (Rossetti et al., 1999a).

In this chapter, we first review the rheological properties of 52/54 EN-type paraffin as an analogue material to model natural deformation; then we describe the experimental apparatus in detail and discuss its usefulness to simulate the evolution of thick orogenic wedges. As an example of the application of this technique, we illustrate results of an experiment in which the paraffin wedge is accreted against a rigid vertical backstop. This experiment is part of a first modeling program, where the influence of variation of both backstop geometries and strain rate have been tested (Rossetti et al., 1999b and 2000).

MODELING THE RHEOLOGY OF OROGENIC WEDGES

Natural prototype

Most rheological models of thick orogenic wedges assume frictional (Coulomb) plasticity at shallow levels and power-law ductile flow at depth (e.g., Pavlis and Bruhn, 1983; Jamieson and Beaumont, 1988; Barr and Dahlen, 1989; Willett et al., 1993; Beaumont et al., 1994). In general terms, water-rich quartz-feldspar rocks represent the dominant lithologies in the deeper parts of orogenic wedges (Pavlis and Bruhn, 1983; Barr and Dahlen, 1989). Consequently, the ductile behavior of mature orogenic wedges should be controlled by wet quartz and feldspar rheology (e.g., see Pavlis and Bruhn, 1983). These minerals, for $T > 1/2 T_m$ (where T_m is the melting temperature), exhibit power-law creep with stress exponent in the range $2 \leq n \leq 3$, and activation energy in the range 150–250 kJ mol^{-1} (cf. Ranalli, 1995). In convergent tectonic settings, average geothermal gradients are normally rather low (around 20 ± 5 °C/km) and thermal diffusivity is typically 10^{-6} m^2 s^{-1} (Pavlis and Bruhn, 1983; Barr and Dahlen, 1989; Ord and Hobbs, 1989). Consequently, the strength envelope of a thick orogenic wedge shows a brittle-ductile transition at ~15–20 km. In the ductile field, for strain rates in the order of 10^{-15}–10^{-13} s^{-1} (e.g., Pfiffner and Ramsay, 1982), the creep strength decreases by two to three orders of magnitude with increasing depth beneath the brittle-ductile transition.

Analogue material

Following Rossetti et al. (1999a), we use commercial 52/54 EN-type paraffin with $T_m = 53 \pm 1$ °C, ρ (density) $= 0.85$ g cm^{-3}, and k (thermal diffusivity) $= 8 \times 10^{-8}$ m^2 s^{-1}. Rheological calibration experiments (Rossetti et al., 1999a) show that this material is a nonlinear fluid ($1.6 \leq n \leq 2.1$) for ho-

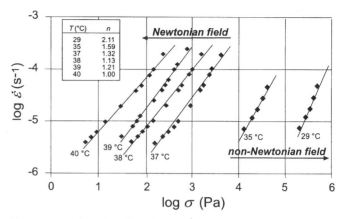

Figure 1. Log-log plot of strain rate ($\dot{\varepsilon}$) versus stress (σ) at different temperatures for chosen paraffin. Best-fit lines providing stress exponent n are also shown (modified after Rossetti et al., 1999a). For temperatures $T > 37$ °C, transition from nonlinear to linear rheology is documented (see inset).

mologous temperatures $T/T_m < 0.70$ and an approximately Newtonian fluid ($1.0 \leq n \leq 1.3$) for $T/T_m \geq 0.70$, i.e., $T = 37$ °C (Fig. 1). Effective viscosity values are strongly temperature dependent, varying by nine orders of magnitude in the investigated temperature range between 30 and 52 °C. In the Newtonian field and under natural gravity conditions, this material fulfills similarity criteria to model crustal ductility, if length and stress are scaled by a factor 10^5–10^6, and time by 10^{11} for a strain rate of 10^{-5} s^{-1} (Rossetti et al., 1999a). Temperature is a controlling factor to obtain mechanical similarity between model and nature. Because of its high activation energy (~900 kJ mol^{-1}) with respect to other laboratory material (e.g., see Davy and Cobbold, 1991), paraffin allows to obtain large variations in creep strength (or viscosity) over a small temperature range. For a temperature change from 37 to 40 °C and the experimental strain rates herein, the creep strength of paraffin drops by about two orders of magnitude, from 500 to 10 Pa.

Figure 2 shows the scaling factors used and compares natural prototype (N) and wedge model (M) strengths. The stress values are correctly scaled to the natural prototype for the ductile portion, while the brittle portion in the model is limited to the upper thermal boundary layer maintained at the temperature of 37 °C. However, the ratio of nature vs. model total strength (shaded areas in Fig. 2) is about 10^{11}, in agreement with the length and stress ratios. For the length ratio $l_N/l_M = 6 \times 10^5$, the characteristic thermal diffusion time, $t_M = l_M^2/k_M$, in the model is ~9 hr, corresponding to a time-scale factor of $t_N/t_M = 3 \times 10^{10}$.

Experimental apparatus

Description. We have built a new specifically designed thermomechanical shortening machine in order to model orogenic wedge evolution. The apparatus is installed in an insulated room kept at constant temperature (37 °C), and is shown in Figure 3. It consists of a 75-cm-long, 40-cm-wide and 50-cm-high Plexiglas box, into which differently colored layers of melted

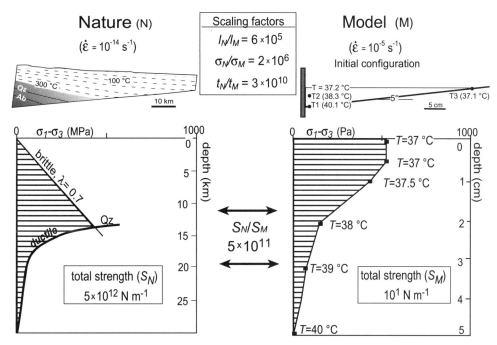

Figure 2. Scaling factors and comparison of strength profiles between natural prototype (left; modified after Barr and Dahlen, 1989) and model paraffin wedge (right) subjected to temperature gradient shown. Ab, albite; Qz, quartz. List of physical parameters: $\dot{\varepsilon}$, strain rate; l, length; σ, stress; t, time; T, temperature; λ, pore fluid pressure ratio; $\sigma_1 - \sigma_3$, differential stress.

paraffin are poured. The paraffin rests on a mobile basal plate, which is thermally conductive (aluminum) and may be inclined at various angles. A computer-controlled resistance heater, set at a constant temperature of 40 °C by a heating system, is positioned below the basal plate. The resistance heater is confined to the rear one-third of the plate, and an insulating panel separates it from the front, which is maintained at room temperature by a computer-controlled cooling system. The resistance heater and the insulating panel are connected to the mobile basal plate, which is displaced backward during the model run. Displacement is applied by means of a screw jack fixed onto the basal plate and connected to a stepping motor. A rigid backstop is placed at the rear of the paraffin wedge. Different backstop configurations, from vertical to low dipping, can be used. Initial model dimensions are usually 5 cm thick, 40–50 cm long, and 40 cm wide.

Movement of the basal plate causes deformation and shortening of the paraffin wedge against the rigid backstop. A displacement transducer measures the incremental shortening, with 0.1 mm precision. Before the model run, lateral shear is reduced by lubrication of the lateral box walls with silicone oil.

Temperature control. Both heating and cooling systems consist of a microprocessor-based programmer-controller (accuracy 0.1% at 25 °C ambient temperature; resolution 0.1 °C) for use with current thermocouples inputs. Temperature control is achieved by means of 10 computer-connected thermocouples (RTD PT 100 type), some of which are inserted into the paraffin wedge at various locations, and some located in the insulated room, to provide a continuous record of both room and model temperatures. The temperature profile inside the paraffin wedge (from 37 to 40 °C from surface to base at the rear), and room temperature show variations of ±0.2 °C. The desired temperature profile inside the paraffin wedge is obtained about 12 hr after paraffin pouring (Fig. 4).

Observation of the model evolution. A grid composed of circular markers is painted onto the surface of the models to evaluate plane finite strain during runs. Model surfaces are photographed at constant time and displacement intervals. At the same time, the model topography profile is monitored with a computer-controlled laser scan machine (resolution of the order of 100 µm) running over the model parallel to the shortening direction. Topography data are recorded as data files to obtain variations of model topography with time. At the end of each run, models are sectioned to examine the development of structures inside the paraffin wedge.

Each model is run at least twice to check the reproducibility of the results.

EXPERIMENTAL RESULTS

We illustrate the evolution of one experiment representative of a large set of two-dimensional experiments (Rossetti et al., 1999b). The boundary conditions of the experiments were: (1) a constant bulk strain rate of 10^{-5} s^{-1}; (2) vertical and rigid backstop; (3) 5° dipping mobile basal plate. The experimental run lasted 10 hr, to 40% shortening, to ensure thermal recovery within the model. This is also confirmed by the experimental check of

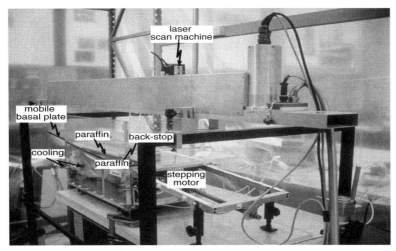
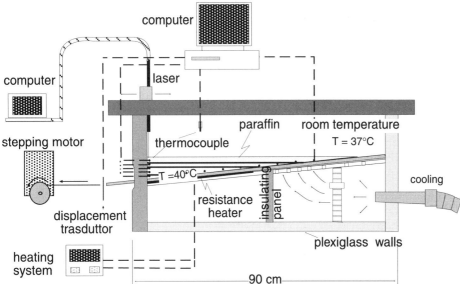

Figure 3. Photograph and schematic drawing of thermomechanical experimental apparatus (see text for further details).

the temperature profile within the model wedges as shown in the inset in Figure 4. The experimental results are described using the parameters L, l, and h, which denote the length of the undeformed top of the wedge, the distance of the deformation front from the backstop, and the maximum elevation of the top surface with respect to the initial topography, respectively (Fig. 5).

The nucleation and growth of three major anticlines characterized the deformation history of the model. These structures were rather cylindrical in shape and sequentially developed from the backstop toward the front (Fig. 5). The first anticline nucleated close to the backstop and grew rapidly during the first 20% shortening ($h = 0.1\ L$); the second began to nucleate at 10% shortening, and the third began to nucleate at 30% shortening. The maximum height of the inner anticline was $h = 0.13L$, at 40% shortening. The sequential development of the anticlinal structures during progressive shortening corresponded to the continuous widening of the deformed area (Fig. 5 and curve f in Fig. 6A). However, the wavelength of each anticline progressively decreases with increasing shortening (Figs. 5 and 6a). The vertical and horizontal motion of four reference points (a, b, c, d in Fig. 5) located on the model surface is illustrated in Figure 6B. The graph shows that the model wedge underwent mainly backward horizontal translation near the front of the wedge (points c and d in Fig. 6B), whereas mainly thickening with vertical movement occured at the rear (points a and b in Fig. 6B), as expected for vertical backstop configuration. The combined effects of uplift and frontward migration of deformation can be adequately described by the variation of the aspect ratio ($A = h/l$) of the model with increasing shortening. It can be seen that the parameter A increases almost linearly during the first 20% shortening and then remains nearly constant around the value of 0.25 (Fig. 6C). Calculations of the value A for previous Coulomb rheology models, both numerical (Willett, 1992) and sandbox (Mulugeta, 1988), show similar trends (Fig. 6C).

To evaluate the role of the thermal gradient on the evolution of the model wedge, we performed a set of experiments at constant temperature and with the same boundary and geometrical conditions as before. We chose a temperature of 38 °C, which

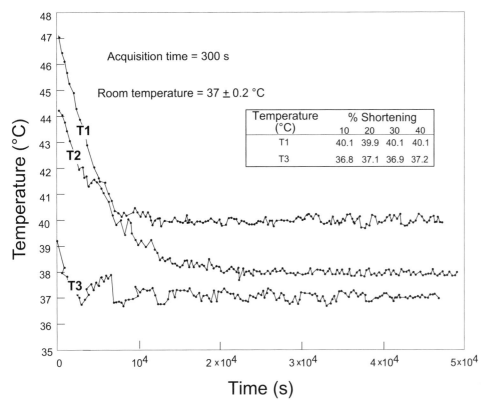

Figure 4. Equilibrium temperature profiles (reached at time $t > 10^4$ s) within model wedge before deformation. Inset shows temperatures at bottom (T1) and at top (T3) surfaces of model for incremental shortening steps (see Fig. 2 for positions of reference points).

provides a total strength for the model wedge of 25 N m^{-1}, comparable with that of the previous experiment. The results of this experiment show a progressive thickening at the rear of wedge, without the sequential formation of the antiformal structures and basal decollement. This means that frontal material cannot be underthrust and incorporated at the toe of the wedge and that homogeneous shortening accommodates thickening. The comparison between these two experiments underlines the role of the thermal gradient within the wedge, generating strength layering from top to bottom and from front to rear of the wedge. Therefore, its use appears fundamental to achieve both the rheological similarities between model and nature as well as to obtain the propagation of the compressional front and accretion of new material at the toe of the wedge.

DISCUSSION AND CONCLUSIONS

The apparatus and material discussed in this chapter allowed for the first time modeling of the ductile strength variations within an orogenic wedge, including the effects of thermal recovery on the model strength.

The application of these results to natural cases, however, should take into account that: (1) the brittle portion of the paraffin wedge is limited to the upper thermal boundary layer set at the room temperature (Fig. 2); (2), paraffin in the experimental temperature range is a Newtonian material, whereas laboratory data indicate that flow of crustal materials usually follows a power-law creep, except for very low grain sizes, as in mylonitic shear zones (e.g., see Ranalli, 1995); consequently, the deformation of paraffin is expected to be more diffuse than that of a crustal material, where strain-softening processes predominate (e.g., see Wejermars and Schmeling, 1986); (3) isostatic compensation, flexural readjustments, and erosion during wedge growth have not been considered. The last two points can explain the inadequacy of this modeling material in reproducing localized deformation (i.e., thrust surfaces) and the excessive thickness of the wedge, respectively.

Despite the limitations noted here, our models correctly simulate the depth decrease in strength in the ductile portion of the crust, taking into account the proper scaling in terms of total strength. In this sense, this approach can be particularly useful to understand the large-scale evolution of mature (ductile-dominated) orogenic wedges. The modeling technique is mostly addressed to the evolution of the wedge as a whole, which results from the interplay between the gravity forces and the basal traction, here described by the aspect ratio (A) of the wedge. In particular, we observe that the growth of the experimental viscous wedges tends to a self-similar configuration (i.e., constant A), suggesting that the wedge has entered the stability field (sensu Dahlen, 1990) with respect to its internal strength and applied strain rate. This result implies that for each strain rate there is a typical equilibrium wedge shape, and that variation in strain rate can induce a rapid reshaping of the wedge, and possibly a change in deformation style in different parts of the wedge (Rossetti et al., 1999b).

In the model wedge, the material flows backward from the toe to the rear of the wedge, where an upward path is imposed

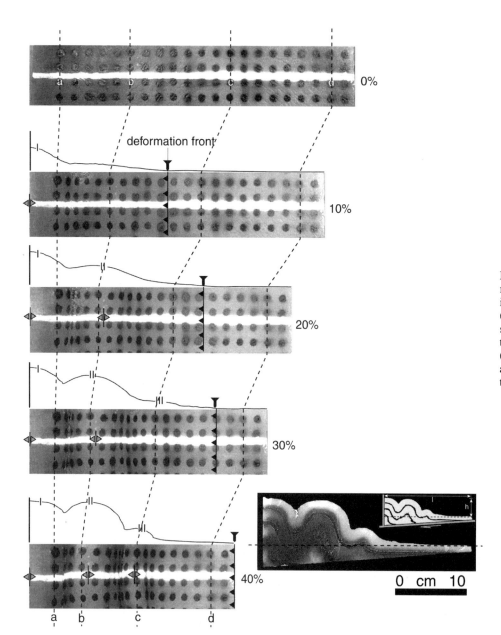

Figure 5. Sequential photographs of model surface for incremental shortening steps, and related model topography (continuous lines). Photograph of model sectioned in middle at end of deformation run is also shown. Roman numerals (I, II, III) refer to sequentially formed anticines; *a, b, c, d* denote points, evolution of which is described in text.

by the vertical rigid backstop. Qualitatively, this behavior is analogous to that described in the viscous flow models of Cowan and Silling (1978).

Further developments of this technique are in progress and will be focused on the particle flow paths and velocity fields inside the wedge, incorporating erosion and variation in strain rate.

ACKNOWLEDGMENTS

ENEL SpA supported and S. D'Offizi encouraged this research. Laboratory experiments have been performed at ISMES SpA and the Analogue Laboratory of University of Roma Tre. Rossetti is grateful to M. Luiselli for his important help during laboratory work. The work by Ranalli is supported by a grant from the Natural Sciences and Engineering Research Council of Canada. Reviews by R. Pfiffer and an anonymous reader helped improve the chapter.

REFERENCES CITED

Barr, T.D., and Dahlen, F.A., 1989, Brittle frictional mountain building 2. Thermal structure and heat budget: Journal of Geophysical Research, v. 94, p. 3923–3947.

Beaumont, C., Fullsack, P., and Hamilton, J., 1994, Styles of crustal deformation in compressional orogens caused by subduction of the underlying lithosphere: Tectonophysics, v. 232, p.119–132.

Chapple, W.M., 1978, Mechanics of thin-skinned fold-and-thrust belt: Geological Society of America Bulletin, v. 89, p. 1181–1198.

Cobbold, P.R., and Jackson, M.P.A., 1992, Gum rosin (colophony): A suitable material for thermomechanical modelling of lithosphere: Tectonophysics, v. 210, p. 255–271.

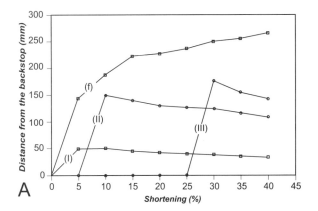
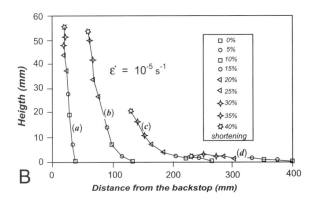
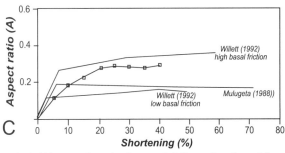

Figure 6. A: Distance from backstop of deformation front (f) and of hinge lines between major anticlines (I, II, III) as function of shortening. B: Trajectories of surface points initially at positions a, b, c, d in Figure 5. C: Aspect ratio (A) as function of shortening in this experiment (squares), compared with Coulomb numerical (Willett, 1992) and sandbox (Mulugeta, 1988) models.

Cowan, D.A., and Silling, R.M., 1978, A dynamic scaled model of accretion at trenches and its implications for the tectonic evolution of subduction complexes: Journal of Geophysical Research, v. 83, p. 5389–5396.

Dahlen, F.A., 1990, Critical taper model of fold-and-thrust belts and accretionary wedges: Annual Review of Earth and Planetary Sciences, v. 18, p. 55–99.

Davis, D., Suppe, J., and Dahlen, F.A., 1983, Mechanics of fold-and-thrust belts and accretionary wedges: Journal of Geophysical Research, v. 88, p. 1153–1172.

Davy, P., and Cobbold, P.R., 1991, Experiments on shortening of a 4-layer model of the continental lithosphere: Tectonophysics, v. 188, p. 1–25.

Jamieson, R.A., and Beaumont, C., 1988, Orogeny and metamorphism: A model for deformation and pressure-temperature-time paths with application to the central and southern Appalachians: Tectonics, v. 7, p. 417–445.

Koyi, H., 1995, Mode of internal deformation in sand wedges: Journal of Structural Geology, v. 17, p. 293–300.

Lallemand, S., Schnurle, P., and Malavieille, J., 1994, Coulomb theory applied to accretionary and non accretionary wedges: Possible causes for tectonic erosion and/or frontal accretion: Journal of Geophysical Research, v. 7, p. 12033–12055.

Liu, H., McClay, K.R., and Powell, D., 1992, Physical models of thrust wedges, in McClay, K.R., ed., Thrust tectonics: London, Chapman & Hall, p. 71–81.

Malavieille, J., 1984, Modélisation expérimentale des chevauchements imbriqués: Application aux chaînes des montagnes: Bulletin de la Société Géologique de France, v. 7, p. 129–138.

Molnar, P., and Lyon-Caen, H., 1988, Some simple physical aspects of the support, structure, and evolution of mountain belts, in Clark, S.P., Jr., Burchfiel, B.C., and Suppe, J., eds., Processes in continental lithospheric deformation: Geological Society of America Special Paper 218, p. 179–207.

Mulugeta, G., 1988, Modelling the geometry of Coulomb thrust wedges: Journal of Structural Geology, v. 10, p. 847–859.

Ord, A., and Hobbs, B.E., 1989, The strength of the continental crust, detachment zones and the development of plastic instabilities: Tectonophysics, v. 158, p. 269–289.

Pavlis, T.L., and Bruhn, R.L., 1983, Deep-seated flow as mechanism for the uplift of broad fore arc ridges and its role in the exposition of high P/T metamorphic terranes: Tectonics, v. 2, p. 473–497.

Pfiffner, O.A., and Ramsay, J.G., 1982, Constraints on geological strain rates: Arguments from finite strain states of naturally, deformed rocks: Journal of Geophysical Research, v. 87, p. 311–321.

Platt, J.P., 1986, Dynamics of orogenic wedges and the uplift of high-pressure metamorphic rocks: Geological Society of America Bulletin, v. 97, p. 1037–1053.

Ranalli, G., 1995, Rheology of the Earth (second edition): London, Chapman & Hall, 413 p.

Rossetti, F., Ranalli, G., and Faccenna, C., 1999a, The rheology of paraffin as an analogue material for viscous crustal deformation: Journal of Structural Geology, v. 21, p. 413–417.

Rossetti, F., Faccenna, C., Ranalli, G., Storti, F., and Funiciello, R., 1999b, Modelling the evolution of viscous orogenic wedges using paraffin as an analogue material: European Union of Geosciences, 10th Meeting, Strasbourg, France: Journal of Conference Abstracts, v. 4, p. 607.

Rossetti, F., Faccenna, C., Ranalli, G., and Storti, F., 2000, Convergence rate–dependent growth of experimental viscous orogenic wedges: Earth and Planetary Science Letters, v. 178, p. 367–372.

Stockmal, G.S., 1983, Modeling of large-scale accretionary wedge formation: Journal of Geophysical Research, v. 88, p. 8271–8287.

Storti, F., and McClay, K., 1995, The influence of sedimentation on the growth of thrust wedges in analogue models: Geology, v. 23, p. 999–1003.

Wang, W.H., and Davis, D.M., 1996, Sandbox model simulation of forearc evolution and non-critical wedges: Journal of Geophysical Research, v. 101, p. 11329–11339.

Weijermars, R., and Schmeling, H., 1986, Scaling of Newtonian fluid dynamics without inertia for quantitative modelling of rock flow due to gravity (including the concept of rheological symilarity): Physics of the Earth and Planetary Interiors, v. 43, p. 316–330.

Willett, S., 1992, Dynamic and kinematic growth and change of a Coulomb thrust wedge, in McClay, K.R., ed., Thrust tectonics: London, Chapman & Hall, p. 19–31.

Willett, S., Beaumont, C., and Fullsack, P., 1993, Mechanical models for the tectonics of doubly vergent compressional orogens: Geology, v. 21, p. 371–374.

MANUSCRIPT ACCEPTED BY THE SOCIETY APRIL 12, 2000

Flow and fracturing of clay: Analogue experiments in bulk pure shear

Fernando O. Marques*
Departamento de Geologia, Faculdade Ciências, Universidade de Lisboa, Edifício C2, Piso 5, 1700 Lisboa, Portugal

ABSTRACT

To simulate the behavior of rocks capable of simultaneous flattening and fracturing, we used clay. When this material was deformed under bulk pure shear, two families of conjugate faults formed initially at an angle of ±30°–35° to the direction of maximum compressive stress (σ_1). With continuing deformation, two general situations developed. (1) Faults were rectilinear and conjugate sets showed identical development with small relative rotations. (2) One of the conjugate sets prevailed, the earlier faults rotated significantly about the axis defined by the intersection between conjugate faults, and duplexes developed. Because the experiments took place at constant volume, flattening, relative rotations, alternating kinematics, and intersection of conjugate faults must solve space problems. Commonly, minor faults developed that terminated at the main conjugate fault; nevertheless, they showed a significant amount of slip. This can be analyzed in terms of the rotation of faults and/or flattening of clay. Flattening was far from being homogeneous across the model, especially when there was considerable rotation of blocks about axes parallel to conjugate fault intersection; then, significant domains showed very little strain, despite the 40% shortening. Stress/strain graphs, at constant strain rate, show that clay initially underwent strain hardening, and, after the maximum of σ_1 was reached, strain softening slowly developed as faults grew and spread throughout the model. In the absence of considerable rotations, and despite brittle failure, flow patterns of an entire model did not deviate significantly from that expected for plastic flow in pure shear, with the exception of relatively small domains. The used clay proved to be an excellent rock analogue.

INTRODUCTION

The evolution of faulting in the continental crust is still not fully understood, especially concerning fault shape, fault hierarchy, fault rotation, slip along faults, strain distribution, and flow patterns. To try and answer these questions, rock samples and analogue materials have been tested (e.g., Reches and Dieterich, 1983; Oertel, 1965). When using natural rocks, total strains can be only a few percent (e.g., Reches and Dieterich, 1983), and thus many of the questions cannot be answered. In addition, such experiments were performed at room temperature (brittle behavior), and our aim was to simulate the behavior of rocks capable of simultaneous flattening and fracture. Cloos (1955) and Oertel (1965) used clay as a rock analogue, but their results were not conclusive regarding the preceding questions.

This lack of information was the main incentive for our investigation. We chose clay as the analogue material because it contains fluids, is capable of simultaneous flattening and fracturing (according to Freund, 1974, this is an essential requirement for the strain by faulting to occur), and is suitable as a model material for large rock bodies subject to geological deformation, as shown by Hubbert (1937) and Cloos (1955). We chose pure shear with controlled side stress, because it approximates the conditions expected in a continental crust subject to shortening. Our results bear little similarity with published results (Cloos, 1955; Oertel, 1965; Hoeppener et al., 1969), and

*E-mail: fmarques@fc.ul.pt

Marques, F.O., 2001, Flow and fracturing of clay: Analogue experiments in bulk pure shear, *in* Koyi, H.A., and Mancktelow, N.S., eds., Tectonic Modeling: A Volume in Honor of Hans Ramberg: Boulder, Colorado, Geological Society of America Memoir 193, p. 261–270.

this can be the consequence of, for example, using different types of clay, with differences in water content, and/or using different machines (with significantly different specifications) to deform clay by bulk pure shear.

The ultimate objective of this experimental work is to help understand the fault patterns observed in eroded mountain belts, and their significance, by giving an extraordinary visualization of strain by flow and faulting in an analogue material and by bringing new insights to the understanding of fault development in large crustal areas. In this chapter we present the preliminary results of a study still in progress. They are in part descriptive and interpretative, because some logistical problems still remain to be solved: e.g., the used strain markers (grid of squares and circles), essential to the studies of flow and strain distribution, were in great part disrupted by fractures and faults.

EXPERIMENTAL PROCEDURE

The analogue material used for the modeling is molding clay, with ~26% water content. Therefore, we assume that fluid pressure can play a role in the development of fractures. X-ray diffractometry (Philips PW1710 with a PW1820/00 goniometer) revealed that the used clay is composed of smectite, chlorite, quartz, illite, and calcite. Volume changes by compaction and/or water loss could not be detected by measurements of volume of models before and after deformation.

The pure shear rig we used to perform the experiments is the one described in Mancktelow (1988). Throughout the text we refer to the walls that are pushed closer as top walls (where compression is exerted), and the ones that are pulled apart as side walls (where constant side stress is monitored) (Fig. 1). The finite bulk strain ellipsoid has the longest principal axis (X) horizontal and parallel to the top walls; the shortest principal axis (Z) is also horizontal but parallel to the side walls; and the intermediate principal axis is vertical (Y, with constant length, thus guaranteeing plane strain).

The experiments were performed in the Experimental Tectonics Laboratory of the Laboratório de Tectonofísica e Tectónica Experimental (LATTEX), University of Lisbon, under the following experimental conditions: temperature = 21 °C, strain rate constant, = $6.00\,E - 5\,s^{-1}$, maximum shortening percentage ~40%, and initial size of the model 290 × 70 × 60 mm. Most experiments were performed with a constant side stress of about 30 kPa, but we also investigated the effects of lower (down to ~10 kPa) and higher (to 50 kPa) values of this parameter. A grid of squares with inscribed circles was carved on top of the models to analyze strain (Fig. 2). After thorough lubrication of all confining walls with Vaseline, the parallelepipedic cake of clay was put into the rig with the longest axis parallel to the σ_1 direction, and then it was covered with a thick glass (for constant vertical dimension).

EXPERIMENTAL RESULTS

In the early stages of compression (shortening <10%), clay began to flatten and showed very small fractures like open gashes, subparallel to the direction of σ_1. The later formation of faults was commonly preceded by, and resulted from, en echelon fracturing and their coalescence, but this was not always the case. Other faults were born as small fractures that, with continuing deformation, progressively increased size by propagation at the ends.

At the early stages of fracturing, two sets of conjugate faults formed at an angle of about 60°–70°, with σ_1 as acute bisectrix. With continuing deformation, two main typical situations were observed: (1) rectilinear faults and conjugate sets with identical development and small relative rotations (less common) (Fig. 3); (2) one of the conjugate sets prevailed, the earlier faults (and blocks) rotated significantly about axes defined by intersection of conjugate faults, sometimes with a sigmoidal shape, and duplexes developed (common) (Fig. 4). The intersection be-

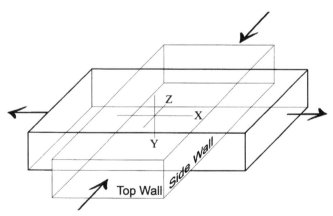

Figure 1. Schematic representation of clay model, moving walls, and principal axes of bulk finite strain ellipsoid. X—longest axis, Y—intermediate axis, and Z—shortest axis.

Figure 2. Photo of XZ plane of initial stage, common to all experiments.

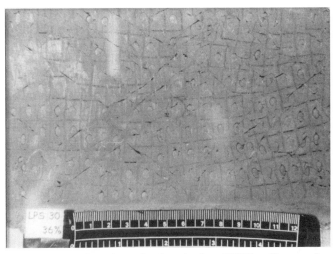

Figure 3. Photo of XZ plane of end result of run LPS30 to illustrate similar development of conjugate fault sets and rectilinear geometry of faults.

tween conjugate sets was typically a conflict zone, more or less complex depending on equal or unequal development of conjugate sets and timing of intersections.

Fault geometry

Fault geometry was very simple: they were subvertical, (at least) initially rectilinear, and oriented at ±30°–35° to σ_1 (conjugate sets at 60°–70°). At later stages, some faults rotated and bent significantly.

Some main faults showed sigmoidal shapes, as illustrated in Figure 4. The central section of the fault gradually rotated, clockwise in the sinistral faults and counterclockwise in the dextral, and the overall shape became sigmoidal. In some cases the terminations of faults asymptotically curved toward the side walls of the rig (which acted as mechanical barriers) or toward the main conjugate fault. According to Ramsay and Huber (1987), the increase in the acute angle between conjugate sets can be the result of flattening of the entire mass. This would explain rotations of faults along their entire length, but not the sigmoidal shape of some faults. Freund (1974) presented a theoretical view on fault terminations (Fig. 8 of Freund, 1974) and stated that sigmoidal shape of faults is one of the ways that faults can terminate without disruption of the faulted area boundaries. In our opinion, these views seem complementary and a satisfactory, but not a sufficient, explanation for the results we obtained experimentally. Here we add two other probable explanations, one that is a geometric consequence of pure shear and another that takes into account the mechanical behavior of the faults. (1) If a fault grows fast in the early stages and terminates at opposite corners of a deforming rectangular mass, then late in the experiment this line will undergo considerable shortening (to 20% if the fault is, e.g., a diagonal of the rectangles of Fig. 5, B and E). The dextral fault most to the right in Figure 5 underwent shortening and rotation, and became curved, i.e., it was deformed by buckling (which implies the occurrence of continuous heterogeneous deformation within the model). (2) Strain hardening of the faults should lead to their blockage, and, from the moment they become inactive, they should behave as passive markers, i.e., rotate progressively into parallelism with the XY plane with continuing deformation.

In addition to the sigmoidal shape presented by some faults in late stages of the experiments, some faults showed obvious curvatures close to the intersection with the dominant conjugate fault, which they did not cut (Fig. 6). If we did not have access to the stress field (known orientation of the applied $\sigma 1$ or deducted from strain ellipses) or to the kinematics of faulting (dislocated markers), one could think that the referred fault curvature was a drag effect and would draw a wrong conclusion on fault kinematics. What is the meaning of this curvature? Three explanations are possible: (1) termination of fault by bending, as proposed by Freund (1974); (2) curving of the older fault (acting as a passive marker) by rotation of the dominant conjugate; and (3) a combined effect of the previous two.

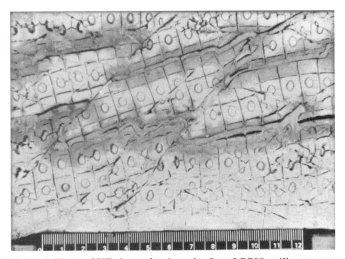

Figure 4. Photo of XZ plane of end result of run LPS09 to illustrate rotation of faults and blocks, which leads to sigmoidal shape of major faults, high slip in faults even close to terminations, heterogeneous distribution of strain, and domains with very low strain. Line markers were initially parallel to photo edges. Open fractures are result of desiccation of clay after experiment.

Fault terminations

In his theoretical approach, Freund (1974) contended that transcurrent faults can terminate, without disrupting the boundaries of the faulted area, in two ways: (1) by splay faulting at the ends, or (2) by bending the fault's ends (or, conversely, by rotating the middle section of the faults). Both situations were observed in our experiments (Fig. 6), either close to the walls of the rig, or close to the main conjugate fault.

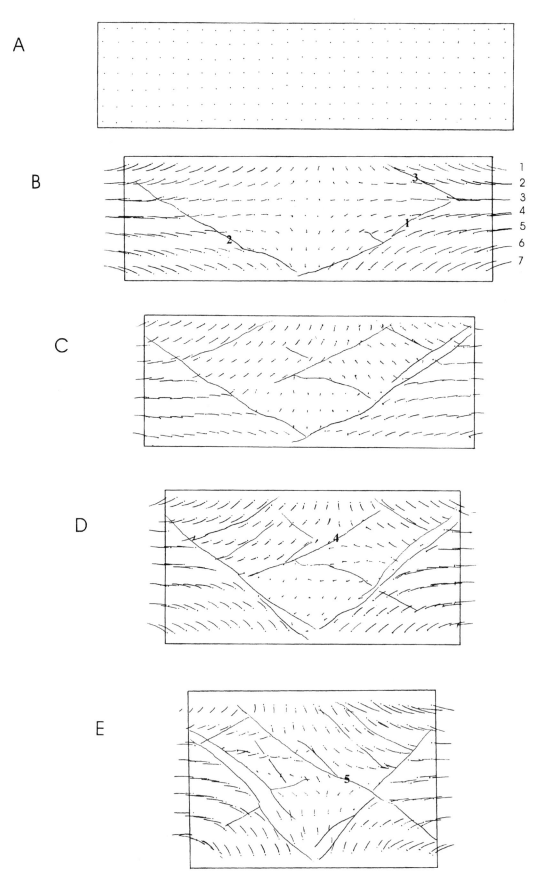

Figure 5. Flow patterns from run LPS26. Dots mark positions of centers of ellipses at considered shortening percentage. Therefore, lines mark trajectory of material points between shortening intervals. A: Undeformed state. B: After 13% shortening. C: After 20% shortening. D: After 30% shortening. E: After 40% shortening.

Figure 6. Photo of XZ plane of end result of run LPS32 to illustrate: (1) fault terminations by bending (top middle to left corner of photo) and by splaying (middle, bottom left, and right corners, and top right corner of photo); (2) similar development of conjugate fault sets and rectilinear geometry of faults; (3) false drag folds in sinistral fault at top middle to left corner. Photo taken with low-angle lighting.

Flow path

Although the initial marker grids were strongly disrupted by faulting, we could still construct flow paths of one experiment, in which rotation of marker lines was one of the least, by following the trajectories of circle and/or ellipse centers. The results are shown in Figure 5.

In Figure 5A we plotted centers of circles of the undeformed model. After the first 13% of shortening (Fig. 5B), the behavior of clay deviated only slightly from what is theoretically expected for plastic flow in pure shear, despite the generation of the first conjugate faults, which grew synchronously and at an angle of ~55°. After 20% shortening (Fig. 5C), flow departed from that expected for plastic materials under pure shear, mostly because in the lozenge domain in the middle bottom part of the model the flow (and strain) was practically absent. Also note the symmetry of flow in the two triangular domains in the lower left and right corners, and tendency of flow to become parallel to the faults in these domains. Fault 3 stopped growing at the conjugate. After 30% shortening (Fig. 5D) the behavior was similar to the previous interval, but more pronounced. After 40% shortening (Fig. 5E) the conjugate faults 1 and 2 were at an angle of ~80° (which implies a rotation of ~25°) and there was a change in the behavior of the lozenge domain in the middle bottom part of the model, which then presented internal strain. Between 30% and 40% shortening, fault 1 became inactive and was cut by the conjugate fault 5.

When we look at the Figure 5 in more detail, flow paths close to faults can give us some indication on the mechanical behavior of faults. Theoretically, flow lines 1, 2, and 3 should move upward, 4 should move in a straight fashion because it is in the middle of the model, and 5, 6, and 7 should move downward. However, we observe the following. (1) In Figure 5B, flow line 4 moves downward, instead of keeping straight, especially in the block to the right of fault 1, which could denote easy slip along this fault (strain softening). (2) In Figure 5, C and D, flow lines from 4 to 7 in opposite sides of faults 1 and 2 do not show flow continuity and are at a great angle, which could also reveal strain softening. (3) There was practically no flow in the lozenge domain in the middle bottom of Figure 5 until 30% shortening, because strain was easier to accommodate by slip along the limiting faults then by clay flattening, which is indicative of strain softening in the faults. (4) After 30% shortening this situation was reversed (it is easier to accommodate strain by clay flow then by fault slip), revealing that strain hardening took place in faults. (5) In Figure 5E there is an obvious difference of flow between the triangular domain to the right of faults 1 and 5, and the domain to the left of fault 2; in this block flow lines 3 to 7 continued to move downward, which was not the case with flow lines 3 to 5 in the triangular block between faults 1 and 5. Note that faults 1 and 2 shared an identical history (synchronous development and dominance until ~30% shortening) and rotated significantly and in identical amounts; however, after ~30% shortening, fault 1 was active until 40% shortening, but fault 2 became inactive and was cut and displaced by the conjugate fault 5. As mentioned herein, rotation to an unfavorable position relative to σ_1 cannot justify the behavior of fault 1, and the explanation must be sought in its mechanical behavior; flow lines in the block to the right of this fault in Figure 5, B, C, and D seem to indicate easy slip along the fault, probably related with strain softening, and flow lines in Figure 5E seem to indicate strain hardening and blockage. Then, it was easier to form a new conjugate to accommodate continuing strain. Flow after 30% shortening underwent some changes (deviations from theoretical predictions if faulting was not present) in the vicinity of the newly formed fault 5.

Fault hierarchy

Hierarchic development of faults was diverse. In some cases, conjugate faults grew synchronously and approached each other, and alternating kinematics and propagation of the conjugate sets, with generation of many minor faults (top middle of Fig. 7A), made the intersection. However, in later stages one of them always prevailed, cutting and displacing the other conjugate set (Fig. 8). In other cases, the two conjugate sets grew and approached each other, and one of them prevailed (at least for a while, because later in the experiment the situation could be reversed). When conjugate faults developed at different times during the experiment, two opposite situations occurred: the growth of the younger conjugate stopped at the older conjugate that continued to prevail, or simply cut through the preexisting conjugate and prevailed from then on (Fig. 9). What ruled hierarchy, and why and when did one of the conjugate sets prevail over the other are questions to be asked. In some situations the answer seems

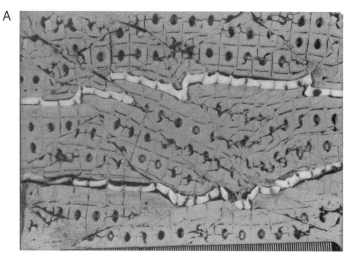

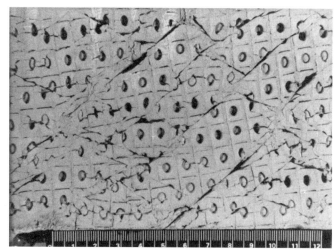

Figure 8. Photo of XZ plane of end result of run LPS13. Prevailing dextral conjugate fault set cuts and displaces earlier formed sinistral conjugate set.

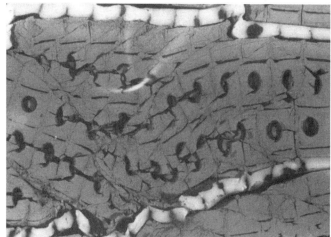

Figure 7. A: Photo of XZ plane of end result of run LPS18/02 to illustrate: (1) sigmoidal shape of main faults; (2) high rotation of blocks with internal low strain (center of photo); (3) intersection of faults cutting each other, with formation of minor faults (middle top of photo). B: Close up of middle-right area of photo to show minor faults that terminate at main conjugate fault, but show significant slip, and rotation by bookshelf mechanism. Open fractures are result of desiccation of clay after experiment and lighter layer is plasticine marker.

simple: relatively early rotation of faults in the experiment (either along its entire length or only at its central section—sigmoidal shape) leads to a progressively less favorable position relative to stresses exerted on their surfaces. Normal stress becomes progressively higher and shear stress progressively lower, leading to inhibition of slip. It is easier to accommodate deformation by developing younger conjugate sets that will cut through, and displace, the older rotated fault.

Observation of Figure 4 and the discussion herein of Figure 5 show that significant rotation to unfavorable positions cannot be the only explanation for nonsynchronous development of conjugate faults: in spite of significant rotation, the dextral conjugate did not block and prevailed until 40% shortening. Then, other possible explanations for hierarchical and differential behavior of faults during their evolution in the experiments must be sought in their mechanical behavior, i.e., whether they go through strain softening or strain hardening (or both, depending on time), and could it be the result of faster strain softening in the prevailing conjugate. The answer cannot be found in the stress/strain graphs because they do not discriminate the behavior of individual faults. What we can predict, in general terms, is that (1) faults undergoing strain softening have no apparent reason to become inactive throughout their history; and (2) faults undergoing strain hardening should tend to undergo blockage, and tend to passive marker behavior the moment they become inactive.

Often, sets of minor faults formed as conjugate sets to an older major fault, but never crossed it, and even showed a significant amount of slip (Fig. 7B). This can be analyzed in terms of the rotation of faults and/or flattening of clay in the surroundings. It is important to note that, although one of the conjugate sets was younger than the other, both conjugate sets were active at the same time, i.e., from the moment of formation of the minor conjugate set. Otherwise, these would have cut through the older fault and displaced it, which was not the case. In some cases, the minor faults showed very little rotation, and, to account for slip along minor faults, clay flattened in the acute domain between conjugate sets. In other cases (Fig. 7B), the original ~65° between faults gradually increased (to 90°) by rotation of the minor faults, and considerable slip can be justified by a bookshelf mechanism (Mandl, 1987). If, along with rotation, there is also flattening of clay in the acute domain between conjugate sets, the amount of slip can be even greater.

Rotation of structural elements

Relative rotations about axes parallel to conjugate intersections were commonly observed in faults and blocks limited by faults and/or mechanical contacts (plasticine and/or walls of rig).

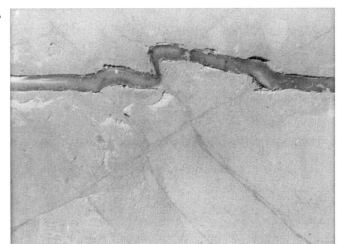

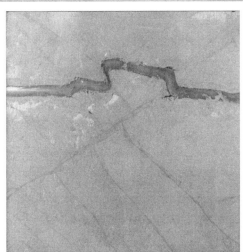

Figure 9. Photo of XZ plane of end result of run LPS03. Rotation in entire length of sinistral faults leads to their blockage, and consequent easier generation of younger dextral conjugate fault that intersects and displaces older faults. A: Initially, intersection has more plastic character with development of sigmoidal shape in older faults close to intersection zone. B: Later, younger fault displaces older, and geometric arrangement is that of drag folds associated with fault. Dark layer is passive marker. Photo is rotated 90° clockwise from A.

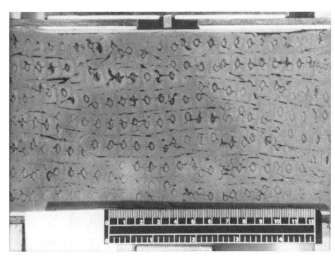

Figure 10. Photo of XZ plane at 32% shortening of run LPS27. Considerable clockwise rotation (~15°–20°) of major left-lateral fault to right of model was accomplished mostly by evenly distributed flattening of clay.

Rotations of faults were achieved in two distinct ways: by flattening of clay (little rotation of marker lines and homogeneous flattening of clay throughout the model; Fig. 10), or by rigid rotations (very heterogeneous and little flattening, and high rotation of marker lines; Fig. 4). A third way can be a mixture of both mechanisms. Rigid rotation of blocks of considerable dimension (to as much as 20° from the original position) was typically associated with significant rotation of limiting faults and with very low values of strain of clay (Fig. 4). Considerable block rotations and concentration of strain in major faults mostly accommodated deformation.

Rotation of blocks between parallel faults was accomplished by a bookshelf mechanism (Fig. 7B). This could also accommodate part of the rotation of major faults.

Slip along faults

The used marker grids proved to be too coarse for a thorough quantitative analysis of slip. However, a qualitative evaluation of slip revealed that great offsets were generally observed at the central part of major faults, but also occasionally very close to the ends when these terminated against side walls that worked as decollements (Fig. 4). The amount of slip was also greatly dependent on rigid block rotation (as can be seen in Fig. 4), on the amount of differential flattening between blocks of a fault, and on the amount of rotation of faults. These last two were generally responsible for significant offsets in minor faults that terminated against the conjugate fault. Differential flattening was also responsible for considerable offsets at terminations of faults that ended at the top walls (where slip should be close to null).

Strain distribution

If deformation applied at the top walls of the model could be homogeneously transmitted to the entire clay model, and if clay deformed by homogeneous plastic flow (no disruption or brittle failure), all finite strain ellipses would have identical axial ratios, the principal shortest axis would be parallel to σ_1, and principal axes of finite strain ellipses would be parallel to the principal axes of the instantaneous ellipse (coaxial deformation). Our experiments departed from this ideal situation because none of these conditions were fullfilled (mostly because the used clay failed by fracture), and the result can be observed on the heterogeneous transformations of the original grid of squares and circles (e.g., Fig. 4).

With the used coarse marker grids, many circles and ellipses were destroyed (cut and displaced by faults). However, from observation of the surviving finite strain ellipses that developed from the original circular markers, we can say, in general terms, the following. (1) Distribution of strain is very heterogeneous from model to model, even when initial conditions seem to be similar. (2) In most cases flattening is far from being homogeneous across the model (Fig. 4), especially when there is considerable rotation of blocks; then, significant domains show very little strain, despite the final 40% shortening (Fig. 11). (3) In other cases strain of clay is more evenly distributed (Fig. 10). (4) Strain distribution seems to vary with varying side stress—the greater the side stress, the more homogeneous the flattening distribution (probably due to better transmission of bulk strain into the clay model). (5) In the early stages of compression, strain tends to concentrate closer to the top walls, and conjugate faults are originated here and grow toward the center of the model. (6) In experiments with little relative rotation, the X axes of the finite strain ellipses are roughly perpendicular to the direction of σ_1, but in experiments with considerable rotations, X axes of finite strain ellipses are not orthogonal to the σ_1 direction applied at the top walls. (7) There seems to be no obvious drag associated with fault movement, which could be detected by curved distribution of principal axes of ellipses; this could be indicative of low friction on fault surfaces.

Domains that underwent very little strain occurred in three typical situations: (1) inside strongly rotated blocks, where bulk shortening was not accommodated by flattening, but by rigid rotation (Fig. 4); (2) inside little or nonrotated lozenge-shaped blocks limited by faults that take up most of the deformation (Fig. 11); and (3) inside nonrotated triangular domains limited by two conjugate faults and the side wall of the rig.

Graphical representations

The analysis of the σ_1 versus percent shortening (stress/strain) graphs, at constant strain rate, shows an early rapid increase of σ_1 to a peak of variable value. The shape of this initial curve in the various experiments under similar conditions is not always identical (Fig. 12), but they all mean that the early stages of deformation of the clay are accompanied by strain hardening (increase in σ_1 at constant strain rate). The peak value of σ_1 is dependent on the chosen value of side stress (Fig. 12), and on small variations in the intrinsic properties of the used clay (e.g., water content) when experimental conditions are identical. After the maximum of σ_1 is reached, strain softening slowly develops, deduced from the negative slope of the curve. A couple of cases show a horizontal flat line after the peak of σ_1 (no strain softening or hardening), and a few others show a broad curve concave upward (strain softening after peak σ_1, and then some strain hardening) (Fig. 12B).

From the analyses of the stress/strain graphs, at constant strain rate, we become aware only of the behavior of the entire model, which is the sum of the behavior of the clay and faults. The initial strain hardening can be due to the elastic deformation of clay, because there are no faults. However, from the peak σ_1 onward, the stress drop should be related to fracture and/or fault formation. Coincidence of fault initiation with peak stress in our experiments suggests that faulting triggers strain softening. This should be equivalent to the drastic stress drop observed in experiments with natural rocks (e.g., Reches and Dieterich, 1983).

Another important feature of the stress/strain graph, is the episodic stress variations. Our observations suggest that this be related to the way slip occurs along major faults. Stress drops correspond to major slip episodes, which means that slip is not constant and homogeneous. Slip along faults in our experiments looks like the movement of a block being pulled by a spring over a hard surface: it moves in jerks (stick slip). Confirmation of this mechanism in our experiments will be possible by future improvement of the software that controls graph construction.

Variations in the value of side stress induce direct proportional variations in the values of σ_1; the greater the side stress, the greater the σ_1 (Fig. 12, C and D).

CONCLUSIONS

Clay accommodates strain by flattening and brittle failure, with formation of conjugate faults at $\pm 30°–35°$ to σ_1.

Conjugate faults were always in the shortening field of the bulk strain ellipsoid. Then, they could behave in two different ways: when growth was not restrained, they grew to the limits of the area under stress; when growth was restrained, they rotated and buckled. This is an alternative explanation for the curved geometry of some faults.

Fault terminations were commonly realized by splaying and/or bending, in order to dissipate slip toward the fault ends without disrupting the boundaries of the faulted area.

Faults do not have to cut their conjugate fault while both are active; even though they can present considerable slip, it is mostly accommodated by rotation and/or differential flattening of clay.

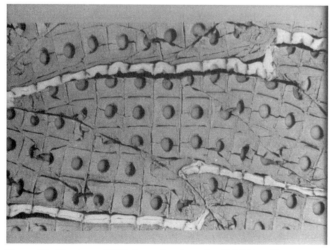

Figure 11. Close-up of XZ plane of end result of LPS19 to illustrate domains with very low strain and strong curvature of faults. Open fractures are result of desiccation of clay after experiment.

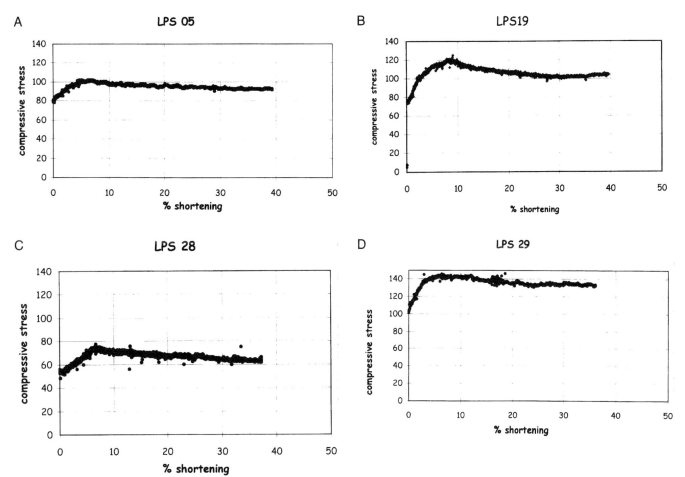

Figure 12. Most typical σ_1 vs. percent shortening graphs of experiments. A: Initial strain hardening to ~100 kPa, and then strain softening, at constant ~30 kPa of side stress. B: Initial strain hardening to ~120 kPa, then strain softening, and finally some strain hardening again, at ~30 kPa of side stress. C: Graphical pattern similar to that in A, but peak σ_1 is lower because of lower side stress of ~10 kPa. D: Graphical pattern similar to that in B, but peak σ_1 is higher because of higher side stress of ~50 kPa.

Distribution of strain was usually very heterogeneous across a model (especially when there was considerable rotation of blocks), and also from model to model, even when initial conditions seemed to be similar. Domains of very little strain occurred in three typical situations: (1) inside strongly rotated blocks; (2) inside little or nonrotated lozenge-shaped blocks limited by faults; and (3) inside nonrotated triangular domains limited by two conjugate faults and the side wall of the rig.

Block rotations and flow paths show that when faults undergo strain softening it is easier to accommodate strain by slip along faults than by flow of clay. Hierarchic development of faults also seems to be mainly dominated by their mechanical behavior: the easier the slip along a fault (strain softening), the greater the probability of dominance over the conjugate set. Strain hardening and blockage can result from significant rotation of a fault.

Clay models reacted initially by strain hardening, but from ~10% shortening onward faults developed and strain softening was triggered.

Faults and blocks limited by faults and/or mechanical contacts rotated significantly about the axes defined by the intersections between conjugate faults. As concluded by Cobbold et al. (1989) from their sand experiments, rotation of blocks occur at all scales in the models, and this could be of great importance in geological studies that must deal with rotations.

The clay proved to be an excellent rock analogue.

ACKNOWLEDGMENTS

We greatly acknowledge the experimental work carried out by Luísa P. Ribeiro. This chapter benefited from discussions with J. Cabral, and also with colleagues during EUG10. X-ray diffractometry was carried out by Patrícia Conceição in the Chemistry Dep. of FCUL. Experiments were carried out in the Experimental Tectonics Laboratory of LATTEX, a Research Unit funded by PLURIANUAL (125/N/92).

REFERENCES CITED

Cloos, E., 1955, Experimental analysis of fracture patterns: Geological Society of America Bulletin, v. 66, p. 241–256.

Cobbold, P.R., Brun, J.P., Davy, P., Fiquet, G., Basile, C., and Gapais, D., 1989, Some experiments on block rotation in the brittle upper crust, in Kissel, C., and Laj, C., eds., Paleomagnetic rotations and continental deformation: NATO ASI Series C, Volume 254, Dordrecht, Kluwer Academic Publishers, p. 145–155.

Freund, R., 1974, Kinematics of transform and transcurrent faults: Tectonophysics, v. 93, p. 93–134.

Hoeppener, R., Kalthoff, E., and Schrader, P., 1969, Zur physikalischen Tektonik: Bruchbildung bei verschiedenen Deformationen im Experiment: Geologische Rundschau, v. 59, p. 179–193.

Hubbert, M.K., 1937, Theory of scale models as applied to the study of geologic structures: Geological Society of America Bulletin, v. 48, p. 1459–1520.

Mancktelow, N.S., 1988, An automated machine for pure shear deformation of analogue materials in plane strain: Journal of Structural Geology, v. 10, p. 101–108.

Mandl, G., 1987, Tectonic deformation by rotating parallel faults: The "bookshelf" mechanism: Tectonophysics, v. 141, p. 277–316.

Oertel, G., 1965, The mechanisms of faulting in clay experiments: Tectonophysics, v. 2, p. 343–393.

Ramsay, J.G., and Huber, M., 1987, The techniques of modern structural geology. Volume 2: Folds and fractures: London, Academic Press Inc., 700 p.

Reches, Z., and Dieterich, J.H., 1983, Faulting of rocks in three-dimensional strain fields: I. Failure of rocks in polyaxial, servo-control experiments: Tectonophysics, v. 95, p. 111–132.

MANUSCRIPT ACCEPTED BY THE SOCIETY APRIL 12, 2000

Index

[Italic page numbers indicate major references]

A

accretionary wedge, *124*
  analog model, *125*
  Messianian, 127
Aegean microplate, 127
Africa, 149
African plate, 127
Alapaevsk terrane, 202
Alberta Foothills, 180
alluvial successions, Pyrenees, 113
Alps, basal decollement, 203
amalgamation, segment, 207, 208, *218*
analog material
  barite, 211
  clay, 53, 261, 262, 268, 269
  colophony, 248
  glass microbead, 154
  Granucol, 22, 23, 24, 25, 26
  paraffin, 235, 245, 248, *254*, 255
  plaster of Paris, 208, 215
  plasticine, 103, 116, 117
  plastilina, 116
  polydimethylsiloxane, 181, 238
  putty, 235
  quartz sand, 154, 223
  Rhodorsil Gomme, 239
  silicone putty, 103, 104, 107, 111, 116, 117, 223
  temperature sensitive, 246, 253
analysis, four dimensional, 179, 187
anhydrite, Cardona Formation, 113
anisotropic growth, 39, 40
anticline
  deformation, 157
  model, 256
anticrack, 29
antiform, 54, 58, 61, 67
Apennines thrust belt, 174
apophysis, oblique, 11, 16
apparatus
  deformation, 235, *236*, 245, *246*
  thermomechanical, *253*
Apulian
  carbonates, 174
  foreland, 174
  platform, 173
Archean terrain, vii
Artés Formation, 113
Arthur Holmes Medal, vii
Arthur L. Day Medal, vii
asperity bifurcation, 208, 209, 210, 211, 214, 215, 216, 219, 297
asymmetry, 66
  type A, 61, 64, 66
  type B, 61, 64

B

back rotation. *see* rotation
backkink, 194, 199, 202. *see also* kink band
  dip, 199
  model, 195, 196, 197
  propagation, 194
backshear, 194, 196, 199, 202, 204
  asymmetry, 203
  dip, 205
  imbrication, 201
backslope
  dip, 205
  gravitation stability, 199
  gravity collapse, 202
  slumping, 191
  stability, 201, 205
backthrust, 161, 164, 170, 183, 186, 189, 202. *see also* fault, thrust
Baltic Sea, 202
Baltic shield, 202
  seismic profile, 204
Barbados Ridge complex, 124
Barbastro Formation, 113, 114
barite, analog material, 211
basal
  decollement, 192, 193, 198, 201, 202, 203
  detachment, 112, 164, 179, 183
  friction, 158, 180, 186, 193
basement ramp, structure, 163
Basin and Range Province, 221
Betics, 112
bifurcation
  asperity, 207, 208, 209, 210, 211, 214, 215, *216*, 219
  tip line, 207, 208, 210, *215*, 216, 219
Biot-Ramberg equation, 96
bookshelf mechanism, 267
Bothnian Bay, 202
boudin, *101*, 102, 103
  Pinós Cardona anticline, 115
  plasticine, 107
  rotation, 102, *104*, 108, 109
boudinage, vii
  chocolate tablet, 17
boundary
  conditions, 168, 169, 176, 180, 222, 255
  constraints, 154
  work, 34
British Caledonides, 202
brittle domain, 179, 180, 186, 187
  modeling, *183*
brittle failure, 94, 268
brittle-ductile transition, 254
brittle-viscous domain, 179, 180, 186, 187
  modeling, 179, *183*
buckle fold, *54*, 58, 64, 66, 67, 79, 94, 117
buckling, 51, 52, 89, 263
buoyancy, 124
Byerlee relationship, 223

C

Cabo Frio, 133
  basin, 148
  fault zone, 138, 140
  region, 143
Calabro-Lucano region, model, 174
Calabro-Lucano Southern Apennines, *172*
camera, thermal infrared, 250
Campano-Lucano Southern Apennines fold and thrust belt, 174
Campos basin
  Brazil, 133, 138, 147, 148, 231
  faulting, 150
  seismic profile, 133
  structure, 134, *149*
carbonate platform, 138
Cardener River, 116
Cardona anticline, 113
  age, 114
Cardona diapir, 111, *112*, *114*, *118*
  age, *119*
Cardona Formation, *112*
Cardona-Pinós fold, 111
Castelltallat Formation, 113, 114
Catalan Coastal Ranges, 114
Celsius Medal, vii
centrifuge, 117, 118
Centro-Iberian autochthon, transpression, 19
chemical potential, *33*
chlorite, 102, 262
clastic loading, 138
clay, analog material, 53, 261, 262, 268, 269
cleavage, 17, 18, 29
coalescence
  segment, 219
  tip line, 207, 215, 219
cohesion, 120
  rock, 124, 193
collapse
  gravity, 198, 199
colophony, analog material, 248
compaction
  differential, 181
  lateral, 198
compression, 116, 126, 174
  Pyrenean, 112
conglomerate, Solsona Formation, 114
conic section, 7
conservation
  of mass, 34, *35*
  of momentum, *34*
continental
  collision, 191, 192
  indentation, 192
  margin, 133
  plate, 192
  suturing, 192
contraction
  diapir, 128
  longitudinal, 65
corundum powder, 181, 182
Coulomb
  criterion of failure, 181, 223, 253
  plasticity, 23, 192, 253
  slip, 23, 27
crack, 29
creep, 90, 92

crust
  faulting, 261
  orogenic, 192
CT volume scanning, 187

D

Deborah number, 74
decollement, 123, 124, 125, 127, 129, 267. *see also* detachment
  basal, 124, 192, 193, 198, 201, 202, 203
  Pyrenean triangle zone, 113
decoupling, mechanical, 174
Deevo fault, 202
deformation, 207, 231, 261, 266
  apparatus, 235, *236,* 245, *246*
  brittle, 193
  bulk, 67
  Calabro-Lucano region, 174
  coaxial, 16, 52, 65
  disharmonic, 160
  fold and thrust belts, 253
  four dimensional, 179, 181, 187, 208
  geometry, 162
  instantaneous, *6*
  kinematics, *155,* 172, 176
  lineation, 51, 52, 56, 63, 64
  migration, 256
  orogenic wedge, 253
  partitioning, 176
  patterns, 193
  piggy-back, 158, 168, 169, 174, 176
  progression, 237
  propagation, 245, 246
  rotation, *9*
  three stage, 58
  transpression, viii, *1, 7, 16,* 236
  transtension, viii, 1, *7, 16,* 17, 236
  two stage, 53
  types, *7,* 16, *18*
  viscous, 29, 34
density
  contrast, 124
  upper crust, 193
deposition, syntectonic, 159, 169, 172
detachment, 120, 123, 180, 230. *see also* decollement
  basal, 112, 164, 179, 183
diapir, ix, *123,* 132,
  collapse, 133
  contraction, 128
  Eocene, 112
  erosional trigger, *116, 118*
  location, 128
  model, 118
  mud, 123, *124,* 127
  Priabonian, 112
  Pyrenees, *112*
  rise, 148
  salt, *111,* 131, 133, 140, 149
differential loading, 118
diffusion, 29, 30, 31, 32
diffusive loss, *30*
dilatancy, 27, 239

disharmony
  mechanical, 176
  structural, 168
displacement vector, 19, 43
dissipation, *30,* 32, 33, 34
dolomite, 102
dolostone, 173
domain
  brittle, 179, 180, 182, 186, 187
  brittle-viscous, 179, 180, 183, 186, 187
  external, 173, 174
  tectono stratigraphic, *173*
  triangular, 268, 269
double eye pattern, sheath folds, 108
double mushroom pattern, sheath folds, 108
Dow Corning
  compound DC3179, 92
  silicone, 116, 117
downlap, 138
ductile strength, thermal variation, 257
duplex, 207, 208, 211, 212, 214, 219, 261, 262
  collapse, 214
  extensional, 215

E

East African rifts, 231
East European craton, 202
Ebro basin, 114, 118, 119
El Guix anticline, 113, 114, 120
Elle
  numerical model, 41
  simulations, *44*
ellipse, equation, 14
equation
  Biot-Ramberg, 96
  conic section, 13
  constitutive, *37*
  diffusion, 31
  dilatancy number, 239
  dissipation, *30*
  ellipse, 14
  extrusion number, 239
  rate of displacement, *2*
  steady-state creep, 92
  strain ellipse, *4*
  strain rate, *5, 30,* 31
  volume change number, 239
  vorticity vector, *239*
erosional trigger, diapir, *116*
Espirito Santo basin, 148
Eurasian plate, 127
exploration, hydrocarbon, 149, 154, 174, 207
extension, 65, 116, 132, 133, 140. *see also* transtension
  Late Jurassic, 112
  overburden, 149
external domain, 173, 174
extremum principle, 30
extrusion number, equation, 239
eye pattern, sheath folds, 108, 109

F

failure, brittle, 94, 268
fan, listric, 208
fault, 155, 161
  amalgamation horses, 218
  antithetic, 131, 132, 133, *134,* 138, 143, 147, 148, 149, 150, 211, 218, 219
  arch shaped, 163
  asperity bifurcation, 207, 208, 209, 210, 211, 214, 215, 216, 219
  backthrust, 192, 198
  box, 208
  Campos basin, 150
  conjugate, 261, 263, 265, 266, 268, 269
  crust, 261
  detachment, 222
  development, *219,* 265
  en echelon, 208, 215
  extensional, *209, 215*
  floor, 208, 211
  growth, 215
  hinterland, 158
  listric, 221, 225
  master, 219, 221, 225, 230, 231, *231,* 232
  normal, 133, 138, 222, 230, 232
  planar, 221
  propagation, 172
  roof, 208
  segment linkage, 210
  sigmoidal, 263
  splay, 208, 209, 211, 212, 214, 218, 263
  strike slip, ix, 21
  synthetic, 131, 138, 147, 148, 149, 211, 218, 219
  system, 21
  termination, 263, 268
  thrust, 162, 164, 179, 183
  tip-line bifurcation, 207, 208, 210, 215, 216, 219
finite element, viii, 35
  grid, 78
flattening, 51, 261, 268
  differential, 267, 268
floor fault, 208, 211
flow, *265*
  model, 258
  monoclinic, *239,* 242
  plastic, 265
  shear box, *240*
  strain hardening, 90
  strain softening, 90
fluid
  migration, 124
  overpressure, 124, 125
flux, 31, 35, 37
foam texture, 39, 40, 44
fold, 102
  amplification rate, *95*
  amplitude, 75
  angular, 90, 95, 96, 97
  asymmetry, 52, 125, 155
  box, 116, 125, 126
  buckle, *54,* 58, 64, 66, 67, 79, 94, 117
  buckling, 51, 52, 89, 263

chevron, 90, 93
disharmonic, *61*
distribution, 70, *75,* 109
fault bend, 183
fault propagation, 123
faulted, 155, 162, 168
flattening, 52, 261, 267, 268
flexural, 65
fractal distribution, *69,* 70, 75, 82
growth rate, 84, 90, *94*
limbs, 104, 105, 108
modeling, *89*
open, 54
ptygmatic, vii
shapes, 90
sheath, viii, 101, 102, 104, 105, 107, 108, 109, 115
sinusoidal, 93, 95, 96, 97, 98
tubular, 104
wavelength, 70, 75, 90
wavelength:thickness ratio, 58, *76,* 97
fold and thrust belt, 179
    Calabro-Lucano Southern Apennines, 172, 173
    deformation, 253
    kinematics, 253
    propagation, 172
fold axis
    angle with hinge, 58
    angle with lineation, 51, 52, *58,* 63
fold shape, 93
    rheology, *93*
foliation, 45, 107
    mylonitic, 102
forekink, 194, 201, 202. *see also* kink band
    dip, 204
    model, 195, 197
    multiple, 196
foreland, 160
    domain, 174
    ramp, 163, 164, 169, 170, 172, 174
    structure, 162
foreshear, 202, 204
    asymmetry, 203
    dip, 205
    imbrication, 201
foreslope
    dip, 205
    slumping, 205
    stability, 201
forethrust, 202
    wedge, 126
Formations
    Artés, 113
    Barbastro, 113
    Castelltallat, 113, 114
    Solsona, 113
    Súria, 113, 114
forsterite, 102
four-dimensional analysis, 179, 181, 187, 208
Fourier transform, 70
fractal, 69
    self-affine, *70,* 75, 82
    self-similar, 70

friction
    basal, 172
    internal, 202
    wedge, 123
Front Ranges, Canadian Rocky Mountains, 180
frontal ramp, 180, 186. *see also* ramp

G

Gabon basin, 140, 148
Gaussian normal probability distribution, 70
glaciers, vii
gliding, gravity, 147
graben, 131, 149
    salt evacuation, 132, 149
grain
    growth, 39, *40,* 44, 47
    shape, 39
    size, *40*
grain boundary, 45
    energy, *43*
    migration, 39, *40*
    mobility, 43, 44
Granucol, analog material, 22, 23, 24, 25, 26
gravity collapse, 202
growth rate, 39, 93
    fold, 90
Gulf of Mexico
    pseudoclinoform, 138
    sedimentary basins, 131, 140, 148
    structure, *149*
gypsum, Barbastro Formation, 113

H

halite, Cardona Formation, 113
halokinesis, 131, 133, 148, 149
Hans Ramberg Tectonic Laboratory, 193
heat flow, 225, 232
Hellenic nappes, 127, 129
Hellenides, 112
hinge
    angle with fold axis, 58
    sharpness, 90
hinterland, 161
    anticline, 164
    deformation, 174
    domain, 174
    fault, 158
    thrust, 125, 129
    wedge, 123, 126, 130
horse, 183, 207, 208, 211, 212, 214, 216, 219
    asperity bifurcation, 218, 219
    internal shearing, 219
    splay, *218*
hydrocarbon, 174, 187
    evaluation, 153
    exploration, 149, 154, 174, 207
    migration, 149
hydrofracturing, 124
hyperbola, 7, *14, 15*
    rectangular, 6

I

illite, 262
imbrication, 160, 162, 164, 165, 168, 194, 198
    shear, 198, 201
    thrust, 183, 186
inclusion
    rotated, 101
    viscous, 32
indentation, continental, 192
indenter, 204
    dip, *199*
    effective, ix, 192, 193, 194, 195, 196, 198, 199, 202, 203, 204
    geometry, 195
    rigid, 191, 192, 193, 194, 197, 202
    shape, 192
intergranular processes, 29
internal domain, 173
isostasy, vii, 193, 257

J

Japan, 202
Jura Mountains, 180

K

kinematics, 261
    alternating, 265
    anomalous, 176
    compression, 153, 154
    deformation, *155,* 172, 176
    evolution, 181
    fold and thrust belt, 253
    model, 168, 187
    orogenic wedge, 253
    piggy-back, 159, 170
    structure, 176
    vorticity number, 235
kink band, 191, *193,* 197, 201.
    see also backkink; forekink
    conjugate, 125
    nucleation, 195
    reverse, 197
    sand thickness, *198*
    spacing, 197, 198, 204
Kvarkush
    anticline, 202
    frontal thrust, 202
Kwanza basin, 133, 148
    structure, 134

L

l'Estany anticline, 113
lacustrine system, Pyrenees, 113
lateral ramp, 179, 180, 186, 189
layering, mechanical, 222
limestone, 173
    Castelltallat Formation, 113, 114
    platform, 148
lineation, 59, 61, 65, 66, 102
    angle with fold axis, 51, 52, *58,* 63

lineation (*continued*)
  asymmetrical, 67
  deformation, 51, 52, 56, 63, 64
  disharmonic fold, *61*
  pattern, *52, 58, 64,* 67
  sigmoidal, 62
  switch, 18
  U pattern, 51, *59,* 64, 65, 66, 67
linkage
  deformation, 207
  segment, 219
lithosphere
  stratification, 222
  strength profile, *222,* 223
loading, 230
  differential, 124
  tectonic, 124
lozenge, 265, 269

M

M folds, 61, 67
marble, 102
marginal domain, 173, 174
marl, Súria Formation, 114
mass transfer, 29, *30*
master fault, 209, 211, 218, 219
Maxwell viscoelastic model, 74
mechanical hopper, 22
Mediterranean Ridge, 123, 127, *127,* 128
mélange, 115
Messinian evaporites, 127
Mica fish, *241*
microbead, analog material, 154, 160, 162
microlaminate, 111, 116
migration
  deformation, 256
  hydrocarbon, 149
model
  boundary layer, 39
  Calabro-Lucano region, 174
  finite element, *29,* 69
  kinematic, 187
  non-Newtonian, 53, 90
  numerical, 231
  rotation, *242*
  scale, 23
  thermotectonic, 250
Moho, 203
Mohr-Coulomb
  criterion of failure, 124, 125 (*see also* Coulomb, criterion of failure)
  envelope, 124
  material, 193
  theory, 21
monocline, 140
Murge region, 174
mushroom pattern, sheath folds, 108, 109

N

Nagsuggtoqidian orogen, vii
necking instability, *226,* 232

Newtonian fluid, silicone putty, 116
Newtonian layers, perturbation, 226
Nile Rifts, 231
North Sea, 140
North West Shelf, Australia, 231
number of kinks, indenter dips, *199*
numerical model, Elle, 41

O

octachloropropane, viii, 39, 40
olivine, 222
orogen, 202
Orphan Basin, Canada, 231
Ossa-Morena Zone, 102
overburden thickness, Cardona formation, 113
overpressure, 129
overthrusting, model, 248

P

paleopiezometry, 44, 47
paraffin, 91
  analog material, 235, 245, 248, *254*
particle path, viii, 3, *10,* 16, 17
  apophyses, 4
partitioning, deformation, 176
PatMatch, computer program, 240
patterns
  double eye, 108
  double mushroom, 108
  eye, 108
  mushroom, 108, 109
Perspex simple shear rig, 103
perturbation
  amplitude, 78, 80, 82, 84
  distribution, *70,* 77, 80
piercing, Cardona diapir, 120, 121
piggy-back deformation, 158, 168, 169, 174, 176
Pinós-Cardona anticline, 113, 114, 115, 117, 120, 121
plane-strain critical taper, 253
plaster of Paris, 91
  analogue material, 208, 215
plasticine
  analog material, 103, 116, 117
  boudin, 104, 107
plastilina, analog material, 116
plate
  continental, 192
  tectonics, viii
pluton, vii
Poisson's ratio, 69, 74, 81
polydimethylsiloxane, analog material, 181, 238
polythene sheet, 63
  tracing lineations, 52, 54
pop-up structure, 179, 183, 186, 189, 194
pore pressure, 124
Pozosphere microballoon, 223
pressure vessel, *92*
  uniaxial, 90
principal displacement zone, 23, 27

probability distribution, 81
progradation, clastic, 131, *138,* 140, 143, 147, 148, 149
pseudoclinoform, Gulf of Mexico, 138
pseudodownlap, 131, 140, 148
  Santos basin, 140
pseudoturtleback, 131, 133, 148
putty, analog material, 235
Pyrenean triangle zone, 112, 113
Pyrenees, 112

Q

quartz, 222
  in clay, 262

R

radiolarites, 173
ramp, 180, 186
  deformation, 207
  dip, 199, 204
  foreland, 155
  imbrication, 192
  lateral, 179, 186, 189
  oblique, 179, 180, 189
  rigid, 199
reflection
  Mediterranean Ridge, 127
  profile, 202
reflector, 138
refraction, Mediterranean Ridge, 127
relay structure, 215
rheology
  orogenic wedge, 254
  thermal imaging, 247
Rhodorsil Gomme, analog material, 239
riders, 214
Riedel shear, 24
rift
  basin, 221, 225, 226, 231, 232
  system, 221
  valley, vii
rifting, 224, 231
rock, 222
roll-overs, 149
roof fault, 208
rotation, 24, 51, 58, 63, 64, 65, 66, 101, 104, 147, 195, 199, 261, 262, 265, 266, 268
  block, 131, 132, 133, *140,* 147, 150, 267, 268, 269
  boudin, 102, 103, *104,* 107, 109
  fault, 266
  model, *242*
  particle, 102
  rate, 243
  rigid, 268
  stress axes, 230

S

S folds, 61, 67
salt
  barrier, 150

basin, 133
dissolution, 133, 134, 148
evacuation, 134, 148
expulsion, 138
extrusion, 115
flow, 148
mobilization, 140, 143, 149
reflector, 143
residual, 138
structure, vii
tectonics, ix, 116
tongue, 132
Salt Range, Pakistan, 180
Sanaüja anticline, 113, 114
sand
analog material, 22
quartz, 181, 182, 193
wedge, 23, 24
sand thickness, kinkband, *198*
sand, analog material, 154, 223
sandstone, 173
Solsona Formation, 114
Súria Formation, 114
Santos basin, 133, 138, 140, 148
pseudodownlap, 140
structure, 140, *149*
scaling, 23, 245, 253, *253, 254*, 257
analog, 223
factor, 27
stress, *248*
time, *247*
velocity, *247*
viscosity, *249*
sedge, geometry, 202
sedimentation, rate, 231
segment
amalgamation, *218*
coalescence, 219
linkage, 207, 208, 215, 219
splaying, 219
shale, 173
shear, 25, 26
basal, 140, 221, 230, 231, 232
box, 22, 53, 58, 61, 63, 89, 90, 91, 103
tests, *93*
ductile, 66
gradient, 232
imbrication, 198, 200, 201, 202, 204
lens, 24, 25
pure, viii, 74, 102, 235, 243, 261, 265
Riedel, 23, 24
simple, viii, 1, 2, 16, 23, 51, 53, 54, 60, 63, 65, 66, 67, 101, 102, 104, 107, 108, 235, 240, 243
zone, 1, 19, 21
sheath fold. *see* fold, sheath
shortening, 69, 84, 90, 98, 129, 140, 163, 193
bulk, 113, 197
layer parallel, *79*, 113
machine, 254
model, 182
Siemens Somatom Plus 4 spiral X-ray computed tomographer, 181

Sierra Nevada, stretching, 18
Sierras Marginales thrust, 114
silicone
analog material, 53, 58, 61, 63, 103, 104, 107, 111, 116, 117, 125, 126, 223
Newtonian fluid, 116
siltstone, Barbastro Formation, 113
Skelefte arc, 203
slumping, 202
backslope, 191
smectite, 262
Solsona Formation, 113
South Atlantic basin, 131, 138, 140, 148
Southern Apennines, 153
hydrocarbon exploration, 174
mechanical stratigraphy, 176
thrust belt, 174
spiral X-ray computed tomography, viii, *179*
improvements, 180
splaying, 26
coalescence, 219
deformation, 207
segment, 207, 208, *217,* 219
stability
slope, 201
stereographic projection, 52
strain, 32, 59, 63, 214, 268
bending, 97
coaxial, 1, 2
contractional, 54
creep, 74
distribution, 93, *97,* 267, *268, 269*
elastic, 74
ellipse, *4,* 11, 16, 267
ellipsoid, 1, 5, 17, 39, 47, 54, 56, 58, 65, 66, 67, 262, 268
extensional, 226
gradient, *97*
hardening, viii, 95, 96, 98, 261, 263, 266, 268, 269
heterogeneous, 19
history, 17
homogeneous, 56
layer parallel, 51, 58, 64, 66, 67
longitudinal, 51, 54, 58, 64, 67
normal, 90
partitioning, 204
plane, 240
rate, 1, 2, *5, 10,* 14, 16, 19, 31, 37, 74, 90, 125, 208, 235, 243, 246, 249
shear, 105
softening, viii, 94, 96, 98, 257, 261, 265, 266, 268
tangential, 90
viscous, 74
stratification
lithosphere, 222
rheological, 221
stratigraphy, mechanical, 173, 174
strength profile, lithosphere, *222,* 223
stress
axes, 230
balance model, 126
deformation apparauts, 247
field, 27, 29, 32, 35

lithostatic, 120
maximum principle, 32
mean, 29, 30, 120
scaling, *248*
state, 34
stretching, 15, 17, 64
Sierra Nevada, 18
structure
geometry, 176
kinematics, 176
pop-up, 179, 183, 186, 189, 194
Santos basin, 140, 148
small scale, 235
stylolites, 29
subduction zone, 203
subsidence, 147
thermal, 132, 133
subthrust, 161
surface energy, 39
Súria anticline, 113, 114, 120, 121
age, 114
Súria Formation, 113, 114
suturing, continental, 192
Svecofennides, 191, 192, 202
sylvinite, Cardona Formation, 113
syncline, peripheral, 149
synforms, 54, 58, 61, 64, 67

T

Tagil oceanic and volcanic arc complex, 202
talc, 102
tear faults, 180
tectonics
diapirs, 130
thick skinned, 172
thin skinned, 154, 172
temperature
control, 255
deformation apparatus, *247*
tension gashes, 29
tensor, 33
thermal gradient, 245, 256
thermal imaging, rheology, 247
thermal variation, ductile strength, 257
thermomechanical, apparatus, *253*
thrust. *see also* fault, thrust
conjugate, 116
forward, 183, 186
frontal, 161, 183, 187
oblique, 170
thrust belt, 153, 154, 169, 176
architecture, 176
structural evolution, 154
thrust fault, conjugate. *see* fault, conjugate
thrust plane, 125
thrust system, evolution, 155
thrusting
imbricate, 186
out of sequence, 179, 189
time, scaling, *247*
tip line
bifurcation (*see* bifurcation, tip line)
coalescence, 219
transfer mechanics, 176

transfer zone, 179, *180,* 186, 189
  modeling, *183*
transpression, *1, 7, 16,* 236
  Centro-Iberian autochthon, 19
  deformation, viii
transtension, *1, 7, 16,* 17, 236
  deformation, viii
triangular domain, 268, 269
turtleback, 149

U

U pattern, lineation, 51, 61, 62, 64, 65, 66, 67
underthrust, 124
Urals, 191, 192, 202
  seismic profile, 204

V

Variscan fold belt, 102
Vaseline, 91
velocity
  discontinuity, 192, 194
  scaling, *247*

Vilanova de l'Aguda anticline, 113, 120, 121
viscosity
  contrast, 102, 103
  ratio, 53
  scaling, *249*
  temperature sensitive, 245, 254
vitrinite, 113
volcanic arc complex, 202
volume change number, equation, 239
vorticity number, 1, *13,* 16
  kinematic, 235
vorticity vector, 1, 5, 6, 19, 51, 65
  equation, *239*

W

wafer, 31, 32, 35
wavelength:thickness ratio, folds, 58, *76,* 97
wedge
  asymmetric, 195
  compressional, 159
  deformation, 253
  evolution, 257

  hinterland, 123, 126, 130
  kinematics, 253
  orogenic, 205, 253, 254
  rheology, 254
  two sided, 200, 204
  Ural orogenic, 202
  width, 197
Wollaston Medal, vii
wollastonite, 102
Wulff construction, 43

Y

Young's moduli, 74

Z

Z folds, 61, 67
Zagros, 112